INTERMEDIATE 1

Maths

Brian J Logan

Consultant Editor: **Robert Barclay**

Acknowledgements

The Publishers would like to thank the following for permission to reproduce copyright material:

Photo credits

page 1 and page 47 © Martin Cassell (mpc) / Alamy; page 4 © Dominic Burke / Alamy; page 11 © Steven May / Alamy; page 22 John Mitchell; page 28 © David Robertson / Alamy; page 70 Tim Ockenden/PA Archive/PA Photos; page 61 Michael Heinsen/ Stone/ Getty Images; page 67 © Jonathan Souza - Fotolia.com; page 71 www.purestockX.com; page 83 and page 108 © DLILLC/Corbis; page 88 © Stock Images / Alamy; page 100 © POPPERFOTO / Alamy; page 103 © Trevor Smith / Alamy; page 121 © Paul Doyle / Alamy; page 135 © Peter T Lovatt / Alamy; page 153 © FADI ALASSAAD/Reuters/Corbis; page 137 © Jochen Tack / Alamy; page 147 © ccMacroshots / Alamy; page 168 (top) © peter dazeley / Alamy; page 159 © Yadid Levy / Alamy; page 165 © Roger Bamber / Alamy; page 177 and page 236 © ImageState / Alamy; page 191 Walt Seng/ Nonstock/ Jupiter Images; page 212 © Chris Howes/Wild Places Photography / Alamy; page 226 NASA; page 227 © IAU/Martin Kommesser/Handout/epa/Corbis; page 228 ©iStockphoto.com/Dino Ablakovic; page 231 © PHOTOTAKE Inc. / Alamy; page 244 © Steven May / Alamy; page 249 ©iStockphoto.com/ Christopher O Driscoll; page 241 and page 258 STP Network; page 267 Redferns Music Picture Library / Alamy; page 283 © Map Resources; page 288 © MAPS.com/CORBIS; page 285 © Rick Parsons - Fotolia.com; page 301 © photoman0 – Fotolia.com; page 324 © Daniel dal Zennaro/epa/Corbis; page 318 © Powered by Light/Alan Spencer / Alamy; page 326 © mflippo – Fotolia.com; pages 12, 36, 37, 65, 114, 125, 168 (bottom), 172, 182, 210, 297 Stuart Burns.

Every effort has been made to trace all copyright holders, but if any have been inadvertently overlooked the Publishers will be pleased to make the necessary arrangements at the first opportunity.

If the CD is missing from this package please contact us on 0141 848 1609 or at hoddergibson@hodder.co.uk, advising where and when you purchased the book.

Orders: please contact Bookpoint Ltd, 130 Milton Park, Abingdon, Oxon OX14 4SB. Telephone: (44) 01235 827720. Fax: (44) 01235 400454. Lines are open 9.00–5.00, Monday to Saturday, with a 24-hour message answering service. Visit our website at www.hoddereducation.co.uk. Hodder Gibson can be contacted direct on: Tel: 0141 848 1609; Fax: 0141 889 6315; email: hoddergibson@hodder.co.uk

First published in 2007 by
Hodder Gibson, an imprint of Hodder Education,
and a member of the Hodder Headline Group,
An Hachette Livre UK Company,
2a Christie Street
Paisley PA1 1NB

Impression number	5 4 3 2 1
Year	2010 2009 2008 2007

Cover photo Buzz pictures / Alamy
Illustrations by Richard Duszczak, Cartoon Studio
Typeset in Bembo 13pt by Pantek Arts Ltd, Maidstone, Kent
Printed in Great Britain by CPI Bath.

A catalogue record for this title is available from the British Library

ISBN-13: 978-0-340-93923-9

Contents

Introduction

This book covers the complete course for Intermediate 1 Mathematics (Units 1, 2, 3 and Applications of Mathematics), a National Qualification of the Scottish Qualifications Authority.

All students will study Units 1 and 2 and these are explained in Chapters 1–9. If you are studying Unit 3 you will also need look at Chapters 10–13, and if you are studying Applications of Mathematics you will also need to look at Chapters 14–17.

As well as the assessments that you will do in class, there will also be a final exam. This book contains a lot of exam-style questions which will help you prepare for the exam. The exam contains calculator and non-calculator papers and you will see that there are examples of both types of question throughout the book. You should check each exercise carefully to see whether a calculator is allowed or not.

At the end of each Unit you will find three 'End of Unit Tests'. One test is non-calculator, while the other two allow the use of a calculator. The third test covers the more difficult A/B content of the course. Your teacher may ask you to complete these tests either in class or at home for extra practice.

The Summary at the end of each Chapter has a short list of the topics covered in the Chapter, including details of formulae, worked examples and useful notes. You may find it helpful to use these Summaries when you are revising and they may help you to organise your revision.

There is a revision exercise near the end of each Chapter which covers the main points of each topic. These will help you when you are preparing for the assessments you will do in class and for the exam.

Finally, at the end of book, there is a Practice Exam Paper which contains both a non-calculator and a calculator paper. This is set at the same level of difficulty as the final exam and will give you a good idea of what to expect. The marks awarded for each question are given in brackets.

I hope you find this book helpful during your Intermediate 1 Maths course.

Good luck.

Brian J Logan

UNIT 1

Chapter 1 **Basic Calculations**

Chapter 2 **Basic Geometry**

Chapter 3 **Expressions and Formulae**

Chapter 4 **Everyday Calculations: Earnings and Spending**

End of Unit Tests

1 Basic Calculations

Basic calculations are part of our everyday lives, and you will often have to do them without the aid of a calculator. How much Income Tax will I have to pay on £5000 when the tax rate is 20%? What is $\frac{9}{20}$ as a decimal? What is $\frac{3}{8}$ of 56? What is £20·15 ÷ 12 to the nearest penny? These are often the sorts of calculations which other people will expect *you* to be able to do, usually when they themselves are struggling! Do you feel confident?

Whole Numbers

In this section, and without a calculator, you should be able to

- **add, subtract, multiply, and divide whole numbers**
- **multiply together two 2-digit numbers such as 27 × 11**
- **multiply together three numbers such as 7 × 5 × 4**
- **carry out calculations in the correct order.**

Order of Operations

What is the answer to 5 + 2 × 3?

You might think the answer is 21 because 5 + 2 = 7, and then 7 × 3 = 21. This would be wrong. Calculations must be done in the **correct order**. **Always** deal with muliplication and division **before** addition and subtraction, in a calcuation. So in the calculation above, we must multiply 2 × 3 first (= 6) and then add 5. The correct answer is 11.

Example Calculate $3 \times 50 - 20 \div 2$.

Remember to do the multiplication and division before the subtraction.

Solution $3 \times 50 - 20 \div 2$

$= 150 - 10$

$= 140$.

Exercise 1

Do not use a calculator

1 Work out the answers to the following:

(a) $425 + 314$ (b) $357 + 669$ (c) $584 - 352$ (d) $252 - 175$

(e) 312×4 (f) 256×7 (g) $84 \div 6$ (h) $531 \div 9$.

2 Work out the answers to the following:

(a) 36×11 (b) 43×12 (c) 28×12 (d) 52×13.

3 Work out the answers to the following:

(a) $6 + 5 \times 2$ (b) $20 - 10 \div 2$ (c) $12 + 24 \div 6$ (d) $18 - 6 \div 3$.

4 Work out the answers to the following:

Hint: When multiplying **three** numbers together, you can multiply them **in any order**, so look for the easiest order. For example, in $4 \times 7 \times 25$, it would be easiest to multiply 4×25 (= 100), first, then 100×7 (= 700).

(a) $6 \times 4 \times 5$ (b) $50 \times 7 \times 2$ (c) $25 \times 9 \times 4$

(d) $3 \times 6 \times 23$ (e) $9 \times 6 \times 4$ (f) $25 \times 9 \times 2$.

5 Work out the answers to the following:

(a) $60 + 20 \times 4$ (b) $340 - 70 \times 2$ (c) $250 + 100 \div 2$

(d) $5 \times 25 - 10 \times 10$ (e) $70 \times 3 + 40$ (f) $12 \div 3 + 16 \div 2$

(g) $80 \times 4 - 60 \times 2$ (h) $12 + 18 \times 11$.

Decimals

In this section, and without a calculator, you should be able to

- add and subtract decimal numbers
- multiply and divide decimal numbers by a single digit whole number
- multiply and divide numbers by multiples of 10, 100, 1000
- write decimals as fractions.

Example Work out the answers to:
(a) $3{\cdot}52 \times 400$ (b) $7{\cdot}5 \div 50$.

Solution
(a) $3{\cdot}52 \times 400 = 3{\cdot}52 \times 100 \times 4 = 352 \times 4 = 1408$

(b) $7{\cdot}5 \div 50 = 7{\cdot}5 \div 10 \div 5 = 0.75 \div 5 = 0{\cdot}15$.

Example Write as fractions: (a) $0{\cdot}3$ (b) $0{\cdot}03$ (c) $0{\cdot}003$.

Remember that the first number after the decimal point is in the tenths column, the second number is in the hundredths column, and the third number is in the thousandths column.

Solution (a) $0{\cdot}3 = \frac{3}{10}$ (b) $0{\cdot}03 = \frac{3}{100}$ (c) $0{\cdot}003 = \frac{3}{1000}$.

Exercise 2

Do not use a calculator

1 Work out the answers to the following:

(a) $6{\cdot}32 + 3{\cdot}65$ (b) $8{\cdot}17 - 3{\cdot}08$ (c) $4{\cdot}85 + 3{\cdot}68$ (d) $9{\cdot}23 - 8{\cdot}4$

(e) $8{\cdot}23 + 5{\cdot}2$ (f) $7{\cdot}24 - 3{\cdot}5$ (g) $5{\cdot}36 + 2{\cdot}9$ (h) $3{\cdot}251 + 4{\cdot}73$

(i) $6{\cdot}3 - 3{\cdot}19$ (j) $5{\cdot}29 + 7$.

Exercise 2 continued

Do not use a calculator 

2 Work out the answers to the following:

(a) 7·3 × 5 (b) 16·56 ÷ 4 (c) 4·276 × 3

(d) 0·648 ÷ 8 (e) 3·25 × 8 (f) 2·135 ÷ 7

(g) 14·67 ÷ 9 (h) 2·046 × 9.

3 Work out the answers to the following:

(a) 7·3 × 10 (b) 6·4 ÷ 10 (c) 6·1 × 100

(d) 5·2 ÷ 100 (e) 4·23 × 1000 (f) 9·4 ÷ 1000.

4 Work out the answers to the following:

(a) 4·2 × 300 (b) 3·7 × 2000 (c) 2·15 × 700

(d) 8·42 × 40 (e) 8·4 ÷ 20 (f) 6·5 ÷ 5000

(g) 7·36 ÷ 400 (h) 5·34 ÷ 600.

5 Write the following as fractions:

(a) 0·9 (b) 0·09 (c) 0·009

(d) 0·13 (e) 0·037 (f) 0·191

(g) 0·007.

6 A plank of wood is 6·5 metres long. What length will be left if a joiner cuts off a piece 2·27 metres long?

7 Alan divides £16·47 equally between his three sons. How much money does each son receive?

8 Alice sells ornaments for £17·50 each. How much money will she make if she sells 40 ornaments?

Fractions

In this section, without using a calculator, you should be able to

- work out a fraction of a quantity
- change a fraction to a decimal fraction.

Example Work out $\frac{5}{6}$ of 84.

Method: We work out $\frac{1}{6}$ of 84 first (by dividing 84 by 6).

Then to calculate $\frac{5}{6}$ of 84, we multiply the answer by 5.

Solution $\frac{5}{6}$ of 84 = 84 ÷ 6 × 5 = 14 × 5 = 70.

(**Note** that this type of question would normally be asked in the non-calculator paper in your final exam.)

Example Write $\frac{3}{8}$ as a decimal fraction.

Solution $\frac{3}{8}$ = 3 ÷ 8 = 0·375.

$$8\,\overline{)\,3\cdot{}^{3}0{}^{6}0{}^{4}0}\quad = 0\cdot375$$

Exercise 3

Do not use a calculator

1 Work out the answers to the following:

(a) $\frac{3}{4}$ of 48 (b) $\frac{2}{3}$ of 54 (c) $\frac{5}{7}$ of 84 (d) $\frac{2}{5}$ of 80

(e) $\frac{7}{10}$ of 130 (f) $\frac{5}{8}$ of 144 (g) $\frac{8}{9}$ of 108 (h) $\frac{1}{6}$ of 168

(i) $\frac{4}{7}$ of 112 (j) $\frac{7}{8}$ of 192.

Exercise 3 continued

Do not use a calculator 

2 Write the following fractions as decimal fractions:

(a) $\frac{7}{10}$ (b) $\frac{3}{5}$ (c) $\frac{1}{4}$ (d) $\frac{2}{5}$

(e) $\frac{1}{2}$ (f) $\frac{3}{4}$ (g) $\frac{1}{8}$ (h) $\frac{5}{8}$.

3 Alex earns £62 from his weekly paper round. He saves $\frac{3}{4}$ of his weekly earnings. How much does he save?

4 Mary earns £76 from her part-time job. She gives $\frac{2}{5}$ of her earnings to her mother. How much does she give her mother?

5 There are 1233 pupils on the roll of Limegrove Academy and $\frac{5}{9}$ of the pupils are boys. How many boys are there on the roll?

Rounding

In this section, without using a calculator, you should be able to round calculations to a given degree of accuracy:

- to the nearest whole number
- to the nearest 10, 100, 1000
- to a given number of decimal places.

The attendance at a football match is sometimes rounded to the nearest thousand. Suppose 26 854 spectators attended a match and we were asked to round the attendance to the nearest thousand. Obviously 26 854 is between 26 000 and 27 000, but it is nearer to 27 000.

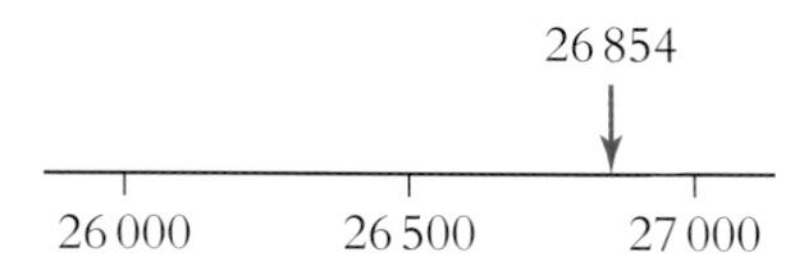

Therefore 26 854 = 27 000 (to the nearest thousand).

If a number is midway, we always round up. For example, 17·5, rounded to the nearest whole number, would be 18.

Exercise 4

1 Round to the nearest 10:

(a) 68 (b) 79 (c) 41

(d) 122 (e) 179 (f) 175

(g) 198 (h) 69·8 (i) 1263.

2 Round to the nearest 100:

(a) 216 (b) 367 (c) 897 (d) 751

(e) 1399 (f) 2741 (g) 601 (h) 7569.

3 Round to the nearest 1000:

(a) 7201 (b) 6050 (c) 5980 (d) 8999

(e) 16 247 (f) 12 345 (g) 26 417 (h) 56 499

(i) 47 500 (j) 50 567.

4 Round to the nearest whole number:

(a) 6·4 (b) 7·5 (c) 17·1

(d) 8·9 (e) 5·72 (f) 15·91

(g) 49·09 (h) 15·6 (i) 74·8.

5 Last Saturday, 52 298 spectators attended a football match. Round this number to the nearest thousand.

6 Carry out the following calculations and round your answers to the nearest 100: (**you may use a calculator**)

(a) 653×24 (b) 52×95 (c) 71×9

(d) $68\,456 \div 5$ (e) $3{\cdot}14 \times 30 \times 30$.

Decimal Places

When we have to measure a quantity to a very high degree of accuracy, we often produce a longer decimal number. (For example, a beam might be measured to be 2·584 m in length.)

Main points

You can tell how many decimal places there are in a number by counting how many figures appear **after** the decimal point.

The number 2·584 has three decimal places. (Three figures (584) appear after the decimal point.)

Consider rounding 2·584 to two decimal places.

- 2·584 lies between 2·58 and 2·59
- look at the figure in the **third** decimal place (2·58<u>4</u>) i.e. 4
- since this figure is **less than 5**, then round down to 2·58.

If the figure in the third decimal place had been **5 or more** then you would have rounded up. For example 2·587 rounds up to 2·59.

Example Round (a) 6·7459 to 1 decimal place
(b) 14·879 632 to 3 decimal places.

Solution (a) 6·7459 lies between 6·7 and 6·8.

6·7<u>4</u>59 rounds down to 6·7 since 4 (the figure in the second decimal place) is less than 5

(b) 14·879 632 lies between 14·879 and 14·880.

14·879 <u>6</u>32 rounds up to 14·880 since 6 (the figure in the fourth decimal place) is more than 5.

Example Calculate £17 ÷ 7.
Give your answer **to the nearest 1p**.

(When you are asked to give an answer to the nearest 1p or the nearest p, this is the same as giving it correct to two decimal places, as there would be two figures after the decimal point to represent pence in the answer.)

Solution

£17 ÷ 7 = £2·428 571 429.

£2·428 571 429 lies between £2·42 and £2·43.

£2.42<u>8</u> 571 429 rounds up to £2·43 since 8 (the figure in the third decimal place) is more than 5.

So £17 ÷ 7 = £2·43 to the nearest 1p.

Exercise 5

1 Round to 1 decimal place:

(a) 6·36 (b) 7·41 (c) 8·79 (d) 16·45
(e) 2·09 (f) 25·11 (g) 7·98 (h) 0·7654.

2 Round to 2 decimal places:

(a) 5·368 (b) 6·743 (c) 18·429 (d) 8·007
(e) 5·789 (f) 6·8123 (g) 2·235 (h) 6·999.

3 Round to 3 decimal places:

(a) 4·7654 (b) 16·5678 (c) 4·0709 (d) 4·1765
(e) 21·345 67 (f) 7·2398.

4 Round to the number of decimal places shown in the brackets:

(a) 8·35 (1) (b) 6·347 (2) (c) 14·983 (1) (d) 5·077 (1)
(e) 231·598 76 (4) (f) 0·761 (2) (g) 18·002 (2) (h) 5·1875 (3)
(i) 12·2457 (1) (j) 56·666 66 (3).

Exercise 5 continued

5 Write the following as decimal fractions, giving your answers to 2 decimal places:

(a) $\frac{1}{3}$ (b) $\frac{2}{3}$ (c) $\frac{5}{6}$ (d) $\frac{5}{9}$ (e) $\frac{3}{7}$ (f) $\frac{7}{8}$.

6 Calculate, giving your answers to the nearest 1p:

(a) £25 ÷ 6 (b) £38·57 × 1·175 (c) £37·97 ÷ 12

(d) £28·35 × 1·15 (e) £200 ÷ 3 (f) £835 ÷ 13.756

(g) £48 ÷ 7 (h) £25 × 1·175.

Percentages

In this section, you should be able to

- write a percentage as a common fraction or decimal
- work out a simple percentage of a quantity without a calculator
- calculate a percentage of a quantity with a calculator
- calculate percentages in a given problem situation.

The most common problem situations involving percentages are Value Added Tax (VAT), Interest from a Bank, and Discounts in a Sale. In these contexts, money is added to or subtracted from an initial sum of money.

Example Write 75% as (a) a decimal (b) a fraction in its simplest form.

Solution

(a) $75\% = \frac{75}{100} = 0{\cdot}75$

(b) $75\% = \frac{75}{100} = \frac{3}{4}$.

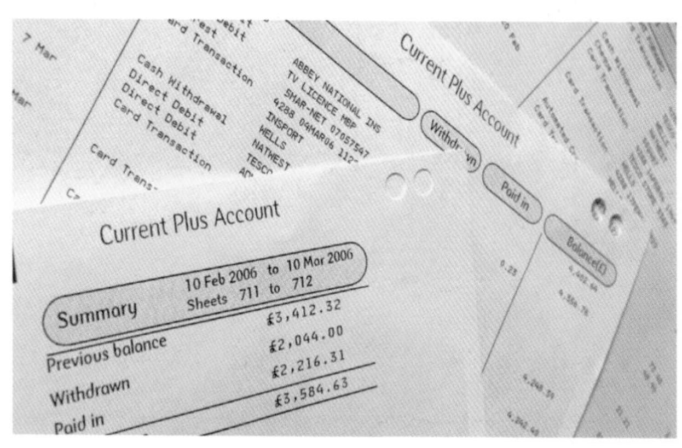

Remember

You should memorise the list of percentages which follows, all of which have been written as fractions

$10\% = \frac{1}{10}$ $20\% = \frac{1}{5}$ $25\% = \frac{1}{4}$ $33\frac{1}{3}\% = \frac{1}{3}$ $50\% = \frac{1}{2}$.

Example Work out, **without using a calculator,**

(a) $33\frac{1}{3}\%$ of £69 (b) 70% of £150

(c) 5% of £150.

Solution (a) $33\frac{1}{3}\%$ of £69 = £69 ÷ 3 = £23

(b) 10% of £150 = £150 ÷ 10 = £15
so 70% of £150 = £15 × 7 = £105

(c) 10% of £150 = £150 ÷ 10 = £15
so 5% of £150 = $\frac{1}{2}$ of £15 = £7·50.

Example Calculate, using a calculator, 24% of £280.

Solution 24% of £280 = $\frac{24}{100}$ × £280 = 0·24 × £280
= £67·2 = £67·20.

Example A piano costs £245 + VAT. If Value Added Tax is charged at 17·5%, find the total cost to the nearest 1p.

Solution VAT = 17·5% of £245 = $\frac{17{\cdot}5}{100}$ × £245
= 0·175 × £245 = £42·875 = £42·88
hence total cost = £245 + £42·88 = £287·88.

Sales Sub Total 30.03
Total Items: 1
Visa/Mastercard/Elec (EFT Tran 30.03
Payment Sub Total 30.03
Change 0.00

VAT Summary

VAT Rate	Ex VAT	VAT	Inc VAT
17.50%	25.56	4.47	30.03

Thank You
Please Call Again

Operated by:
Dolphin Retail Management Ltd

Drive Carefully

The amount of VAT is shown on this receipt

Exercise 6

1 Write each percentage as a decimal:

(a) 25% (b) 63% (c) 60% (d) 9%

(e) 12·5% (f) 99% (g) $87\frac{1}{2}$% (h) 58%.

2 Write each percentage as a fraction in its simplest form:

(a) 10% (b) 30% (c) 70% (d) 40%

(e) 60% (f) $33\frac{1}{3}$% (g) $66\frac{2}{3}$% (h) 5%.

3 Work out the answers to the following, **without using a calculator**:

(a) 30% of £160 (b) 5% of £120 (c) 70% of 230 metres

(d) $33\frac{1}{3}$% of £48 (e) 15% of £80 (f) 75% of £240

(g) 90% of 160 grams (h) $66\frac{2}{3}$% of £60.

4 Calculate:

(a) 36% of £400 (b) 95% of £72

(c) 17·5% of £28 (d) 7% of £5.

5 Calculate 7·5% of £36·87, giving your answer to the nearest 1p.

6 Find the total cost of these items after VAT at 17·5% has been added:

(a) A computer at £680

(b) A garage bill at £75·69 (answer to the nearest 1p)

(c) A hotel bill at £486.

7 Janice gets a bill for £69·60 in a restaurant. Find the total cost of the bill if a 10% service charge is to be added.

8 During a sale, a Department Store offers a 15% discount on the marked price of all items. Find the sale price of a television set with a marked price of £490.

9 Strathclyde Savings Bank offers its customers a yearly interest rate of 3·5% on all savings. How much interest would Dorothy receive in a year if she had savings of £1500 in her account?

Simple Interest

If money is deposited in a bank account, **Interest** will be added as time goes by. Banks offer a rate of interest as a percentage per annum. (Per annum means each year.)

Example A bank offers interest of 4·9% per annum on its *Silver Star Account*. How much interest will Anna receive after a year if she deposits £2500 in this account?

Solution

Interest = 4·9% of £2500 = $\frac{4 \cdot 9}{100}$ × £2500 = 0·049 × £2500

= £122·50.

Take care

For periods of time of less than one year, the amount of interest gathered would be less. For example, in six months you would receive only half of the interest for one year. For other periods less than one year, you must convert the period to a fraction of a year. It is essential you know that

1 year = 12 months = 52 weeks = 365 days.

Example How much interest would Shezad receive if he deposited £500 at an interest rate of 3·6% per annum for a period of 7 months?

Solution

Interest for 1 year

= 3·6% of £500 = $\frac{3 \cdot 6}{100}$ × £500 = 0·036 × £500 = £18

Hence interest for 7 months

= $\frac{7}{12}$ of £18 = £18 ÷ 12 × 7 = £10·50.

Exercise 7

1 A bank offers a rate of interest of 4% per annum. How much interest would you receive from this bank in a year if you deposited:

(a) £200 (b) £2000 (c) £30

(d) £75 000 (e) one million pounds?

2 Calculate the simple interest on:

(a) £480 at a rate of 5% per annum for 6 months

(b) £25 000 at a rate of 4·5% per annum for 5 months

(c) £600 at a rate of 4·8% per annum for 10 months

(d) £500 000 at a rate of 4·92% per annum for 3 months

(e) £40 at a rate of 1·2% per annum for 2 months.

3 Paul invests £7200 in an account at the Scotia Bank. The annual rate of interest is 4·6%. Calculate how much interest Paul should receive after 11 months.

4 Calculate, to the nearest 1p, the simple interest on:

(a) £52 250 at a rate of 4·61% per annum for 1 month

(b) £65·93 at a rate of 3·75% per annum for 9 months

(c) £685 798 at a rate of 5·02% per annum for 4 months.

Expressing One Quantity as a Percentage of Another

In this section, you should be able to express one quantity as a percentage of another. This is achieved by writing one quantity as a fraction of the other, and then muliplying by 100 to convert the fraction to a percentage. The following examples should make this clear.

Example There are thirty pupils in a first year class.
Eighteen are girls.
What percentage of this class is made up by girls?

Solution

Percentage of girls in class

$= \frac{18}{30} \times 100\% = 18 \div 30 \times 100\% = 60\%.$

Example Stewart buys a ticket for a football match for £30 and sells it for £54.
Express his profit as a percentage of the cost price.

Solution

Profit = selling price − cost price = £54 − £30 = £24

Profit as a percentage of cost price $= \frac{24}{30} \times 100\% = 80\%.$

Exercise 8

1 Express these test marks as percentages:

(a) 37 out of 50 (b) 23 out of 25 (c) 28 out of 40 (d) 51 out of 75

(e) 13 out of 20 (f) 28 out of 70 (g) 12 out of 15 (h) 45 out of 60

(i) 82 out of 200 (j) 36 out of 80.

2 There are 1200 pupils at Belltree Secondary School, and 672 are girls.
What percentage of the pupils is made up by girls?

3 Express each of these profits as a percentage of the cost price:

(a) cost price = £25, selling price = £35

(b) cost price = £10, selling price = £19

(c) cost price = 40p, selling price = 60p

(d) cost price = £75, selling price = £78.

Exercise 8 continued

4 In a survey, 125 people were asked to name their favourite flavour of crisps. Of these, 35 said plain was their favourite flavour. What percentage of the group is this?

5 Allison buys an antique vase for £250. She sells it later for £185. Express her loss as a percentage of the cost price.

6 Colin owns sixty golf balls. Nine of them are brand new. What percentage of his golf balls are **not** brand new?

7 What percentage of 75 is 53? Round your answer to the nearest whole number.

8 William scored the following marks in his October tests:

Maths 32 out of 40; English 81%; Geography 50 out of 60; Computing 41 out of 50; Art 19 out of 25.

Put his marks in order, starting with the lowest.

Direct Proportion

In this section, you will see how to solve simple problems involving direct proportion.

Key words and definitions

We say that two quantities are in direct proportion if, as one quantity increases (or decreases), the other quantity increases (or decreases) at the same rate. For example, if we buy a number of similar books, the **cost** is directly proportional to the **number** bought. That is, if two books cost £10, then four of these books would cost £20 (twice as much), and one book would cost £5 (half as much).

Example Seven cans of juice cost £3·85. How much would nine cans of this juice cost?

Solution 7 cans cost £3·85

so 1 can costs £3·85 ÷ 7 = £0·55

and 9 cans cost £0·55 × 9 = £4·95 .

Exercise 9

1 A pineapple costs £0·99. Copy and complete the table below.

Number of pineapples	1	2	3	4	5	6	8	20
Cost (£)	0·99							

2 If eight oranges cost 72p, how much would five oranges cost?

3 Billy pays £8·94 for six similar ties. How much would eleven of these ties cost?

4 Charles can cover 476 miles on the motorway in 8 hours by driving at a steady speed. Travelling at the same steady speed, how far could he travel in 3 hours?

5 Brian earns £207 in 9 hours. How much would you expect him to earn in 24 hours?

6 Sophie's heart beats 35 times in 20 seconds. How many times will it beat in one minute?

7 Pierre can buy 42 euros for £30. How many euros could he buy for £20?

8 Graham spends 5 nights in a hotel at a cost of £240. How much would it cost Anna to stay the night in the same hotel for 3 weeks?

9 Megan can type 500 words in 4 minutes. How long would it take her to type 750 words?

10 100 grams of turkey contain 19·9 grams of protein. How many grams of protein are there in a 5 kilogram turkey?

Exercise 9 continued

11 Cathy earns £168·75 for working 25 hours in a store. How many hours would she need to work to earn £222·75?

12 The cost of a pizza is directly proportional to its area. If a 300 square centimetre pizza costs £3·60, how much would a 750 square centimetre pizza cost?

Exercise 10 – Revision of Chapter 1

Do not use a calculator for questions 1–6

1 Work out the answers to the following:

(a) $8{\cdot}57 - 2{\cdot}8$ (b) $5{\cdot}328 \times 7$ (c) $0{\cdot}558 \div 6$.

2 Work out the answers to the following:

(a) $4{\cdot}5 \times 300$ (b) $7{\cdot}6 \div 20$.

3 Find:

(a) $\frac{4}{9}$ of 198 grams (b) $\frac{2}{5}$ of £76.

4 Work out the answers to the following:

(a) 30% of £260 (b) 15% of 160 kilograms.

5 Write the following in order, starting with the smallest:

32%, 0·31, $\frac{3}{10}$, 0·4, $\frac{37}{100}$.

6 Work out the answers to:

(a) $250 - 3 \times 50$ (b) $6{\cdot}8 + 3{\cdot}1 \times 2$.

7 A television set costs £499 + VAT. If Value Added Tax is charged at 17·5%, find the total cost of the television set to the nearest 1p.

8 A fruit shop has 64 grapefruits and 16 of the grapefruits are pink. What percentage of the grapefruits is represented by pink ones?

Exercise 10 – Revision of Chapter 1 continued

9 Express $\frac{1}{7}$ as a decimal fraction and round your answer to 3 decimal places.

10 The Bank of Tweed offers a special account to its customers.

Amount Invested	Annual Interest Rate
£1 – £10 000	3·6%
£10 001 – £50 000	4·2%
over £50 000	4·8%

Kenneth invests £24 000 in the Bank Of Tweed. How much interest will he receive after 8 months?

11 In a sale, a shop offers a $33\frac{1}{3}$% discount on the marked price of all goods.
How much would Samantha pay for a fridge marked at £249·99?

12 Mr Cowie buys 32 copies of a new mathematics textbook for £255·68.
He likes the book so much that he orders an extra 20 copies.
How much will the extra copies cost?

13 Kenny buys a football programme for £5. He later sells it to a friend for £7.
Express his profit as a percentage of the cost price.

Summary

(Before leaving Chapter 1 you should be confident that you understand each of the following items in this short Summary.)

1 ***Order of Operations***

Remember to deal with × or ÷ before + or –. Hence $50 + 4 \times 10 = 50 + 40 = 90$.

2 ***Decimals*** – Remember your non-calculator skills!

(a) $4{\cdot}25 \times 300 = 4{\cdot}25 \times 100 \times 3 = 425 \times 3 = 1275$

(b) written as fractions, $0{\cdot}1 = \frac{1}{10}$, $0{\cdot}01 = \frac{1}{100}$, $0{\cdot}001 = \frac{1}{1000}$.

Summary continued

3 *Fractions*

To work out $\frac{4}{5}$ of 80, find $\frac{1}{5}$ of 80 (= 80 ÷ 5 = 16), then $\frac{4}{5}$ of 80 = 16 × 4 = 64.

4 *Rounding to a given number of decimal places*

3·657 24 to 2 decimal places:

3·657 24 lies between 3·65 and 3·66.

3·65<u>7</u>24 rounds up to 3·66 since 7 (the figure in the third decimal place) is more than 5.

5 *Percentages* – with and without a calculator:

(a) 60% = 0·6 = $\frac{3}{5}$

(b) Without a calculator, work out 40% of £160.

since 10% of £160 = £16

then 40% of £160 = £16 × 4 = £64.

(c) Using a calculator, 40% of £160 = 0·40 × £160 = £64.

6 *Percentage calculations in context*

(a) If VAT is 17·5%, calculate £299·99 + VAT.

17·5% of £299·99 = 0·175 × £299·99 = £52·498 25 = £52·50 to the nearest 1p

Hence total = £299·99 + £52·50 = £352·49.

(b) Calculate the simple interest on £660 for 5 months at 4·2% per annum.

Interest for 1 year = 4·2% of £660 = 0·042 × £660 = £27·72.

Hence interest for 5 months = $\frac{5}{12}$ × £27·72 = £27·72 ÷ 12 × 5 = £11·55.

7 *One quantity as a percentage of another*

In a test worth 80 marks a student scores 52. What is his percentage score?

Percentage = $\frac{52}{80}$ × 100% = 52 ÷ 80 × 100% = 65%.

Summary continued

8 ***Percentage Profit***

An article costing £80 is sold for £96.

Hence profit = selling price − cost price = £96 − £80 = £16.

Hence profit as percentage of cost price = $\frac{16}{80} \times 100 = 20\%$.

9 ***Direct Proportion***

If 9 magazines cost £34·65, how much would 7 of these magazines cost?

9 magazines cost £34·65,

hence 1 magazine costs £34·65 ÷ 9 = £3·85,

and so 7 magazines cost £3·85 × 7 = £26·95.

2 Basic Geometry

Geometry is the branch of mathematics which is concerned with shapes and sizes. It deals with two-dimensional shapes (such as triangles, squares, pentagons, and circles), and also with three-dimensional shapes (such as cubes, cylinders, and cones). Some knowledge of geometry is vital to those working in areas like design or engineering. When you look back on all of your school mathematics, you may well think of geometry as the most attractive part of the subject.

The Calculation of Area

Area is the term we use to measure the size of a surface. We may need to know the area of a wall before buying wallpaper for it. The amount of turf we buy for a lawn is determined by the area of ground. Sometimes the area is which we are interested has a simple shape. At other times it may be a complex shape.

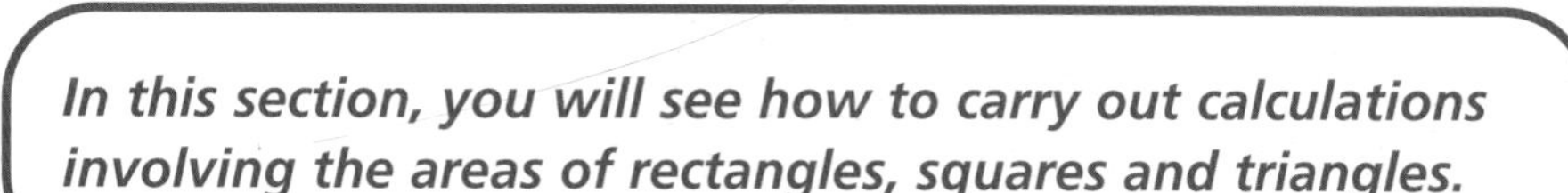

Remember

1 You must know the connections between metric units of length before you approach any problem on area

1 km = 1000 m; 1 m = 100 cm; 1 cm = 10 mm.

2 You should already know the following formulae:

area of a rectangle = length × breadth

area of a square = length × length

area of a triangle = $\frac{1}{2}$ × base × height.

From now on, we shall use *symbols* to write down these three formulae. Thus

area of rectangle: $A = lb$

area of a square: $A = l^2$

area of a triangle: $A = \frac{1}{2}bh$.

Example Calculate the area of the rectangle shown.

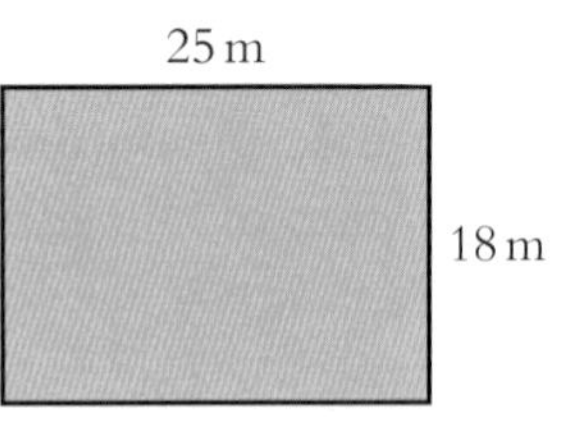

Solution $A = lb$

$= 25 \times 18$

$= 450$

So Area of rectangle = 450 square metres *or* 450 m^2.

Hints and tips

1 If you are given the area of a rectangle, you may be asked to work back to find the length or breadth. For example, if you were told that the area of a rectangle was 450 m^2 and its length was 25 m, you could work out the breadth by dividing. (Breadth = 450 ÷ 25 = 18 m.)

2 Do not mix up **area** with **perimeter**. The *perimeter* of a shape is the distance around its outside. If you were asked to find the perimeter of the above rectangle, you would **add** the lengths of its sides together. (Perimeter = 25 + 25 + 18 + 18 = 86 m.)

Example Calculate the area of the rectangle shown.

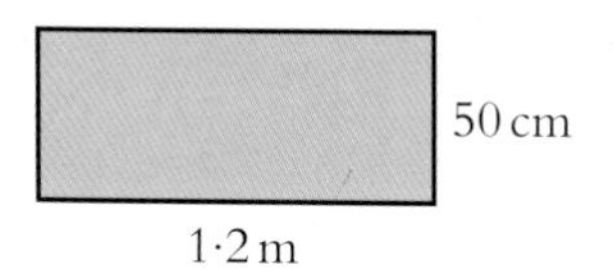

Solution As the units are different, we first change 1·2 m to cm.

Length = 1·2 m = $1{\cdot}2 \times 100$ = 120 cm.

Area, $A = lb = 120 \times 50 = 6000$.

So area of rectangle = 6000 cm^2.

(We could have changed 50 cm to m first. Then 50 cm = $50 \div 100 = 0{\cdot}5$ m, leading to area = $1{\cdot}2 \times 0{\cdot}5 = 0{\cdot}6$ m^2, which is equal to 6000 cm^2.)

Example Calculate the area of the triangle shown.

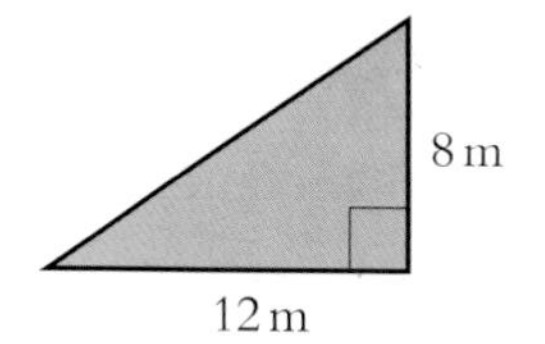

Solution $A = \frac{1}{2}bh = \frac{1}{2} \times 12 \times 8 = 48$

So area of triangle = 48 m^2.

(If you are using a calculator, you can calculate $\frac{1}{2} \times 12 \times 8$ either as $0{\cdot}5 \times 12 \times 8$ or as $12 \times 8 \div 2$. Check too that you could work out the answer without a calculator!)

Example Calculate the area of the triangle shown.

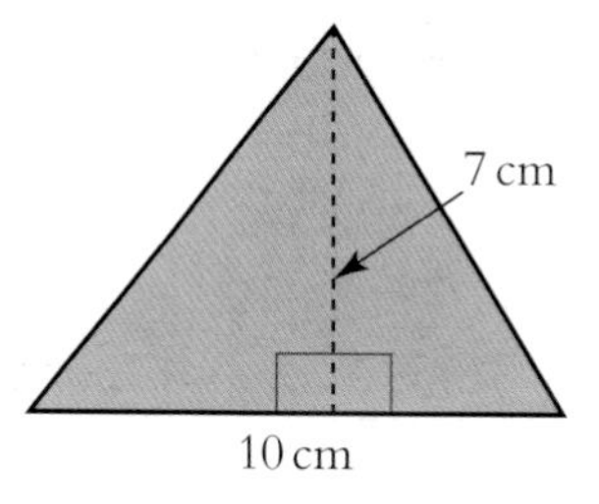

Solution $A = \frac{1}{2}bh = \frac{1}{2} \times 10 \times 7 = 35$

So area of triangle = 35 cm^2.

Exercise 1

1 Work out the area of each of the following four rectangles without using a calculator.

(a) 5 m, 7 m

(b)

(c)

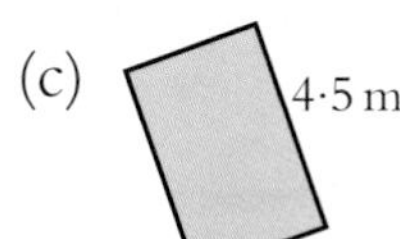

(d)

Exercise 1 continued

2 Work out the area of each of the following three squares.

(a) 13 mm

(b)

(c)

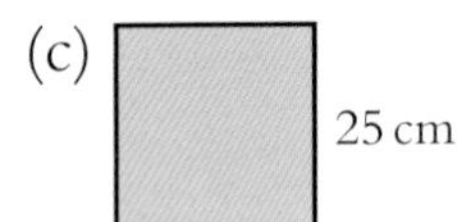

3 A table is 1·5 metres long and 80 centimetres broad.

Find its area in (a) square centimetres

(b) square metres.

4 (a) A rectangle has area 84 cm^2. If its length is 14 cm, calculate its breadth.

(b) A rectangle has area 100 cm^2. If it is 8 cm wide, calculate its length.

5 A farmer keeps his cows in a rectangular field.

The field measures 45 metres by 30 metres.

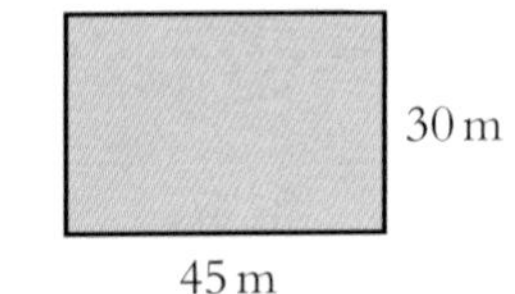

(a) Calculate the *perimeter* of the field.

(b) He puts a new fence around the field.
Find the cost of the fence if fencing costs £2·50 per metre.

6 Helen wishes to fit double-glazing in the windows of her flat.

She has six windows, each rectangular in shape and measuring 1·8 metres by 60 centimetres. Find the total area of **all** the windows in square metres.

7 Find the area of each of the following four triangles.

(a)

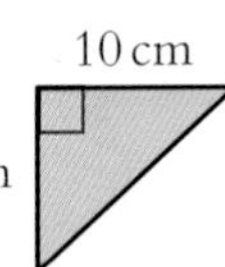

(b) 6 cm 13 cm

(c)

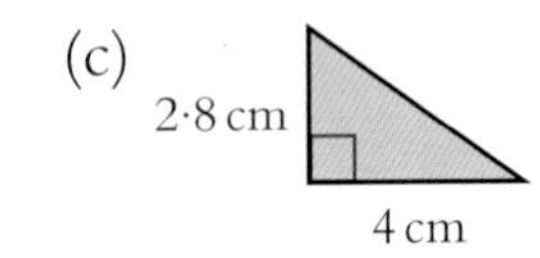

(d) 13 cm 15 cm

Exercise 1 continued

8 Copy and complete the following table for **rectangles**.

Length	Breadth	Area	Perimeter
6 cm	4 cm	24 cm^2	20 cm
7 cm	5 cm		
12 cm	9 cm		
10 cm		80 cm^2	
	7·5 cm	150 cm^2	
15 cm			54 cm

9 A badge in the shape of a triangle has been designed for Clydeview Amateur Football Club.

Calculate the area of the badge.

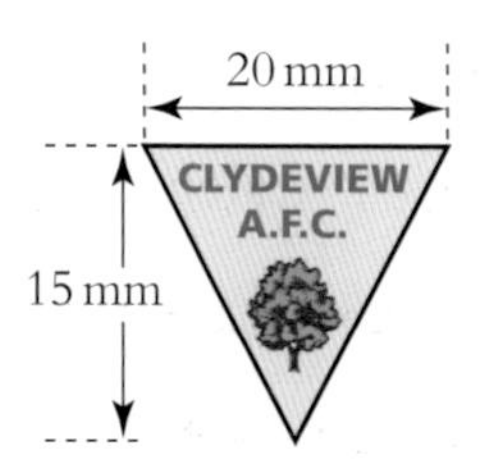

10 Find the area of each of the following four triangles.

(a) 6 cm, 10 cm

(b) 8 m, 11 m

(c)

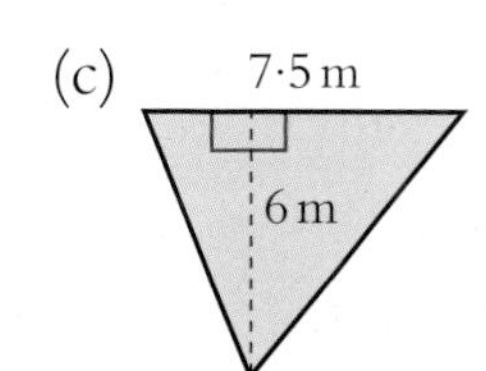

(d)

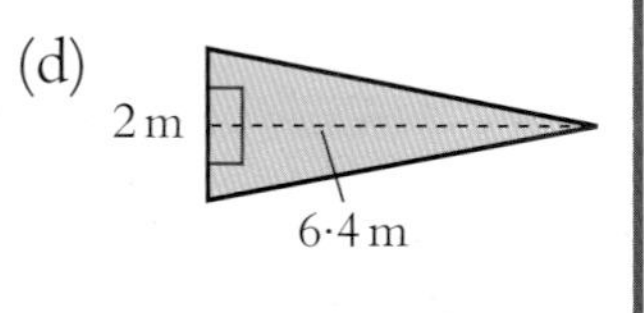

11 Copy and complete the following table for **squares**.

Length of side	Area	Perimeter
8 cm	64 cm^2	32 cm
7 cm		
11 cm		
		20 cm
	144 cm^2	

Exercise 1 continued

12 Mr. Smith has an allotment in the shape of a right-angled triangle.

The area of the allotment is 28 square metres. If the allotment is 8 metres long, as shown, find its width w.

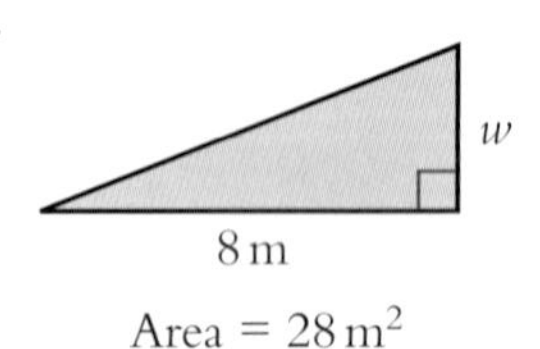

The areas of Composite Shapes

A composite shape is a shape made up of different parts. Its total area can be found by adding the areas of the different parts together. In this section you will calculate the areas of composite shapes which are composed of rectangles and right-angled triangles.

Example The diagram shows a plan of a dining room.

Calculate the area of the dining room.

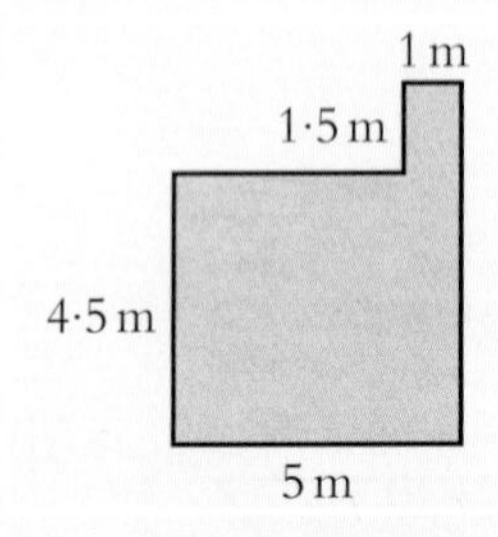

Solution We split the room into two rectangles P and Q, as shown.

1 m
1·5 m Q
4·5 m P
5 m

Area of rectangle P
$= lb = 5 \times 4{\cdot}5 = 22{\cdot}5\ \text{m}^2$

Area of rectangle Q
$= lb = 1{\cdot}5 \times 1 = 1{\cdot}5\ \text{m}^2$

Hence area of room $= (22{\cdot}5 + 1{\cdot}5)\text{m}^2 = 24\ \text{m}^2$.

In some examples, there may be a length missing. You will always be able to find any required length(s) however by **adding** or **subtracting** other lengths given in the diagram.

Example Scott has built a shed in his garden. A side view of the shed is shown in the diagram.

Calculate the area of the side of the shed.

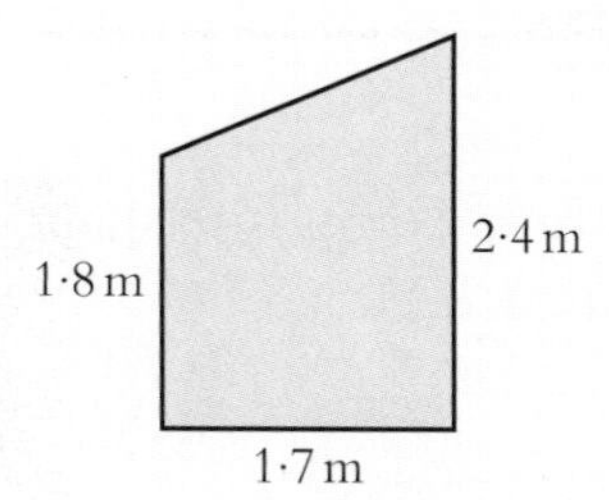

Solution We split the shape into a rectangle P and a right-angled triangle Q as shown.

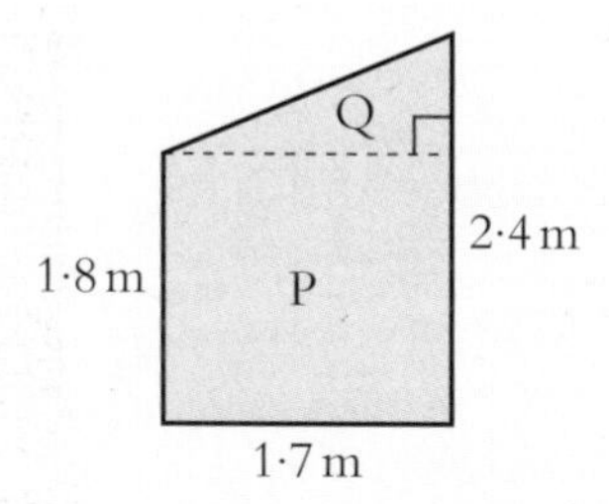

Note that we do not know the height of the right-angled triangle.

However it can be found by **subtraction**:

height of triangle = $(2{\cdot}4 - 1{\cdot}8)$ m = $0{\cdot}6$ m.

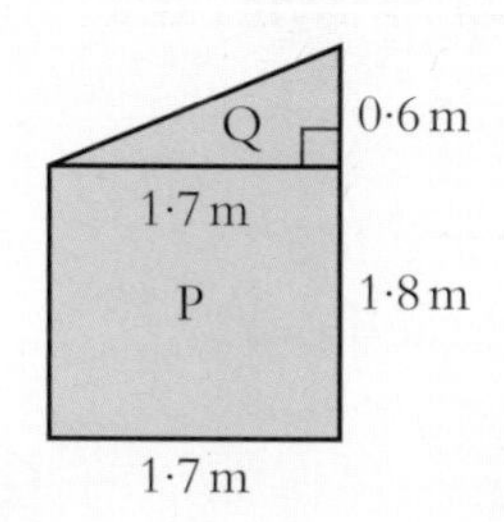

Area of rectangle P = $lb = 1{\cdot}8 \times 1{\cdot}7 = 3{\cdot}06\,\text{m}^2$.

Area of triangle Q = $\frac{1}{2}bh = \frac{1}{2} \times 1{\cdot}7 \times 0{\cdot}6 = 0{\cdot}51\,\text{m}^2$

Hence area of side of shed = $(3{\cdot}06 + 0{\cdot}51)\,\text{m}^2 = 3{\cdot}57\,\text{m}^2$.

Exercise 2

1 Calculate the area of each of the following three composite shapes.

(a)

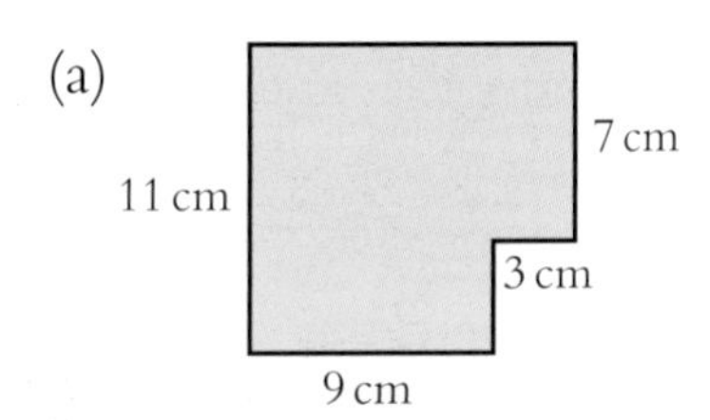

(b)

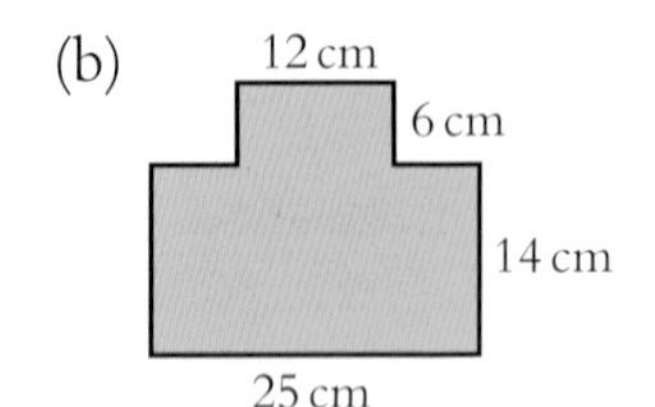

(c)

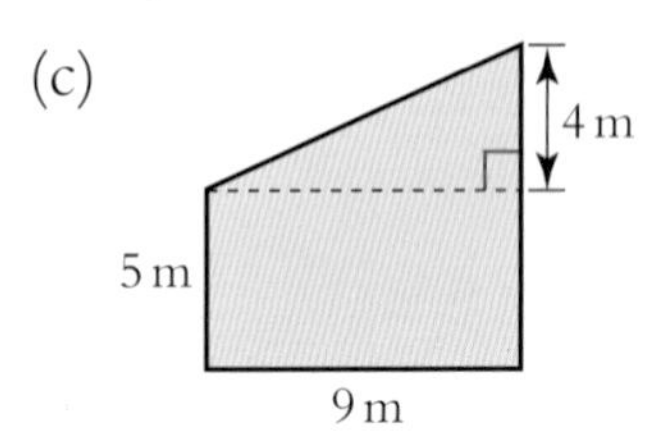

2 Calculate the area of each of the following three composite shapes. (You will have to calculate a 'missing side' in each.)

(a)

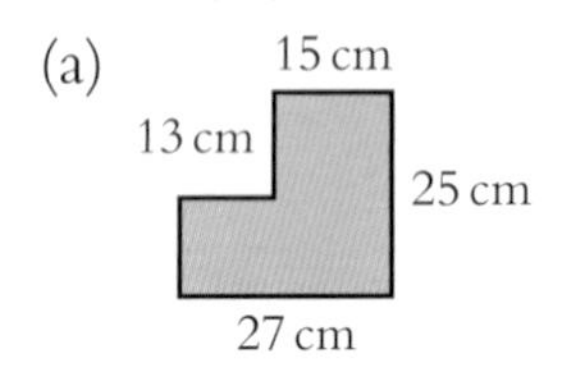

(b)

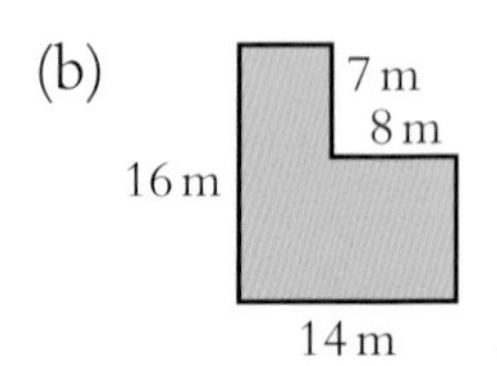

(c)

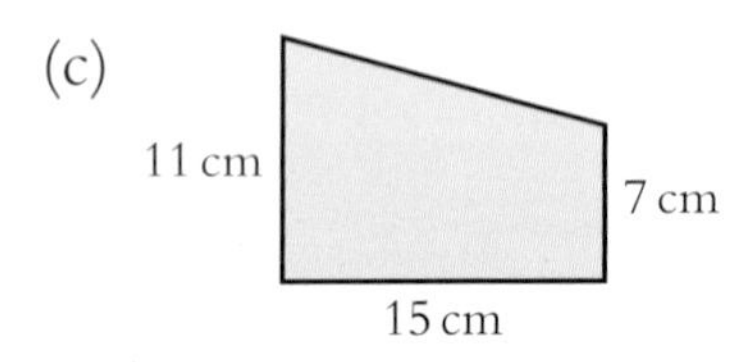

3 Andy is carpeting his bedroom. A plan of the area to be carpeted is as shown.

It will cost £25 per square metre to buy the carpet and have it fitted.

How much will it cost Andy in total?

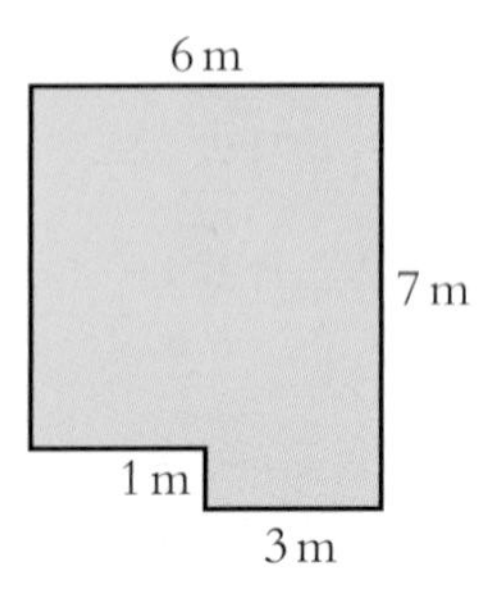

4 Andy decides to have his bathroom floor tiled. A diagram of the bathroom floor is as shown.

(a) Find the area of the bathroom floor.

The tiles Andy will use to cover the floor are square with side 50 centimetres.

(b) Find the area of one tile in square metres.

(c) How many tiles are needed to cover the whole floor?

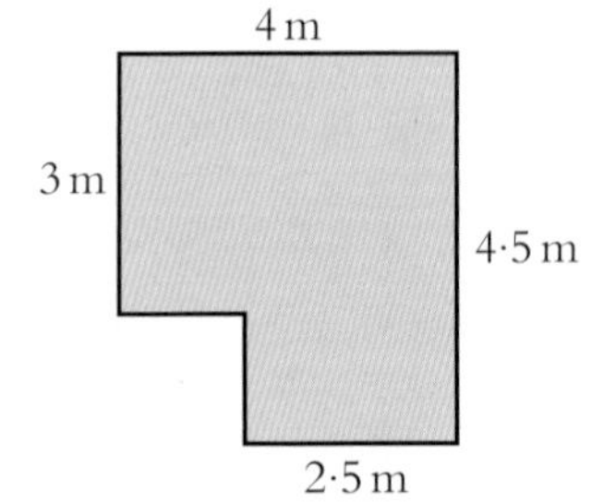

5 The diagram shows the gable end of a building.

Calculate the area of the gable end.

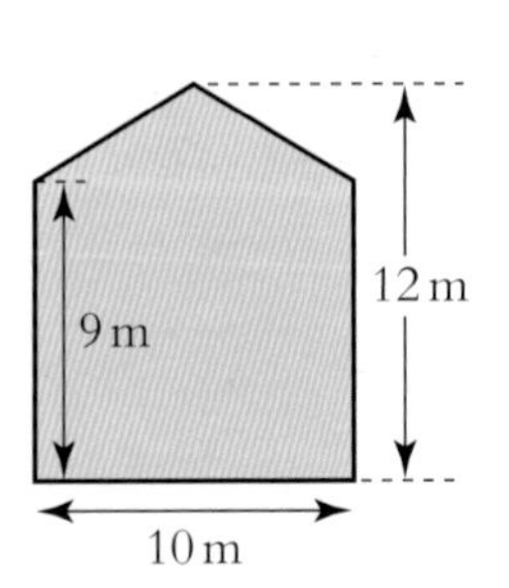

Subtracting Areas

Sometimes, when calculating the area of a composite shape, it is easier to subtract than to add. This is especially true when calculating the area of a border.

Example Anita puts a photograph of her favourite view on a card and places a wooden frame around the photograph as shown.

The photograph measures 15 centimetres by 10 centimetres.

The wooden frame is 5 centimetres wide all around.

Calculate the area of the wooden frame.

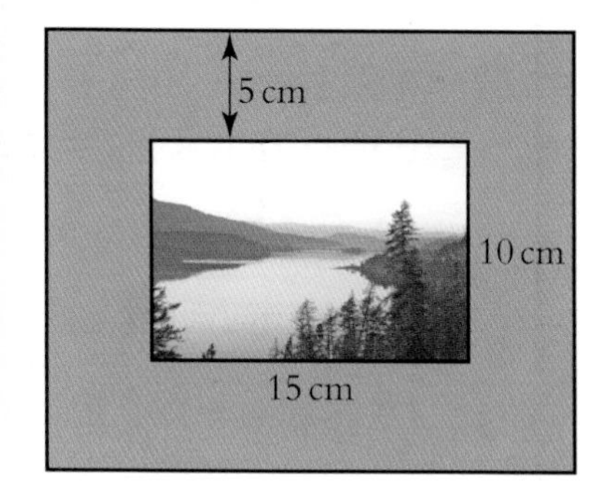

Solution

Length of wooden frame = $(15 + 5 + 5)$ cm = 25 cm.

Breadth of wooden frame = $(10 + 5 + 5)$ cm = 20 cm.

Area of outer rectangle = $lb = 25 \times 20 = 500\ \text{cm}^2$.

Area of inner rectangle (photograph) = $lb = 15 \times 10 = 150\ \text{cm}^2$.

Hence area of wooden frame = $(500 - 150)\ \text{cm}^2 = 350\ \text{cm}^2$.

Exercise 3

1 Find the area of each border (shaded) in the following diagrams.

(a)

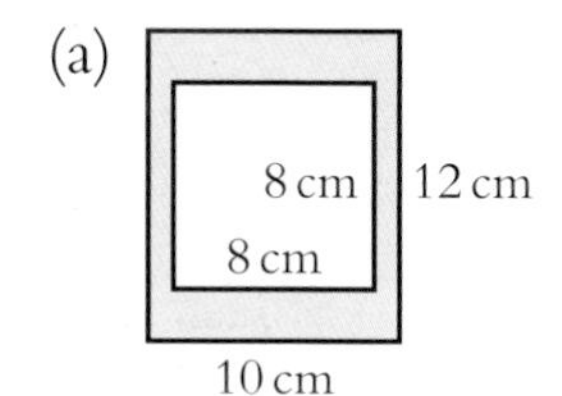

(b)

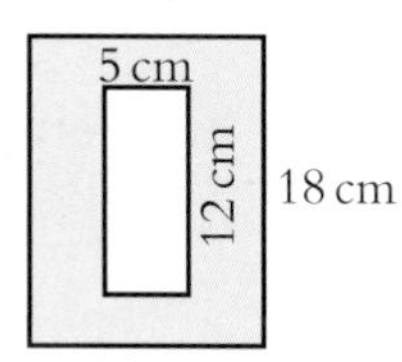

2 John is planning to wallpaper a rectangular wall in his study. He will not paper over the window or door. A sketch of the wall is shown.

Calculate the area to be papered.

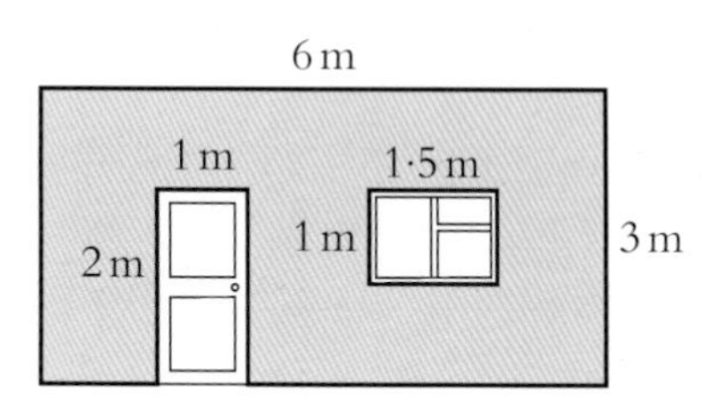

Exercise 3 continued

3 Derek is planning to paint the four walls of his sitting room. Each wall is rectangular.

The measurements are shown in the diagram. (He does not paint the window or door.)

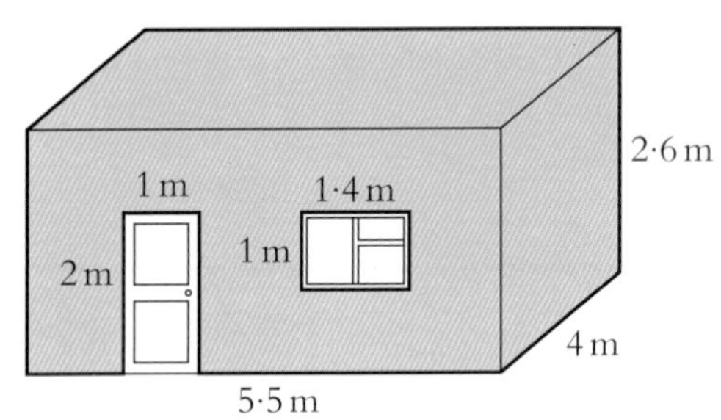

(a) Calculate the total area Derek has to paint.

(b) A tin of paint covers 12 m^2 and costs £9·59.

How much will it cost Derek to paint the sitting room?

4 Albert has ordered decking for the floor of his balcony. The dimensions of the balcony are as shown in the diagram.

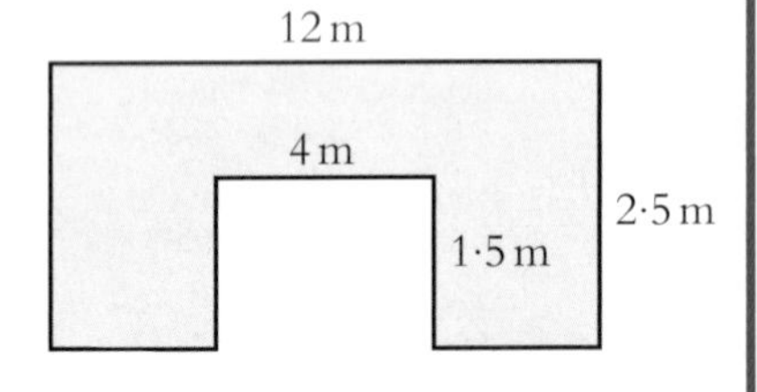

(a) Calculate the area of the decking required.

(b) If it costs £40 per square metre of decking, find the total cost to Albert.

5 The sketch shows a side view of a skip.

Calculate the area of the side of the skip.

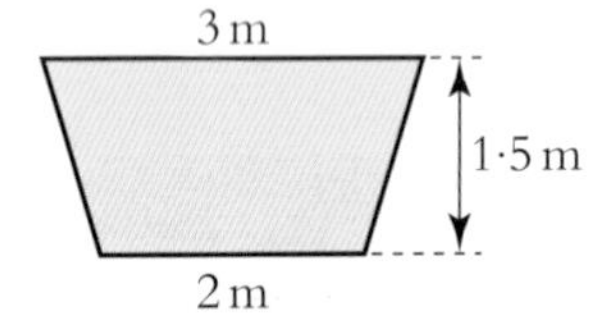

6 The diagram shows the shape and sizes of a dance floor in a hotel.

Calculate the area of the dance floor in two different ways:

(a) by addition

(b) by subtraction.

(Check that you get the same answer each time!)

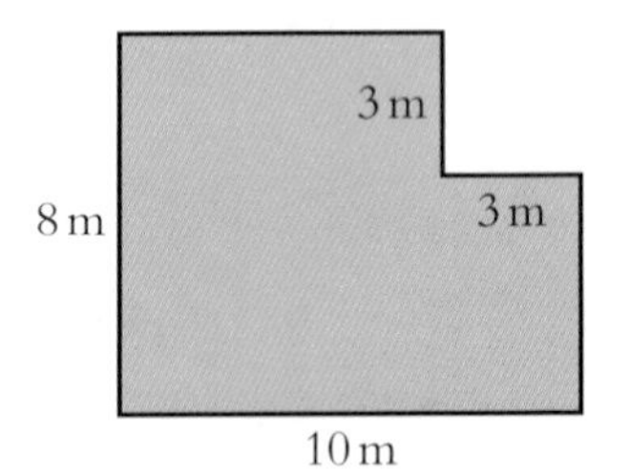

7 Peter's garden is in the shape of a rectangle. It consists of four square flower beds surrounded by a lawn. The measurements are as shown in the diagram.

Calculate the area of the lawn.

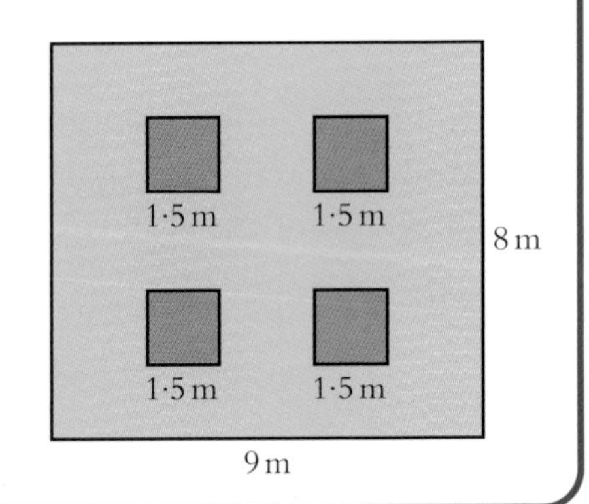

The Calculation of Volume

Volume is the term we use to measure the amount of *space* taken up by an object. The volume of a football is obviouly much larger than that of a golf ball.

In this section, you should be able to calculate the volume of a cube and the volume of a cuboid.

Remember

1 You may already be familiar with these formulae:

volume of a cuboid = length × breadth × height *or* $V = lbh$

volume of a cube = length × length × length *or* $V = l^3$.

2 (the units of volume) 1 cubic centimetre (cm^3) = 1 millilitre (ml)

1 litre (l) = 1000 cm^3.

Example An empty fish tank is in the shape of a cuboid with measurements as shown.

Calculate the volume of the fish tank in litres.

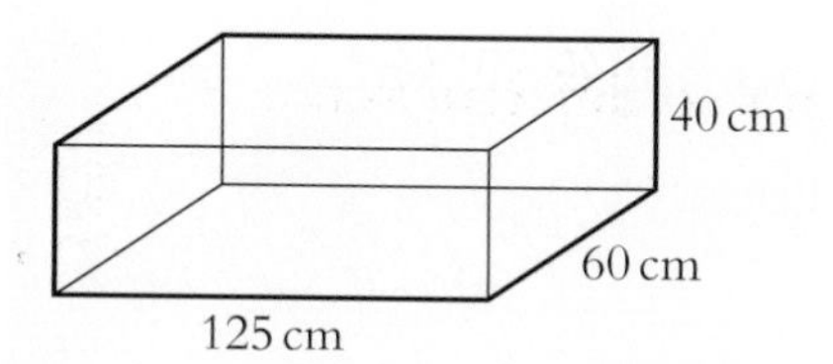

Solution $V = lbh = 125 \times 60 \times 40 = 300\,000\ \text{cm}^3$.

Hence volume in litres = $300\,000 \div 1000 = 300$ litres.

Exercise 4

Do not use a calculator for questions 1 and 2

1 Find the volume of each of the following two cuboids.

(a)

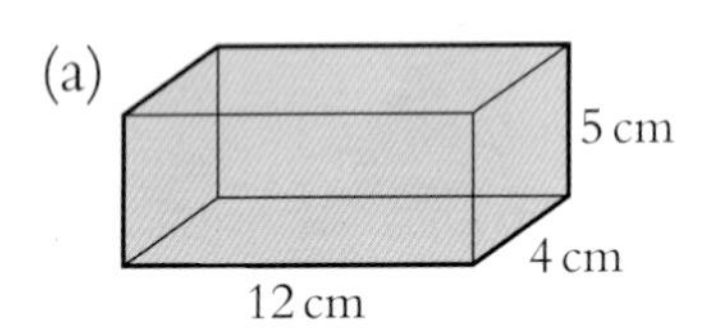

(b)

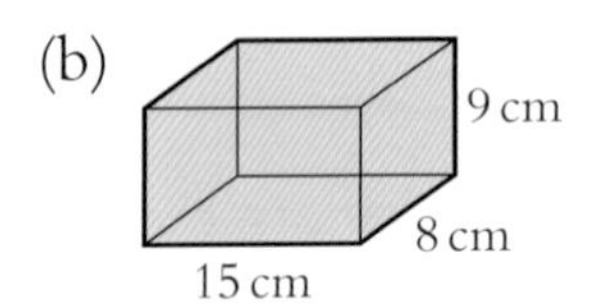

2 Find the volume of each of the two cubes.

(a)

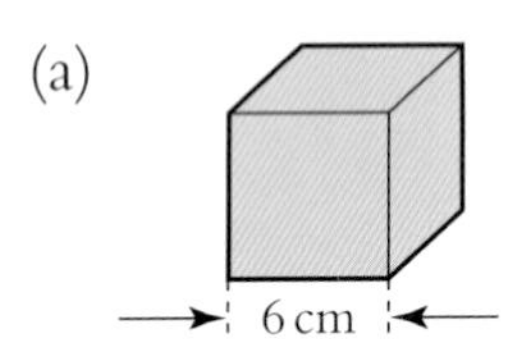

(b)

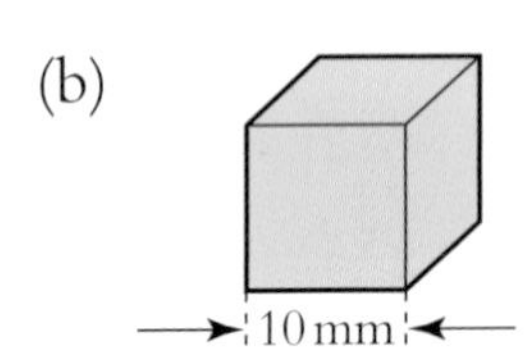

3 Find the volume of cuboids whose dimensions are as follows:

(a) length = 12 cm, breadth = 8 cm, height = 5 cm

(b) length = 18 mm, breadth = 16 mm, height = 15 mm

(c) length = 40 cm, breadth = 35 cm, height = 30 cm (answer in litres)

(d) length = 1·5 m, breadth = 80 cm, height = 50 cm (answer in litres).

4 The container for a 'Jack in the box' toy is in the shape of a cube.

The length of each side in the cube is 25 centimetres.

Find the volume of the container.

Give your answer to the nearest 1000 cubic centimetres.

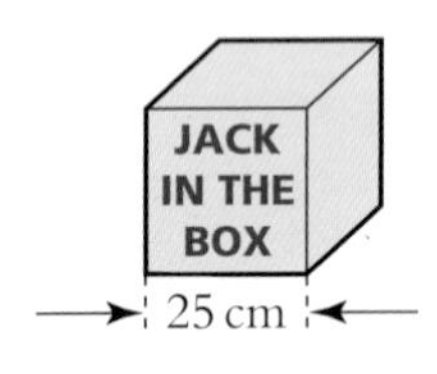

5 (a) Matchboxes are produced in cuboid shape as shown.

Calculate the volume of one matchbox.

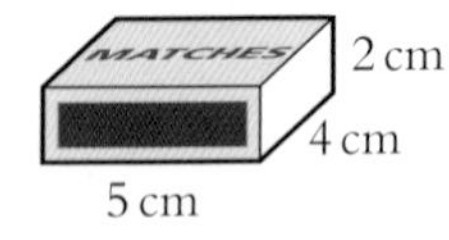

(b) The matchboxes are packed into a large box in the shape of a cube of side 20 centimetres.

What is the volume of this large box?

(c) How many matchboxes will fit into the large box?

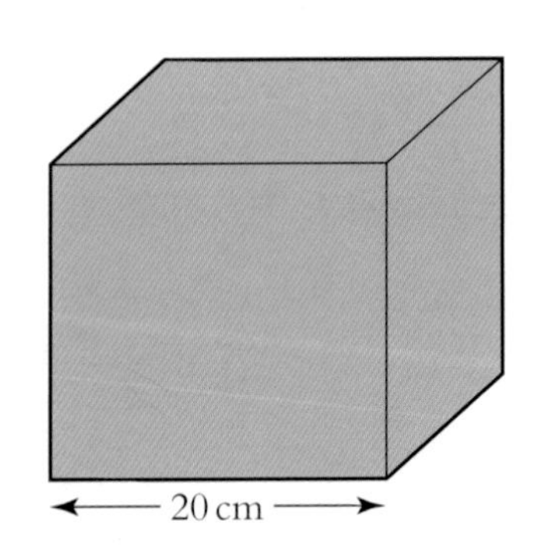

Exercise 4 continued

6 Heather has a set of building cubes. Each cube has side of length 2 centimetres.

(a) Find the volume of one building cube.

(b) The complete set of cubes fit into a large cubical box of side 10 centimetres.

Find the volume of the large box.

(c) How many building cubes does Heather have altogether?

7 Dermot has a biscuit tin in the shape of a cuboid.

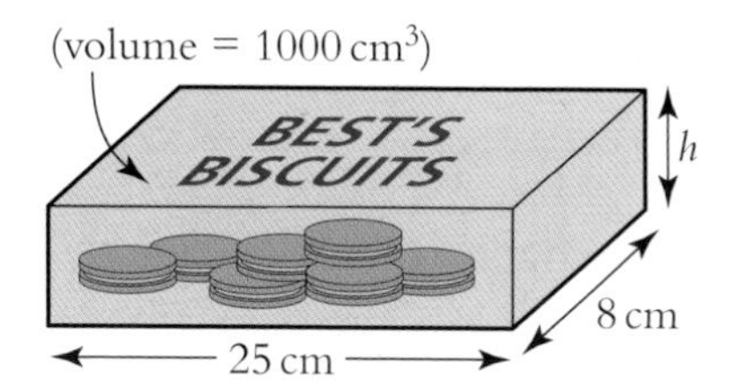

The tin has a volume of 1000 cubic centimetres. If the tin is 25 centimetres long and 8 centimetres broad, calculate its height h.

8 (a) A cuboid has volume 288 cm^3. It is 12 cm long and 4 cm high. Find its breadth.

(b) A vase in the shape of a cuboid has volume 945 ml. It is 15 cm high and 7 cm broad. Find its length.

(c) A cuboid has volume 30 m^3. It is 5 m long and 3 m broad. Find its height.

9 The diagram shows two packets of butter.

Each packet is in the shape of a cuboid.

(a) Find the volume of the packet of *DairyLo* butter.

(b) The *DairyLo* packet costs £1·50.

Find the cost per cubic centimetre of *DairyLo* butter.

(c) Find the volume of the *Country Health* packet.

(d) The *Country Health* packet costs £2·64. Which packet of butter gives better value for money? (Explain clearly the reason for your answer.)

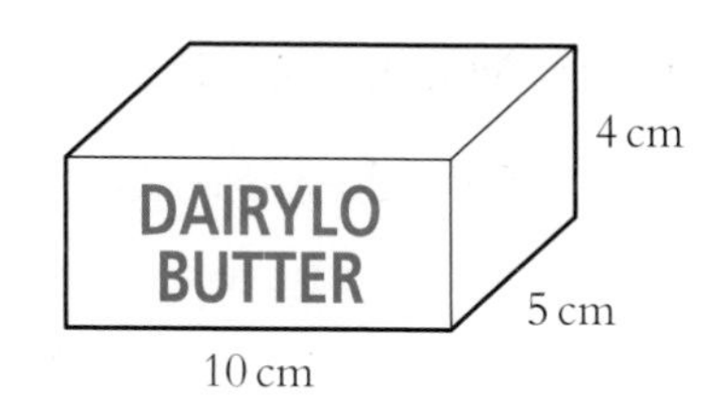

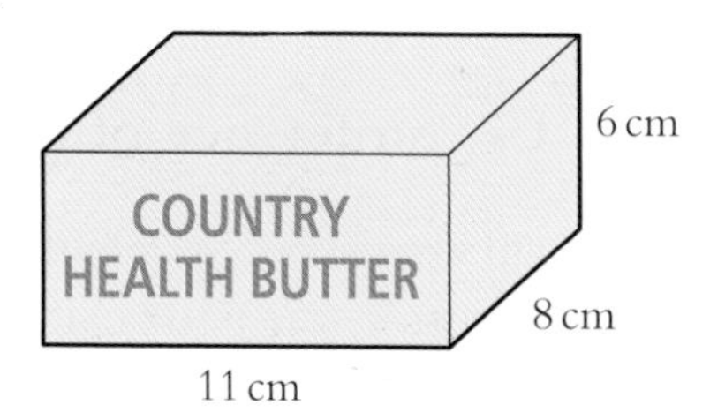

The Circle

The circle is perhaps the most fundamental shape of all, and we see examples of cirlces *many* times each day. In this section we are interested in the calculations of circumference and area of the circle.

The Circumference of a Circle

Key words and definitions

You are already familiar with the words 'radius' and 'diameter' when describing circles.

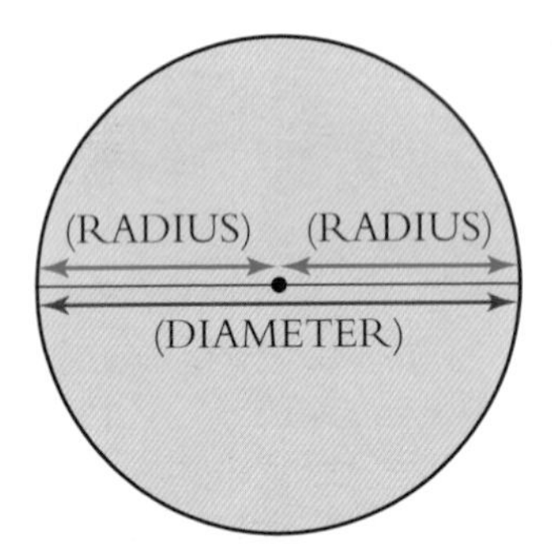

Remember that for every circle **Diameter = 2 × Radius**

The distance around the outside of a shape is usually called the *perimeter* of the shape. However, there is a special name for the distance around the outside of a circle. It is called the **circumference** of the circle.

- You can use a measuring tape to measure the circumference and diameter of some circular objects, e.g. tins, coins. You will find, in every example, that the circumference is just over 3 × the diameter of the circle.
- The **exact** number has been worked out accurately by mathematicians and, because it is such an important number, it has its own symbol, π, called pi.

  There is a button for π on your scientific calculator. Press it and you will find that

  $\pi = 3{\cdot}141\,592\,654.$
- This value is approximate as π continues forever. We usually say that the approximate value of π is 3·14.
- To find the circumference of a circle, we use the formula

  Circumference of a circle = π × diameter or $C = \pi d$

  (If you are asked to calculate the circumference of a circle, use the π button on your calculator. Use $\pi = 3{\cdot}14$ if your calculator does not have a π button or if you are asked to calculate the circumference without using a calculator.)

Example A circle has a diameter of 15 centimetres. Find its circumference.

(Give your answer correct to 1 decimal place).

Solution $C = \pi d$

$= \pi \times 15$

$= 47{\cdot}123\,889\,8$

Hence the circumference = 47·1 cm (to 1 decimal place).

Take care

The formula for circumference is $C = \pi d$. It uses the *diameter* of the circle. So be careful!

If you are given the *radius* of the circle, you must first **double** it to find the diameter before you can use the formula $C = \pi d$.

Example The radius of a circle is 3·8 cm. Find the circumference of the circle.

Solution Since $r = 3{\cdot}8$, then $d = 2 \times 3{\cdot}8 = 7{\cdot}6$.

$C = \pi d = \pi \times 7{\cdot}6 = 23{\cdot}876\,104\,17$

Hence the circumference = 23·9 cm (approximately).

Exercise 5

1 Find the circumference of a circle with diameter:

(a) 10 cm (b) 25 cm (c) 8 mm (d) 4·9 m.

(Give your answers correct to 1 decimal place.)

Exercise 5 continued

2 Find the circumference of a circle with radius:

(a) 6 cm (b) 20 cm (c) 6·5 cm (d) 25 cm.

(Give your answers correct to the nearest centimetre.)

3 The radius of planet Earth is approximately 6400 kilometres.

Find the circumference of the Earth.
(Give your answer to the nearest 100 kilometres.)

4 Find the circumference of each of the following three circles.

(a)

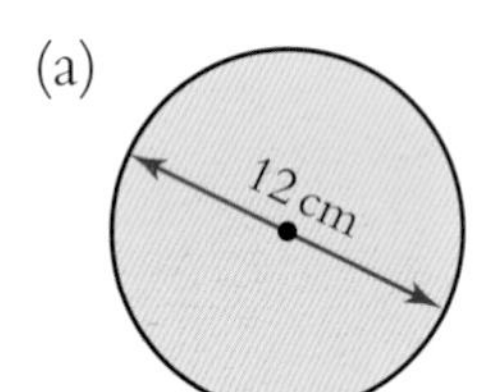

(b)

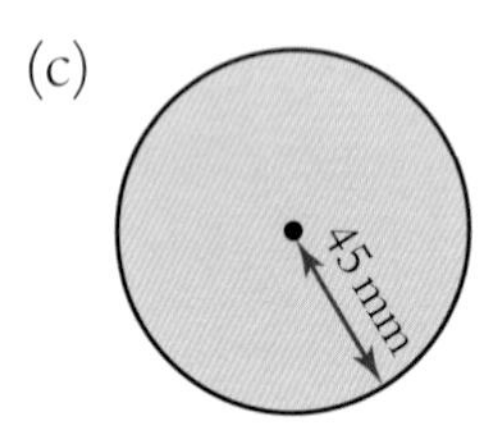

(c)

45 mm

(Give your answers correct to 1 decimal place.)

5 The minute hand on a wall clock is 7 centimetres long. What distance will the tip of the minute hand travel in 1 hour? (Give your answer to the nearest centimetre.)

6 A tyre has diameter 596 millimetres. Find the circumference of the tyre. (Give your answer to the nearest millimetre.)

7 Jock the groundsman has to mark the centre circle on a football pitch.

If the radius of the centre circle is 10 yards, what is the circumference of the circle? (Give your answer correct to 1 decimal place.)

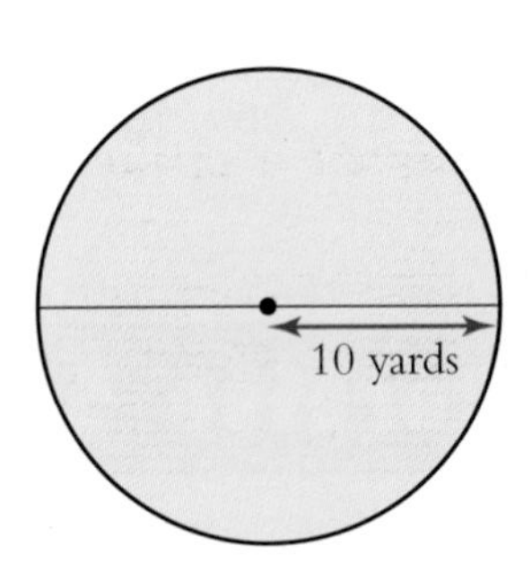

Exercise 5 continued

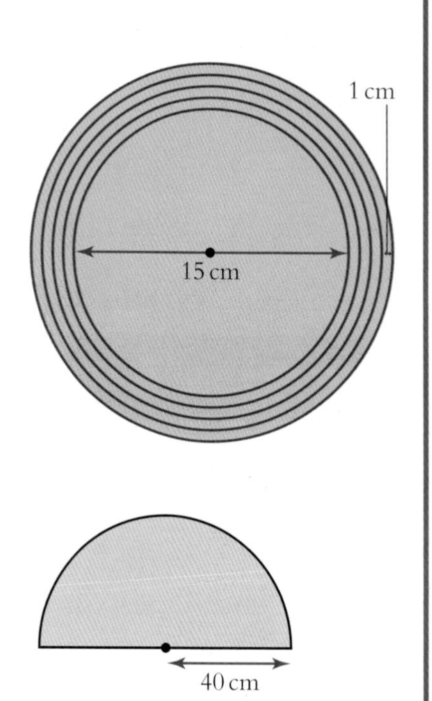

8 The diameter of an oak tree is 15 centimetres.

(a) What is the circumference of the tree?

Each year the radius increases by 1 centimetre.

(b) Calculate the circumference of the tree in

(i) 4 years;

(ii) 10 years.

(Give your answers to the nearest centimetre.)

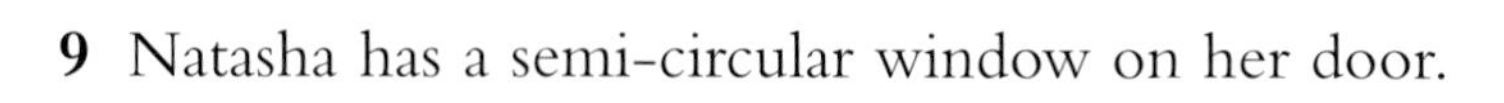

9 Natasha has a semi-circular window on her door.

The window has radius 40 centimetres.

Calculate the **perimeter** of the window. (Give your answer to the nearest centimetre.)

The Area of a Circle

The area of a circle may seem a difficult idea since the circle has no straight edges. However mathematicians have found a formula for calculating the area of a circle.

It also uses the number π, and it is:

Area of a circle: $A = \pi r^2$

Using this formula you should be able to calculate the area of any circle whose radius you know.

Example Calculate the area of a circle of radius 25 centimetres.

Solution $A = \pi r^2 = \pi \times 25^2 = 1963$

Hence the area is $1963\,\text{cm}^2$.

(If however you know the *diameter* of a circle, you must first **halve** it to find the radius before you can find the area of the circle.)

Example Calculate the area of a circle of diameter 35 metres.

Solution Radius, $r = 35 \div 2 = 17{\cdot}5\,\text{m}$

Hence $A = \pi r^2 = \pi \times 17{\cdot}5^2 = 962$

Hence the area of the circle is $962\ \text{m}^2$.

Exercise 6

1 Calculate the area of a circle with radius:

(a) 10 cm (b) 12 cm (c) 6·5 m (d) 20 mm.

(Give your answers correct to 1 decimal place.)

2 Calculate the area of a circle with diameter:

(a) 8 cm (b) 12 cm (c) 14 cm (d) 45 cm.

(Give your answers to the nearest square centimetre.)

3 The radius of the centre circle on a football pitch is 10 yards. Find the area of the centre circle. (Give your answer to the nearest square yard.)

4 Find the area of each of the following three circles.

(a)

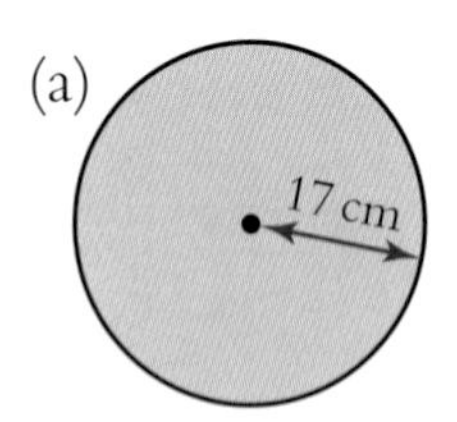

(b)

13·8 cm

(c)

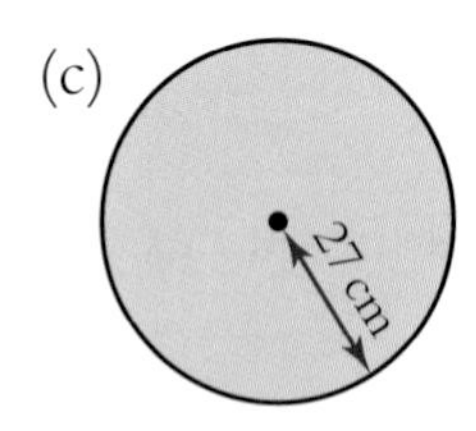

(Give your answers to the nearest square centimetre.)

5 Caitlin has a circular swimming pool in her garden. It has a diameter of 7 metres. Calculate the area of her swimming pool to the nearest square metre.

6 Jacqueline has bought a table. The top is circular with radius 75 centimetres.

Find the area of the top of her table. (Give your answer to the nearest 100 square centimetres.)

Exercise 6 continued

7 Andy uses a circular mirror for shaving. The diameter of the mirror is 15 centimetres. Calculate the area of the mirror to the nearest square centimetre.

8 Louis has a toy railway. It has a circular track with outer radius 42 centimetres.

Calculate (a) the area of the outer circle to the nearest $10\,\text{cm}^2$

(b) the circumference of the outer circle to the nearest cm.

9 The shooting area on a hockey pitch is a **semi-circle** of radius 14·63 metres.

Calculate the area of the shooting area to the nearest square metre.

Some advanced questions

We have already met composite shape problems in which rectangular or square parts have been added or subtracted. Here we extend these problems to include circular parts.

Example A bathroom rug is in the shape of a rectangle with a semi-circle removed as shown in the diagram.

Calculate the area of the rug.

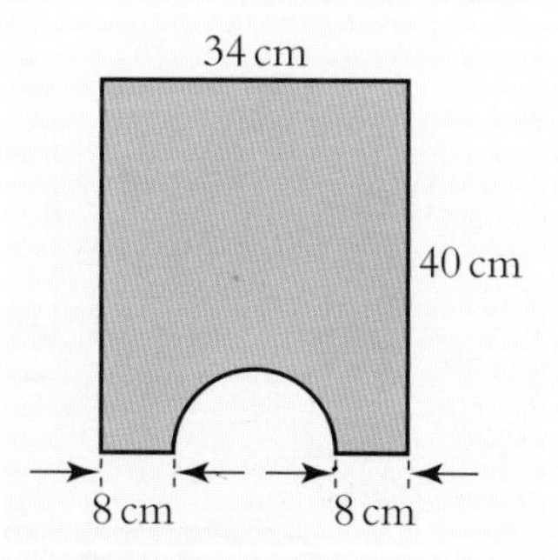

Solution

Area of complete rectangle $A_1 = lb = 40 \times 34 = 1360.$

Diameter of circle $= 34 - 8 - 8 = 18,$

so radius of circle $= 18 \div 2 = 9.$

Area of semi-circle $A_2 = \frac{1}{2}\pi r^2 = 0{\cdot}5 \times \pi \times 9^2 = 127.$

Hence area of rug $= (1360 - 127)\,\text{cm}^2 = 1233\,\text{cm}^2.$

Exercise 7

(Be careful when doing this exercise. Check that you are using the correct measurement in any circle calculations – the radius or the diameter. Watch out too for semi-circles!)

1 A church window is in the shape of a rectangle with a semi-circular top. Its sizes are shown in the diagram.

Calculate the total area of the window.

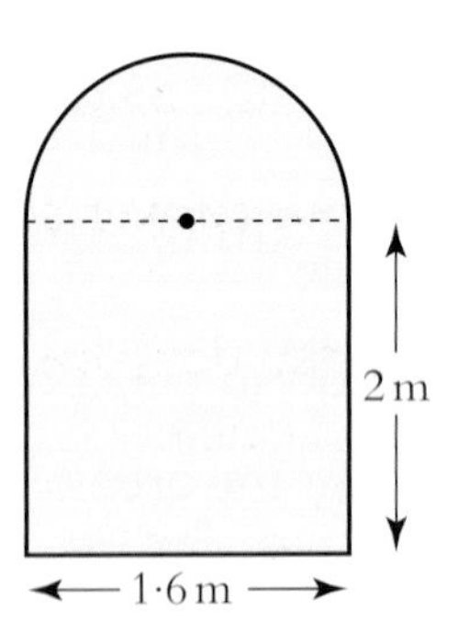

2 Percy has a garden in the shape of a square of side 5 metres. It consists of a circular flower bed of diameter 2 metres surrounded by a grass lawn.

Calculate the area of the grass lawn.

(Give your answer correct to one decimal place.)

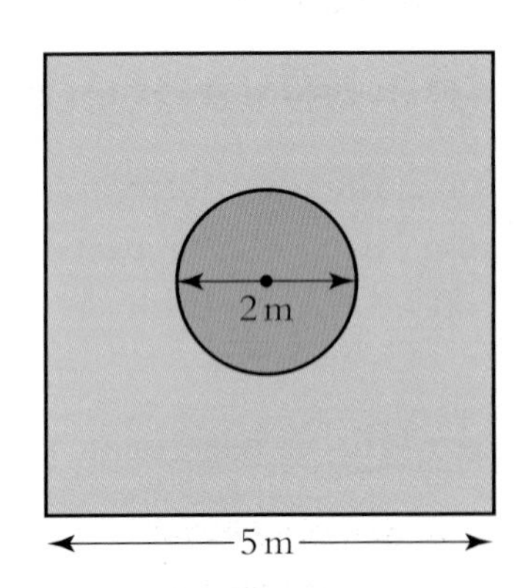

3 A road sign is in the shape of a rectangle with a semi-circle on top. The diagram shows the sizes of the sign.

Find the total area of the road sign. (Give your answer to the nearest square centimetre.)

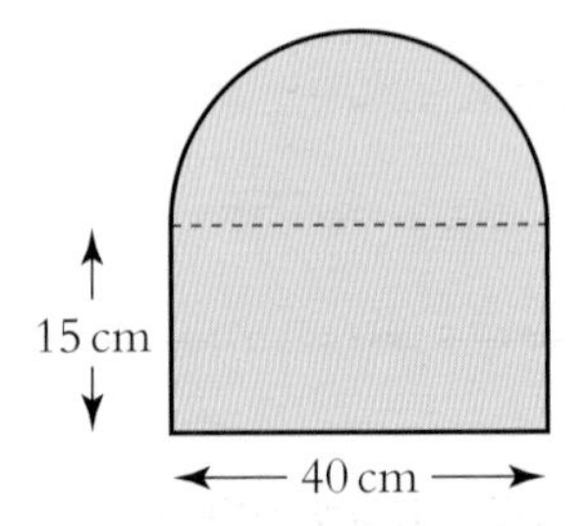

4 The sports field at a school consists of a rectangle 90 metres by 60 metres, with semi-circular ends, as shown in the diagram.

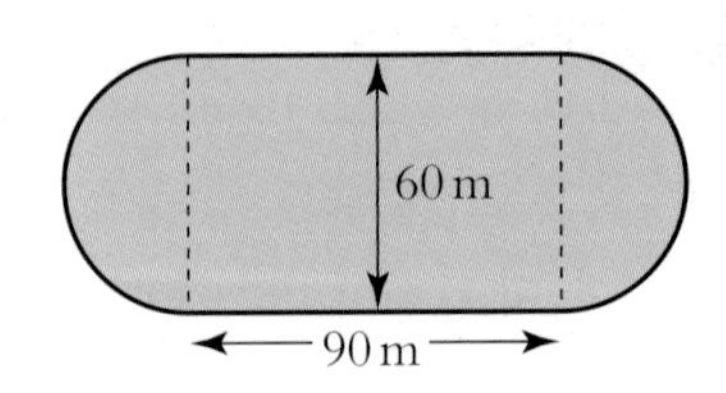

On a training exercise, Darren runs 8 times around the outside of the sports field.

How far does he run? (Give your answer to the nearest ten metres.)

Exercise 7 continued

5 The diagram shows a bridge over a tunnel.

The structure is rectangular in shape and with a semi-circular space for the tunnel.

Calculate the area of the bridge structure.

(Give your answer correct to one decimal place.)

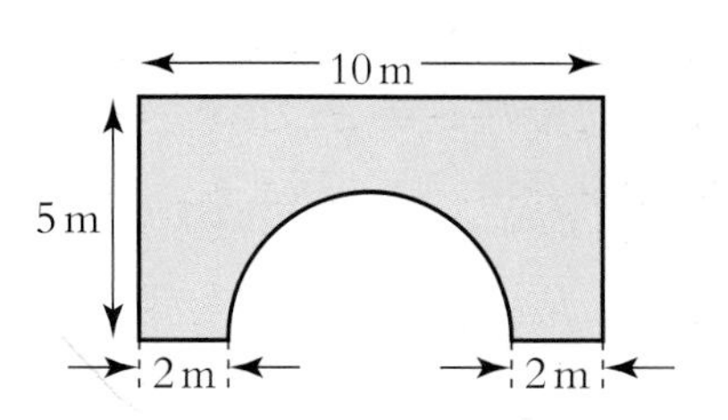

6 A wooden door has a window in the shape of a semi-circle.

The sizes of the door are shown in the diagram.

Find the area of wood on the front of the door.

(Give your answer correct to one decimal place.)

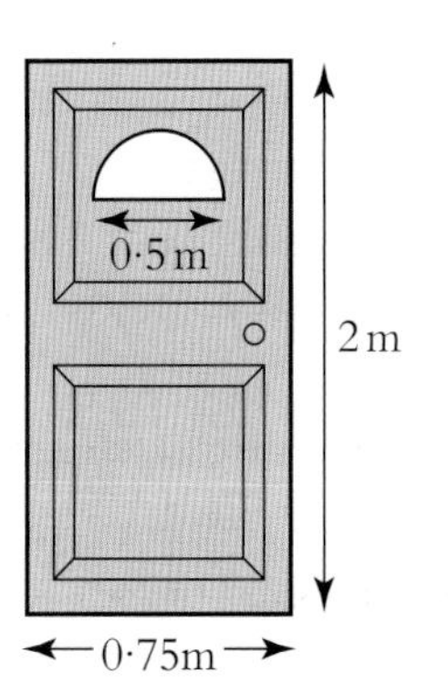

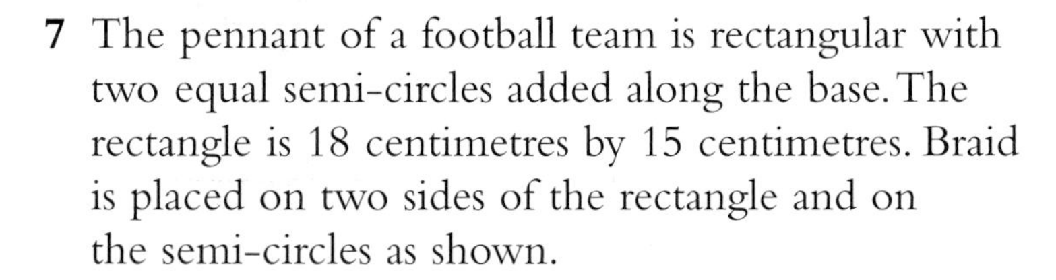

7 The pennant of a football team is rectangular with two equal semi-circles added along the base. The rectangle is 18 centimetres by 15 centimetres. Braid is placed on two sides of the rectangle and on the semi-circles as shown.

Find the length of braid needed for the pennant.

(Give your answer correct to two decimal places.)

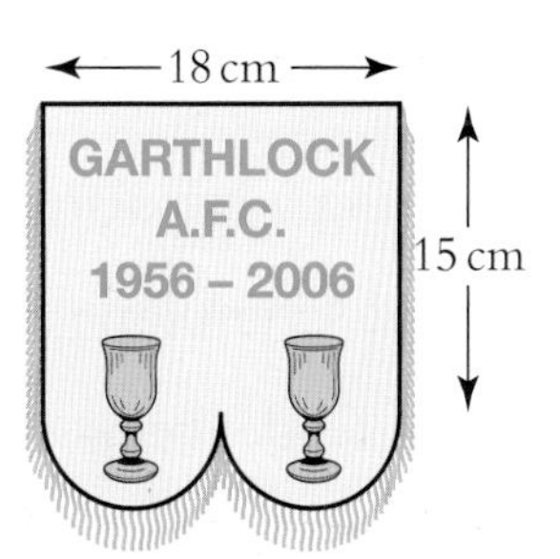

Exercise 8 – Revision of Chapter 2

1 A rectangle is 14 metres long and 10 metres broad.

Calculate (a) its area

(b) its perimeter.

Exercise 8 – Revision of Chapter 2 continued

2 A rectangle has an area of $60\,\text{cm}^2$. If its length is 8 cm, find its breadth.

3 A stamp is in the shape of a square of side 22 millimetres. Find its area.

4 A sketch of Charlie's garden is shown in the diagram.

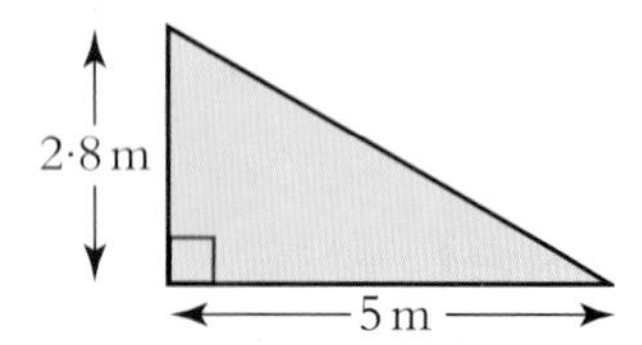

Calculate the area of the garden.

5 By adding or subtracting, find the area of each of the following shapes.

(a)

(10 cm)
(11 cm)
(5 cm)
(19 cm)

(b)

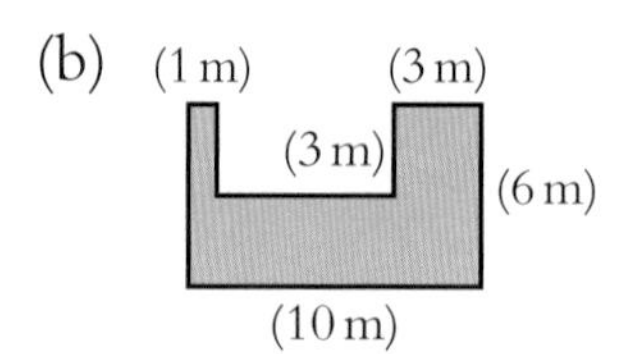

6 Without using a calculator, find the volume of:

(a) a cuboid 13 cm long, 7 cm broad and 3 cm high

(b) a cube of side 7 cm.

7 A cuboid is 50 cm long, 15 cm broad and 32 cm high.

Calculate its volume. (Give your answer in litres.)

8 A cuboid has a volume of $450\,\text{cm}^3$. It is 20 cm long and 5 cm high.
Calculate its breadth.

Exercise 8 – Revision of Chapter 2 continued

9 A circle has diameter 18 cm.

Calculate (a) its circumference, correct to one decimal place

(b) its area, to the nearest square centimetre.

10 A fireplace is in the shape of a rectangle with a semi-circle on top. The sizes are shown in the diagram.

A thin brass edge is to be placed around the fireplace as indicated in the diagram.

Calculate the length of the brass edge. (Give your answer correct to 1 decimal place.)

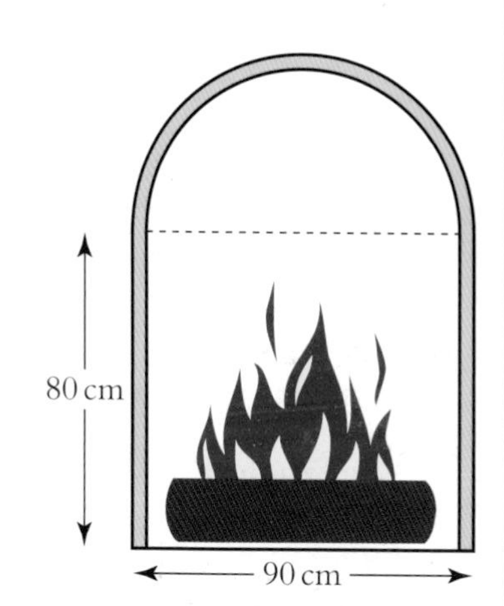

11 The cross-section of a water channel is shown in the diagram.

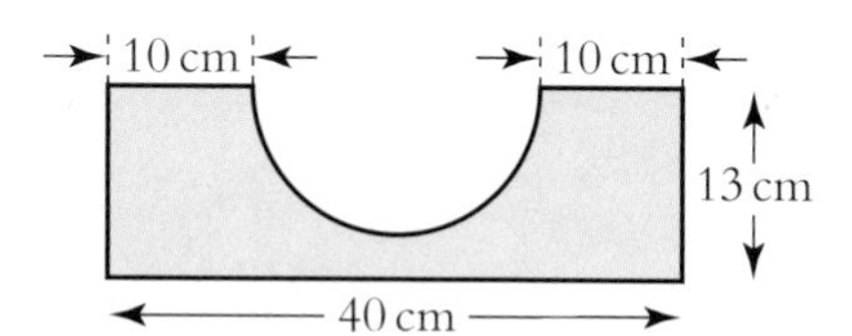

Find the area of the cross-section. (Give your answer to the nearest cm^2.)

12 An octagonal floor is to be covered in tiles. The sizes of the floor are shown in the diagram. Each tile is a square of side 30 centimetres. The tiles may be cut in half.

How many tiles will be needed to cover the octagonal floor?

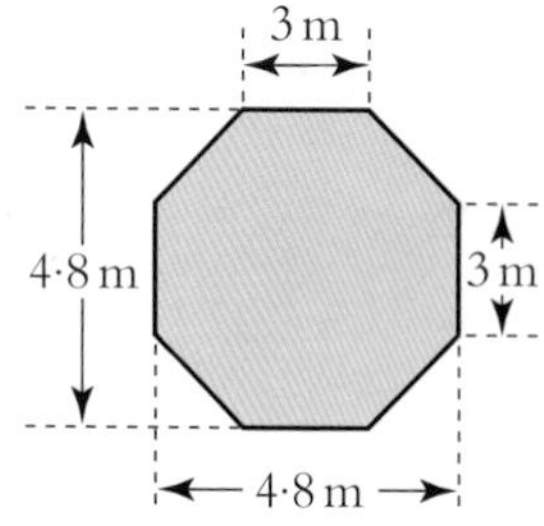

Summary

(You should check that you understand each of the five items in this Summary before moving on. If you find difficulty, repeat some of the questions in the Exercises in this chapter.)

1 ***Basic areas*** (Memorise these important formulae!)

Area of a rectangle $A = lb$

Area of a square $A = l^2$

Area of a triangle $A = \frac{1}{2}bh$

2 ***Composite Shapes*** (The area of a composite shape may be found by adding or subtracting areas.)

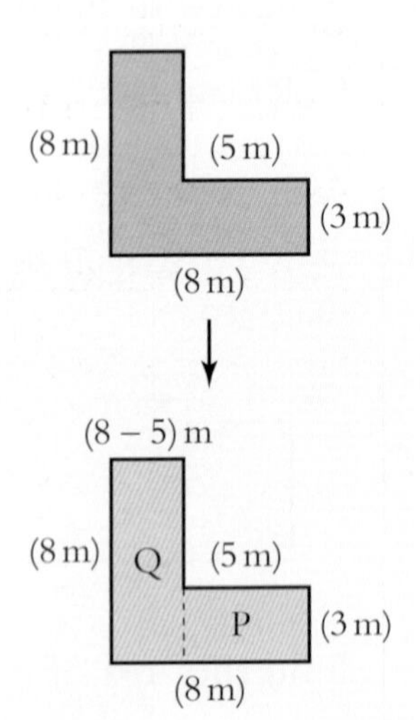

For example, to find the area of this L-shaped room.

Rectangle P: Area $= lb = 5 \times 3 = 15$

Rectangle Q: Area $= lb = 8 \times (8 - 5) = 8 \times 3 = 24$

Area of room $= (15 + 24)\,\text{m}^2 = 39\,\text{m}^2$.

(Check that you are able to get the same answer by 'subtraction'.)

3 ***Basic volumes*** (Memorise these important formulae!)

Volume of a cuboid: $V = lbh$

Volume of a cube: $V = l^3$

Thus to find the volume in litres of a cuboid measuring 45 cm by 28 cm by 20 cm:

$V = lbh = 45 \times 28 \times 20 = 25\,200\,\text{cm}^3$.

Hence volume $= (25\,200 \div 1000)$ litres $= 25{\cdot}2$ litres.

Summary continued

4 You should be able to calculate the circumference and area of a circle using the formulae:

$C = \pi d$ and $A = \pi r^2$

For example, to find the circumference and area of a circle of radius 12 metres:

Circumference $C = \pi d = \pi \times 24 = 75{\cdot}4\,\text{m}$. $(d = 2 \times 12 = 24)$

Area $A = \pi r^2 = \pi \times 12^2 = 452\,\text{m}^2$.

5 You should also be able to do calculations with composite shapes involving rectangles and semi-circles.

For example to find the area of this shape:

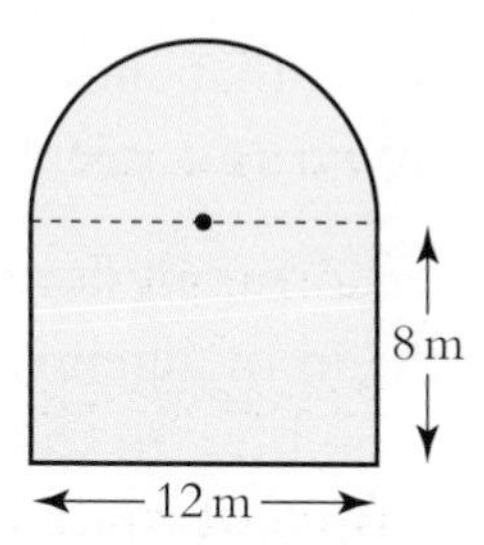

Area of rectangle $= lb = 12 \times 8 = 96$.

Area of semi-circle $= \frac{1}{2}\pi r^2 = 0{\cdot}5 \times \pi \times 6^2 = 56{\cdot}5$.

Hence total area of shape $= 96 + 56{\cdot}5 = 152{\cdot}5\,\text{m}^2$.

There are examples of geometry all around you

3 Expressions and Formulae

While Geometry is often regarded as the most attractive part of a mathematics course, Algebra is sometimes seen as the least attractive.

This is a pity because algebra is a kind of secret language in the same way that shorthand is. By using algebraic symbols we can usually write a problem in the shortest and clearest way!

Expressions and Formulae are only laws expressed in the shorthand of algebraic symbols. Imagine trying to explain to someone how to calculate the volume of a cone if you didn't know the formula
$V = \frac{1}{3}\pi r^2 h$!

Equations are straightforward too, because they are there to show that quantities are equal. A statement such as 'three times a certain number added to five is equal to twenty-six' can be written simply as $3n + 5 = 26$.

Once you become familiar with algebraic symbols and their use, this part of mathematics should become less frightening!

Evaluating Algebraic Expressions

An algebraic expression is a term such as $5a$, or collection of terms, such as $5a + 2b$. You should be able to evaluate an expression, and so you must replace the letters a, b, by numbers. (The letters in an expression are called *variables*.)

Algebraic expressions occur frequently in mathematics. In the last chapter we saw the expressions πd and πr^2. Expressions like these provide us with a general method of solving any problem (in this case involving circumference and area of a circle).

Remember

$5a$ means $5 \times a$; a^2 means $a \times a$; ab means $a \times b$;

$\frac{a}{b}$ means $a \div b$; $2a^2$ means $2 \times a^2 = 2 \times a \times a$.

Example If $e = 5$ and $f = 3$, evaluate:
(a) $4e$ (b) ef (c) $2e^2$.

Solution (a) $4e = 4 \times 5 = 20$

(b) $ef = 5 \times 3 = 15$

(c) $2e^2 = 2 \times e^2 = 2 \times 5 \times 5 = 2 \times 25 = 50$.

Some expressions are more difficult to deal with, and may involve addition, subtraction, multiplication, or division of terms. Other expressions will involve terms in brackets. There is a 'correct order' for evaluating these more difficult expressions and it is as follows:

Technique

The correct order for dealing with algebraic expressions is:

1 Brackets

2 'Of'

3 Multiplication/Division

4 Addition/Subtraction.

You may wish to use the mnemonic 'BOMDAS' to help you remember the correct order.

Example If $e = 5$ and $f = 3$, evaluate:
(a) $4e - 2f$ (b) $7(e + f)$ (c) $e + \frac{2}{3}f$.

Solution (a) $4e - 2f = 4 \times 5 - 2 \times 3 = 20 - 6 = 14$

(b) $7(e + f) = 7 \times (5 + 3) = 7 \times 8 = 56$

(c) $e + \frac{2}{3}f = 5 + \frac{2}{3}$ of $3 = 5 + 2 = 7$.

Exercise 1

Do not use a calculator 

1 If $a = 5$, evaluate:

(a) $3a$ (b) $a + 6$ (c) $7a + 3$

(d) $2a - 4$ (e) a^2 (f) $4 + 3a$.

2 If $m = 10$, evaluate:

(a) $7m$ (b) $3m - 11$ (c) $2m + 4$

(d) $5m - 8$ (e) m^2 (f) $75 - 2m$.

3 If $p = 7$ and $q = 2$, evaluate:

(a) $p + q$ (b) $p - q$ (c) pq

(d) $4p - q$ (e) $4q - p$ (f) $p^2 - q^2$.

4 If $x = 3$ and $y = 1$, evaluate:

(a) $x - y$ (b) xy (c) $5x + y$

(d) $x^2 + y^2$ (e) $4x - 2y$ (f) $12x + 4y$.

5 If $u = 12$ and $v = 11$, evaluate:

(a) $2u + v$ (b) uv (c) $5v - u$

(d) $4u + 3v$ (e) $u^2 - v$ (f) $v^2 - 2u$.

6 If $a = 35$ and $b = 5$, evaluate $\frac{a}{b}$.

7 If $m = 3$ and $n = 2$, evaluate:

(a) $2m^2$ (b) $2n^2$ (c) $5m^2$

(d) $7n^2$ (e) $10m^2$.

8 If $s = 7$ and $t = 3$, evaluate:

(a) $4(s + t)$ (b) $t(2s - 5)$ (c) $3(s - t) + 8$

(d) $s(1 + 2t) - 4$ (e) $\frac{1}{2}(s + t^2)$.

Formulae using Words

A *formula* in mathematics is a rule expressed in symbols. For example $P = IV$ is the formula used to calculate electric power in terms of current and voltage. Knowing values for I and V, we can calculate P by evaluating the expression IV.

In Chapter 2, we used several formulae, which could be expressed either in words or in symbols. For example,

Area of a rectangle = length × breadth or $A = lb$.

When we use a formula to calculate a quantity, the words or symbols are replaced with numbers, as in Exercise 1 of this Chapter. A formula provides a general method which can always be applied to a particular problem.

Example The volume of a triangular prism may be found using the formula

$$\text{Volume} = \frac{1}{2} \text{ of (length} \times \text{breadth} \times \text{height)}.$$

Find the volume of the triangular prism shown in the diagram.

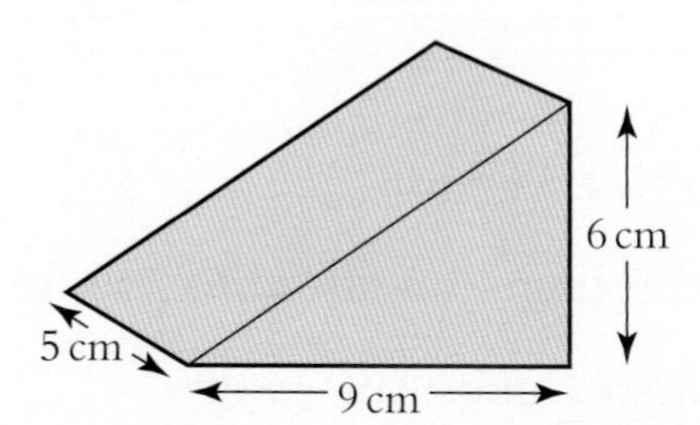

Solution Using BOMDAS, we deal with the bracket first, and then find $\frac{1}{2}$ of the answer.

Hence volume $= \frac{1}{2}$ of $(9 \times 5 \times 6) = \frac{1}{2}$ of $270 = 135$.

Therefore volume $= 135\,\text{cm}^3$.

Exercise 2

Remember 'BOMDAS'!

1 Peaches cost 15 pence each. The cost (in pence) of buying a number of peaches is given by the formula

cost = 15 × number of peaches.

Use the formula to calculate the cost in pence of buying:

(a) 3 peaches (b) 5 peaches (c) 12 peaches (d) 24 peaches.

(e) How many peaches can be bought for £3?

2 The cost of hiring a van for a number of days is found using this rule.

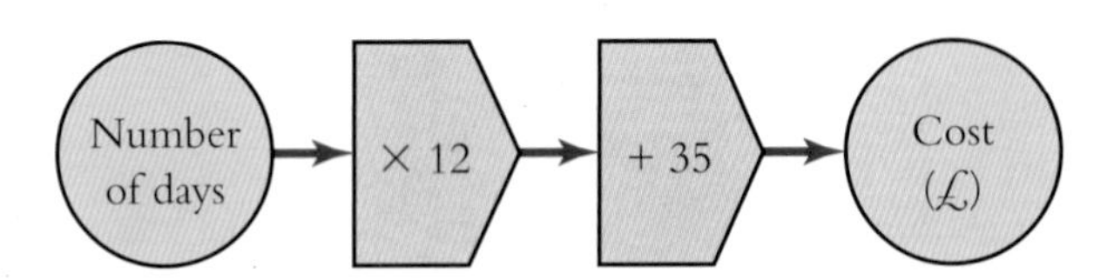

How much would it cost to hire the van for (a) 5 days (b) 12 days?

(c) For how many days did Gary hire the van if he paid £119?

3 A bank has a formula to calculate the price (in pounds) that its customers can afford to pay for a house.

Price in pounds = (4 × Annual salary) + Deposit

Calculate the price that the following customers can afford to pay for a house:

(a) Anne – Annual salary of £24 000 and a Deposit of £8000

(b) Brian – Annual Salary of £28 000 and a Deposit of £9000

(c) Mandy – Annual Salary of £41 500 and a deposit of £13 500.

(d) Calculate the deposit for Joanne who has an agreed price of £150 000 and an annual salary of £35 000.

4 Sam fits blinds in houses. He uses this formula to work out how long a job takes.

Number of minutes = (Number of blinds × 25) + 30

How long does it take Sam to fit:

(a) 1 blind (b) 3 blinds (c) 5 blinds (d) 12 blinds?

(e) How many blinds would Sam fit in 3 hours?

Exercise 2 continued

Remember 'BOMDAS'!

5 Sajid works out his profit when selling cars using the formula:

Profit = Selling Price less Cost Price.

Calculate the profit he makes on each of the following cars:

(a) Selling Price = £5000, Cost Price = £3800

(b) Selling Price = £12 500, Cost Price = £7800

(c) Selling Price = £7775, Cost Price = £6999.

6 The cost of hiring a car can be found using the formula:

Cost = £35 per day plus 5 pence per mile.

Find the cost each day for Michael if he drives:

(a) 100 miles on Sunday

(b) 150 miles on Monday

(c) 80 miles on Tuesday

(d) Michael paid £46 to hire the car on another day. How many miles did he drive that day?

7 The cost (in pounds) of gas to a household is given by the formula:

Cost = 23 + (number of units × 0·13).

Calculate the cost (in pounds) to householders who use:

(a) 100 units (b) 180 units (c) 95 units (d) 215 units.

8 The number of units of alcohol in a drink can be found using the formula:

Number of units = percentage proof × volume in millilitres ÷ 1000.

Calculate the number of units in:

(a) a 330 ml bottle of beer which is 5% proof

(b) a 750 ml bottle of wine which is 12% proof

(c) a 700 ml bottle of whisky which is 40% proof.

Exercise 2 continued

Remember 'BOMDAS'!

9 The volume of a pyramid can be found using the formula:

Volume = $\frac{1}{3}$ of (area of base × height).

The pyramids shown in the diagram both have square bases. Calculate their volumes.

(a)

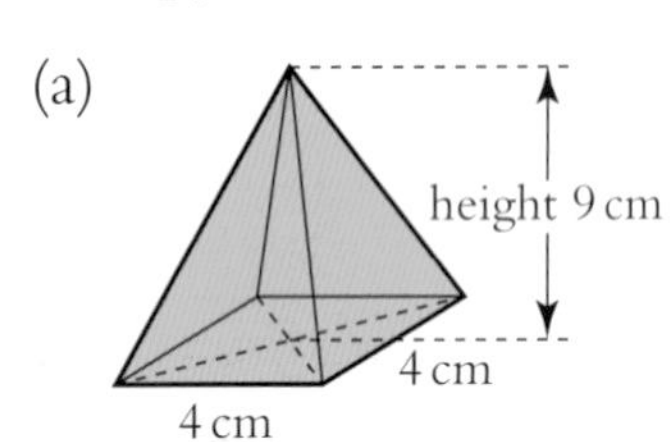

(b)

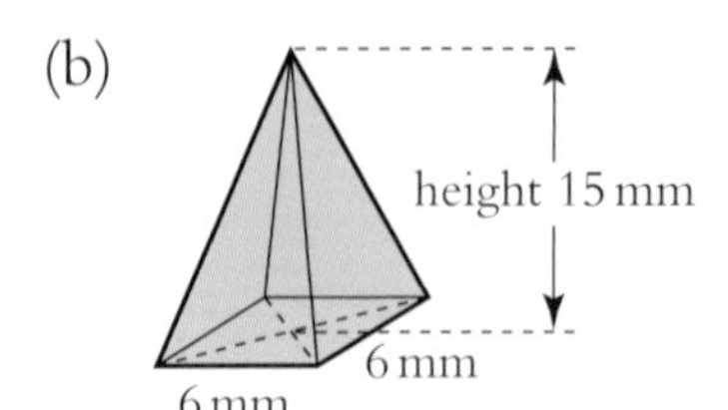

Formulae using Symbols

Formulae are more often given in symbols than in words. Some you have already used. For example $A = \pi r^2$, and $V = lbh$, have allowed you to calculate the area of a circle and the volume of a cuboid.

By replacing the symbols with numbers, we can carry out the evaluation.

Technique

Always remember the correct order of operations (using 'BOMDAS'). If you are substituting difficult numbers and decide to use a calculator, remember too that scientific calculators will do the calculations in the correct order for you.

If your calculator is not a scientific one, then check carefully *your* order of calculations.

Some formulae include the **square root** symbol ($\sqrt{\ }$).

This symbol indicates the opposite process to squaring a number. That is, since $5^2 = 5 \times 5 = 25$, then the square root of 25 is 5 (written as $\sqrt{25} = 5$).

If you are using a formula involving a square root sign, do the calculations covered by the sign first, then press the square root key on your calculator.

Example $D = 4\sqrt{h}$. Calculate D when $h = 49$.

Solution $D = 4 \times \sqrt{49} = 4 \times 7 = 28$.

Exercise 3

Remember 'BOMDAS'!

1 The formula $P = 3s$ can be used to find the perimeter, P, of an equilateral triangle of side s.

Calculate P when s is:

(a) 6 cm (b) 14 cm (c) 5·5 cm (d) 7·6 cm.

2 The formula $P = 3w + 2v$ can be used to find the perimeter of the shape in the diagram.

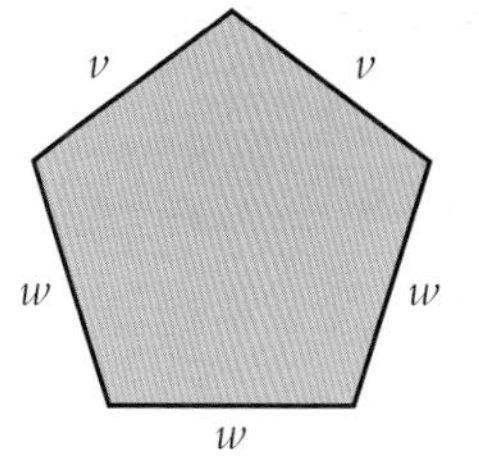

Calculate P when

(a) $w = 5$ and $v = 4$

(b) $w = 4{\cdot}8$ and $v = 3{\cdot}2$.

3 The formula $F = ma$ is used in physics.

Calculate F when

(a) $m = 18$ and $a = 7$

(b) $m = 6{\cdot}8$ and $a = 5$.

4 The cost of buying a radio on hire purchase is given by the formula $C = d + 12p$, where d is the deposit and p the monthly payment.

Calculate the cost of buying a radio on hire purchase when:

(a) the deposit is £25 and the monthly payment £8

(b) the deposit is £19·99 and the monthly payment is £7·99.

Exercise 3 continued

Remember 'BOMDAS'!

5 The mean, M, of three numbers x, y and z, is found using the formula

$M = \dfrac{x + y + z}{3}$.

Using this formula, find the mean of 27, 33 and 48.

6 The following formulae are used in physics.

(a) $s = vt$. Calculate s when $v = 45$ and $t = 3$.

(b) $R = \dfrac{V}{I}$. Calculate R when $V = 24$ and $I = 5$.

(c) $v = u + at$. Calculate v when $u = 20$, $a = 5$ and $t = 12$.

7 The volume of a cuboid is given by the formula $V = lbh$.

Calculate V when $l = 7$, $b = 4$ and $h = 2{\cdot}5$.

8 The perimeter of a rectangle is found using the formula $P = 2(l + b)$.

Calculate P when $l = 3{\cdot}7$ and $b = 2{\cdot}8$.

9 The diameter of a circle is found using the formula $d = \dfrac{C}{\pi}$.

Calculate d when $C = 157$ and $\pi = 3{\cdot}14$.

10 Use the formula $D = 6\sqrt{h}$ to calculate D when $h = 81$.

11 Use the formula $K = 5m^2$ to calculate K when $m = 4$.

12 The formula $h = \dfrac{2A}{b}$ is used to find the height of a triangle.

Calculate h when $A = 56$ and $b = 14$.

13 Use the formula $y = \dfrac{\sqrt{x}}{2}$ to calculate y when $x = 100$.

14 Use the formula $V = \dfrac{L^2}{3}$ to calculate V when $L = 4{\cdot}5$.

Exercise 4 – Revision of Chapter 3

1 Three variables have the values $a = 8$, $b = 3$, $c = 4$.

Find the value of

(a) $3a + 4$ (b) $a + b - c$ (c) $5a - 3b$ (d) $\frac{a}{c}$

(e) $a^2 + b^2$ (f) $4c - 3b$ (g) $\frac{ab}{c}$ (h) $2b^2$

(i) $bc - a$ (j) $ab - 20$ (k) $a(b + c)$ (l) $b^2 + ac$.

2 The cost (in pounds) of hiring a carpet cleaner from a shop is:

Cost = 12 + (number of days × 5).

Calculate the cost of hiring the carpet cleaner for 3 days.

3 A regular polygon is a shape with all its sides the same length and all its angles equal.

The following formula can be used to find the number of sides in such a polygon.

Number of sides in polygon = 360 ÷ (180 – size of any angle in polygon)

How many sides has a polygon in which each angle is 140°?

4 If $L = 5m - 3n$, find L when $m = 7$ and $n = 4$.

5 The volume of a cylinder can be calculated using the formula $V = \pi r^2 h$.

Calculate V when $r = 7$ and $h = 8$.

6 Using the formula $K = \frac{Lm}{N}$, calculate K when $L = 80$, $m = 12$ and $N = 7{\cdot}5$.

7 Using the formula $b = \sqrt{(c^2 - a^2)}$, calculate b when $c = 13$ and $a = 5$.

Summary

(Take time to read through this Summary and repeat any of the Exercises in the Chapter which contain questions with which you are still finding difficulty.)

1 ***Expressions***: (Remember:)

(i) $5y$ means $5 \times y$, y^2 means $y \times y$, yz means $y \times z$

$\frac{y}{z}$ means $y \div z$, $2y^2$ means $2 \times y^2 = 2 \times y \times y$.

(ii) If $a = 9$ and $b = 4$, then $3a - 5b = 3 \times 9 - 5 \times 4 = 27 - 20 = 7$.

2 When evaluating an expression, remember the correct order of operations: '**BOMDAS**':

Brackets, 'Of', Multiplication/Division, Addition/Subtraction.

3 Formulae are sometimes given in words. For example.

Instructions for cooking a chicken are given by the formula:

Cooking time (minutes) = 20 + (weight of chicken in pounds × 25)

How long will it take to cook a 5 pound chicken?

(Cooking time = $20 + (5 \times 25) = 20 + 125 = 145$ minutes.)

4 A formula can be given in symbols as well as in words. For example,

The cost, C pounds, of a special ticket to a swimming pool is given by the formula:

$$C = 25 + 2(n - 6)$$

where n is the number of visits to the pool.

How much will it cost Kirsty to visit the pool 14 times?

($C = 25 + 2 \times (14 - 6) = 25 + 2 \times 8 = 25 + 16 = 41$. It will cost £41.)

4 Everyday Calculations: Earnings and Spending

When, in the not too distant future, you are in employment, you will find yourself doing basic calculations concerning your earnings, and your spending patterns. Later, when you are a householder, your problems will become more complex when you have to budget for such items as Insurance or a car loan in addition to weekly shopping, gas bills, telephone bills, and others. You will eventually become quite 'enthusiastic' about calculations!

In this Chapter we look at several basic calculations involved with earnings and spending.

Earnings

In this section you should be able to perform calculations (with and without a calculator) on different aspects of earning money. This may include pay increases, commission, overtime, and bonuses.

Remember

Percentages are very important in this chapter!

Can you remember how to work out 15% of £720 with and without a calculator?

(a) With a calculator: 15% of £720 = $0{\cdot}15 \times 720$ = £108.

(b) Without a calculator: 10% of £720 = £72, so 5% of £720 = £36. Hence 15% of £720 = £72 + £36 = £108.

Also, some people are paid monthly and others weekly, so you must remember that

1 year = 12 months = 52 weeks = 365 days (366 in a leap year).

Hints and tips

Mistakes are often made in money calculations by mixing up pounds and pence. Always check that your answer is sensible. If for example you calculate that a shop assistant is earning £20 000 per week, you have almost certainly made a mistake. A figure of £200 is more likely!

Pay increases

When you are in employment, it is usual to receive an annual increase in your earnings. This increase is awarded as a percentage of your existing earnings.

Example Matthew earns £14 400 each year in his administrative post with the Scottish Executive. He is to receive an annual increase of 3%.

(a) What will be his new annual earnings?
(b) How much extra will he be earning each month?

Solution (a) 3% of £14 400 = 0·03 × £14 400 = £432

His new earnings are therefore £14 400 + £432.

This is, £14 832.

(b) His extra monthly income is $\frac{£432}{12}$.

That is, £36.

Commission

Commission is the name given to a percentage payment (to sales staff) of the amount of goods sold. It encourages salespersons to sell as much as possible!

Example Elizabeth sells cosmetics in a department store. Her basic salary is £8850 per annum. She is also paid 2% commission on all her sales. Find her total salary for the year if her sales were £72 800.

Solution Basic salary = £8850.

Commission = 2% of sales = 2% of £72 800
= 0·02 × 72 800 = £1456.

Hence total salary = £(8850 + 1456) = £10 306.

Exercise 1

1 Alice earns £16 200 each year. How much does she earn each month?

2 Marcus has an annual salary of £12 500. He is given a pay rise of 3%.

Find his new annual salary.

3 Annie has a pension of £83·50 per week. What is her total pension for the year?

4 Rosemary earns £5·35 per hour. She works a 40-hour week. How much will she earn in a week?

5 Brian is paid £5·80 per hour. His rate of pay is to be increased by 5%.

What will his new hourly rate be?

6 Derek works from 9 am until 2 pm in a café every Friday, Saturday, and Sunday.

How much will he earn for the three days if he is paid £6·20 per hour?

7 Stewart sells programmes outside a football stadium. The programmes cost £3 each. Stewart is paid 4% commission on each programme he sells. How much will he earn if he sells 450 programmes before a match?

Exercise 1 continued

8 Tasnim sells curtains and blinds. Her basic annual salary is £8275 plus 3% commission on all her sales. Calculate her annual salary for the year if she sells goods to the value of £57 000.

9 Neil is a door-to door salesman. His basic monthly salary is £550. He is paid 6% commission on all goods he sells. Find his total monthly salary in March if he sells goods to the value of £35 800.

10 Margo sells cars. She is paid a salary each month of £600 plus 5% commission on all her sales.

Find the value of her sales in a month in which her total salary was £1800.

11 Gordon hands out leaflets advertising a local restaurant. He is paid £1·50 for every customer he attracts to the restaurant. How much he will earn if he attracts 238 people to the restaurant one week?

12 Roberto earns £12 250 per week as a footballer. He is paid a bonus of £1500 every time his team wins a match. In a four week period, his team wins 6 matches.

How much does Roberto earn in this time?

13 Suzie works in a call centre selling double glazing to customers.

Her pay is calculated as follows:

- for every customer agreeing to a home visit she is paid £15
- for every customer who buys double glazing she receives 1·5% commission on the sale.

During one month 65 of Suzie's customers agreed to a home visit.

They ordered £56 000 worth of double glazing.

Calculate Suzie's pay for that month.

14 Edith works in a factory making cardigans. She works a 40-hour week and is paid £6·30 per hour. For each cardigan she makes over her target of 180, she receives a bonus of £1·75.

Calculate her pay for a week when she makes 240 cardigans.

Overtime

Overtime is a way of earning more than a basic wage, by working extra hours in the evening or at the weekend. The most common rates of overtime are *double time* (2 × basic wage rate) and *time and a half* (1·5 × basic wage rate).

To work out the *time and a half* rate, we multiply by 1·5 if a calculator is allowed. For example, if the basic rate is £5·20 per hour, then the time and a half rate is 1·5 × 5·20 = £7·80.

If you are not allowed a calculator, halve £5·20 to get £2·60, and then add £5·20 and £2·60 to get £7·80.

Example **(No calculator allowed)** Tom has a basic rate of pay of £7 per hour. He is paid time and a half for working overtime. During one week he worked 40 hours at the basic rate and 4 hours overtime. How much did Tom earn that week?

Solution Pay at basic rate = 40 × 7 = £280.

Overtime rate of pay = 7·00 + 3·50 = £10·50 per hour.

Pay at overtime rate = 4 × 10·50 = £42.

Hence total pay = 280 + 42 = £322.

Exercise 2

Do not use a calculator for questions 1,2

1 Joe's basic pay rate is £5 per hour. How much will he earn for 4 hours of overtime at double time?

2 Katya has a basic pay rate of £5·60 per hour. How much will she earn for 3 hours of overtime at time and a half?

3 Darren's basic pay rate is £7·20 per hour. Last week he worked 40 hours at the basic rate and 4 hours overtime at time and a half. How much did he earn?

4 Shezad's basic pay rate is £6·80 per hour. He is paid double time if he works overtime.

How much will he earn if he works 30 hours at the basic rate and 5 hours overtime?

Exercise 2 continued

5 Mark earns £5·62 per hour. Find his overtime pay in a month when he works 28 hours of overtime at the time and a half rate.

6 Lynne works in a boutique. Her basic pay rate is £6·50 per hour.

She is paid time and a half for working overtime in the evening and double time for working overtime at the weekend.

In one week she works 35 hours at the basic rate, 4 hours in the evening, and 6 hours at the weekend.

Find her total pay for the week.

7 Philip Parker works in a sports club and earns £5·26 per hour. He is paid time and a half if he works at the weekend.

Philip's timesheet for one week is shown below.

Philip Parker	Hours worked
Monday	8
Tuesday	8
Wednesday	8
Thursday	6
Friday	8
Saturday	5

How much does Philip earn for the week?

Spending

In this section we look at only three aspects of spending money. These are Hire Purchase, Insurance, and buying foreign currency.

Hire Purchase

Although a great deal of spending is done with credit cards, it is still possible to purchase goods though a Hire Purchase agreement.

Hire purchase (HP for short) is a method of buying expensive items by paying only a small part of the cost initially, called the **deposit**, and then paying the rest of the cost by monthly payments called **instalments**.

It is more expensive to buy items with Hire Purchase than to buy them with cash or cheque.

Example Chloe buys a television set on hire purchase. She pays a deposit of £24·99 and then pays monthly instalments of £15·99 for two years.

If the cash price of the television set is £349·99, find the extra cost of hire purchase.

Solution

Deposit = £24·99.

Instalments = 24 × £15·99 = £383·76. (2 years = 24 months)

Total HP cost = £(24·99 + 383·76)
= £408·75.

Hence extra cost of HP = £(408·75 – 349·99) = £58·76.

Exercise 3

Do not use a calculator for questions 1,2

1 Max buys a TV set on hire purchase. Find the total cost if the deposit is £49 followed by 12 payments of £45.

2 Morag buys a cooker on hire purchase by paying a deposit of £57 and then making 12 monthly payments of £32. Find the total cost of the cooker.

3 Alana buys a DVD recorder on hire purchase. She pays a deposit of £80 plus 12 instalments of £17·50. How much does she pay altogether?

4 Sami buys a fridge on hire purchase. The terms require a deposit of £50, followed by 24 monthly instalments of £17. Find the total cost of the fridge.

5 Joan buys a dishwasher on hire purchase. She pays a deposit of £25 followed by 12 monthly payments of £19. If the cash price of the dishwasher is £250, find the extra cost of using hire purchase.

Exercise 3 continued

6 Arthur buys a computer on hire purchase. He pays a deposit of £29·99 followed by 24 monthly payments of £17·75. If the cash price of the computer is £399·99, find the extra cost of buying it on hire purchase.

7 Krzysztof sees this advert in a car saleroom window:

SECOND HAND CAR

Easy Terms

Deposit £500 + 12 payments of £49·50

Or

£999·99 cash

Find the extra cost of buying the car on the 'easy terms'.

8 Bhupinder is thinking of buying a camera. The one he wants can be bought by cash for £199·99, or on hire purchase. The hire purchase terms require a deposit of £9·99, followed by monthly payments of £8·99 for two years. Find the extra cost of buying this camera on hire purchase.

9 Pete sees a car for sale at a cash price of £12 500. He cannot afford this but decides to buy the car on hire purchase. He must pay a deposit, which is 5% of the cash price and then pay 24 monthly instalments of £550.

Find the extra cost of buying the car on hire purchase.

10 Kylie is buying a new bed for her house. The cash price of the bed is £1600.

It can be bought on hire purchase by paying a deposit of 12% of the cash price followed by monthly instalments of £133·33 for one year.

Find the extra cost of using hire purchase.

11 The hire purchase price of a camera is £399.

The terms require a deposit of £75 followed by 9 equal payments.

How much will each payment be?

Insurance

Main points

We take out insurance policies to cover ourselves against accidents, damage, theft, and other unfortunate events. The amount we pay to an insurance company each year is called the **premium**. Insurance companies often give details of their premiums in leaflets. See if you can find some examples in newspapers.

You can insure the building you live in against, for example, fire and flood. You can insure the contents of your home against damage or theft. You can insure your car, you can buy holiday insurance, and you can insure your life.

For buildings and household contents, insurance companies usually quote premiums per £1000 of value.

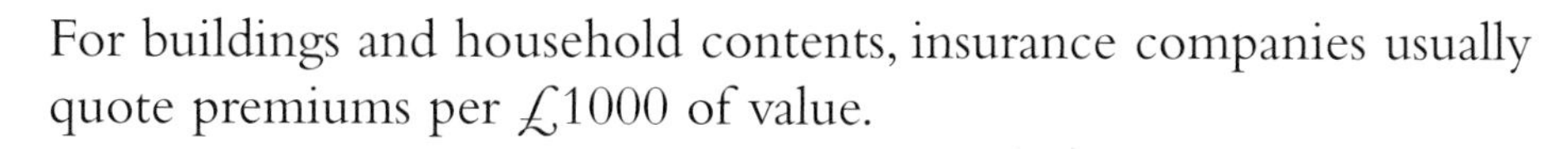

Example Charlie wishes to insure his buildings (worth £165 000) and contents (worth £27 500).
His insurance company charges £2·25 per £1000 for buildings, and £2·80 per £1000 for contents.

What is Charlie's total premium for buildings and contents?

Solution

Buildings:

(value) → (premium)

£1000 ⟶ £2·25

£165 000 ⟶ $\frac{165\,000}{1000} \times 2{\cdot}25 = £371{\cdot}25.$

Contents:

(value) → (premium)

£1000 ⟶ £2·80

£27 500 ⟶ $\frac{27\,500}{1000} \times 2{\cdot}80 = £77.$

Hence total Premium = £371·25 + £77 = £448·25.

Exercise 4

1 Calculate the following premiums:

(a) a £84 000 building insured at £2 per £1000

(b) a £285 000 building insured at £2·20 per £1000

(c) a £125 000 building insured at £1·90 per £1000

(d) £12 000 of contents at £2·30 per £1000

(e) £26 500 of contents at £2·16 per £1000

(f) £18 600 of contents insured at £2·40 per £1000.

2 Find the cost of insuring a building worth £199 900 if the insurance company quotes a premium of £1·80 per £1000.

3 Find the total premium needed to insure a building, worth £89 000, and its contents, worth £11 000, if the insurance company charges £1·60 per £1000 for buildings and £2·30 per £1000 for contents.

4 Isobel wants to insure her buildings (worth £115 000) and contents (worth £14 500). The premiums are £1·60 per £1000 for buildings, and £2 per £1000 for contents.

What is the total premium for Isobel?

5 Ian insures his computer (worth £2800). The insurance company charges £1·35 per £100.

(a) Calculate Ian's premium.

(b) Ian makes a claim when his computer breaks down. The next year the company increases his premium by 5%. Calculate Ian's new premium.

6 Gareth wishes to insure his antiques for £3600. The insurance company charges an annual premium of £1·75 for each £100 insured.

(a) Calculate the annual premium.

(b) Gareth can pay his premium monthly. If he does this he is charged an extra 4%. Calculate the monthly premium.

7 Bobby is quoted a monthly premium of £2·80 per £1000 of cover for life insurance. What will the monthly premium be to insure his life for £45 000?

Exercise 4 continued

8 Karl is quoted a monthly premium of £3·50 per £1000 of cover for life insurance.

If he can afford to pay £70 per month, for how much can he insure his life?

9 The table shows the **monthly premiums** charged by three companies for every £10 000 worth of life insurance cover.

Monthly Insurance Premium (Amount of Cover £10 000 – Age 40 next birthday)		
	Women	Men
Royal & General	£28·50	£34·75
Country Alliance	£24·85	£31·70
Scottish Direct	£27·75	£33·20

(a) John is 39 years old. Find the monthly premium to insure his life for £65 000 with Country Alliance

(b) Samantha, aged 39, changes from Royal & General to Scottish Direct. How much will she save each **year** on cover of £10 000?

10 The table below shows premiums for holiday insurance. All the prices quoted are per person. Children under 12 pay half-price.

	United Kingdom	Europe	Rest of the world
Up to 7 nights	£8·00	£23·50	£52·00
8 to 14 nights	£9·50	£27·60	£61·00
15 to 28 nights	£10·80	£29·40	£70·50

Mr and Mrs Smith and their three children aged 5, 9 and 13, are going on a holiday to Europe for 3 weeks.

Find the cost of holiday insurance for the whole family.

Exercise 4 continued

11 The cost of car insurance is reduced the longer you go without making an insurance claim. A company offers a 40% discount to drivers who have not made a claim for two years.

Davina, whose basic premium is £580, is entitled to this discount. How much will she pay for her car insurance?

12 The basic premium for Bill's car insurance is £920. This is increased by $\frac{1}{4}$ because he is under 20 years of age. He also receives a discount of 30% as he has not made a claim during the last year.

How much does Bill pay to insure his car?

Foreign Currency

When travelling abroad, it is usually necessary to change British currency (pounds sterling) into foreign currency. The amount of foreign currency you receive depends on the exchange rates, which change regularly. These rates appear each day in newspapers, banks, and travel agencies. The following table gives an example of some exchange rates.

Country	Currency	Exchange Rate
Eurozone	Euro	1·41
USA	Dollar	1·81
Canada	Dollar	2·33
Cyprus	Pound	0·81
Japan	Yen	208·41
South Africa	Rand	12·04

Main points

The Eurozone covers Austria, Belgium, Finland, France, Germany, Greece, Ireland, Italy, Luxembourg, Netherlands, Portugal and Spain. All these countries have the Euro as their currency. More countries are expected to join the Eurozone soon.

To change British money (£) to foreign money, we MULTIPLY the number of pounds by the exchange rate.

For example, £50 = 50 × 1·41 euros = 70·5 euros.

To change foreign money to British money (£), we DIVIDE the amount of foreign money by the exchange rate.

For example, 500 US dollars ($500) = 500 ÷ 1·81 pounds = £276·24 (to the nearest penny).

Example Roger is going to Switzerland on holiday. He changes £600 to Swiss francs at a rate of £1 = 2·33 Swiss francs. During his holiday he spends 1300 Swiss francs. When he returns home, he changes the remaining Swiss francs back to pounds. The exchange rate has changed to £1 = 2·29 Swiss francs. How much will he get back?

Solution

Here £600 = 600 × 2·33 Swiss francs = 1398 Swiss francs.

He spends 1300 Swiss francs, so leaving (1398 – 1300) = 98 Swiss francs.

Now 98 Swiss francs = 98 ÷ 2·29 pounds = £42·79 (to the nearest penny).

Exercise 5

Refer to the table of currency conversion where necessary

1 Convert £800 to:

(a) euros (b) US dollars (c) Canadian dollars

(d) Cyprus pounds (e) yen (f) rand

Exercise 5 continued

Refer to the table of currency conversion

2 Pat is going on holiday to Spain. She changes £1250 to euros. How many euros will she receive?

3 Tony changes £750 to Maltese lira. If the exchange rate is £1 = 0·62 lira, how many lira will he receive?

4 Roberta buys perfume at Glasgow Airport for £27·50. She goes on holiday to France. The exchange rate is £1 = 1·40 euros. In France she sees the same perfume on sale for 40 euros. Is the perfume cheaper in Scotland or France?

Explain your answer.

5 Bruno is going to Portugal and wants to change £400 into euros.

Two travel agents offer the following exchange rates.

Azure Travel: £1 = 1·40 euros
(No commission)

Algarve Sun: £1 = 1·43 euros
(2% commission payable)

(a) How many euros would Bruno receive from Azure Travel for £400?

(b) Which travel agent will give Bruno more euros for his £400?

(Show clearly all your working.)

6 (a) Change 362 US dollars to pounds sterling.

(b) Change 4214 rand to pounds sterling.

7 Convert the following amounts of foreign currency to pounds: (Answer to the nearest penny.)

(a) 475 euros
(b) 50 Canadian dollars
(c) 3100 rand
(d) 1 million yen
(e) 1250 US dollars
(f) 85 Cyprus pounds.

8 While on holiday in Greece, George buys a 3-course meal for 17 euros.

How much would this meal have cost in pounds?

9 (a) Convert £60 to Swedish krona given that £1 = 13·5 krona.

(b) The exchange rate changes so that now £60 = 822 Swedish krona.

What is the new exchange rate?

Exercise 5 continued

Refer to the table of currency conversion

10 Patrick is going on a city break to Dublin in Ireland. He changes £450 to euros.

(a) How many euros will he receive if the exchange rate is £1 = 1·38 euros?

(b) While in Dublin he spends 600 euros. How many euros does he have left?

(c) He changes the remaining euros back to pounds when he returns home.
How much will he receive if the exchange rate is now £1 = 1·40 euros?

11 Annabel has planned a trip to Japan. She changes £1500 to Japanese yen. The exchange rate is £1 = 210 yen. During her holiday she spends 300 000 yen. When she returns home she changes the remaining yen back to pounds at a rate of £1 = 208 yen. How much will she receive to the nearest penny?

Spending money – further examples

Examination questions on money are very common since money is such an important part of our everyday lives. In the following exercise there are a variety of different types of question, several involving the completion of tables.

Exercise 6

1 Grant intends to send flowers to his girlfriend. The florist tells him the cost of some different flowers. These are:

Rose £5, Daffodil £3·50, Tulip £3, Carnation £2·50, Violet £4.

Grant wishes to buy **three different** flowers and wants to spend a **minimum** of £11.

The entry in the table shows one way of doing this.

Copy and complete the table to show all the possible ways in which Grant can buy the flowers.

Rose (£5·00)	Daffodil (£3·50)	Tulip (£3·00)	Carnation (£2·50)	Violet (£4·00)	TOTAL COST
✓	✓		✓		£11·00

Exercise 6 continued

2 Martin is making up a box of 12 luxury chocolates for his mother.

He can choose milk chocolates and plain chocolates. He wants more than half of the chocolates he chooses to be milk chocolates.

(a) The entry in the table shows one way of doing this.

Copy and complete this table to show all the possible ways of making up the box.

milk	plain
9	3

(b) Milk chocolates cost £1·50 each and plain chocolates cost £2 each. It costs £3 for the box. If Martin paid a total of £23, how many milk and how many plain chocolates were in the box?

3 Nick is buying a fitted kitchen for his house. The company he is buying from has the following special offer.

> **Special offer**
>
> **Free when you buy our fitted kitchen**
>
> Choose any three different items up to a value of £75
>
> Toaster £10 Kettle £20 Pots £40
>
> Microwave £35 Clock £20

(a) The entry in the following table shows one way in which Nick can choose his items.

Copy and complete the table to show all the possible ways in which Nick can choose his items.

Toaster (£10)	Kettle (£20)	Pots (£40)	Microwave (£35)	Clock (£20)	TOTAL COST
✓	✓		✓		£65

(b) Nick wants all five items. What is the cheapest way to get them if he pays for two of them?

Exercise 6 continued

4 Graham has decided to give up smoking. He smokes 24 cigarettes each day. A packet of cigarettes costs £3·90 and contains 20 cigarettes. How much money will he save in a *year* when he gives up?

5 Ralph is buying clothes at Malone's Menswear.

He sees the following advertisement.

Malone's Menswear

Special Offer

Trousers £55 Shirts £45 Shoes £40

Suits £95 Coats £70

Buy three different items

Free silk tie if you spend over £175

The entry in the table shows one way of getting the free silk tie.

Copy and complete the table to show all the possible ways in which Ralph can get the free silk tie.

Trousers (£55)	Shirt (£45)	Shoes (£40)	Suit (£95)	Coat (£70)	TOTAL COST
	✓		✓	✓	£210

6 Harriett has to take 2 tablets per day to lower her blood pressure.

(a) How many tablets will Harriett take in a year?

(b) Tablets come in boxes of 40. How many boxes will she use in one year?

(c) How much will this cost Harriett if each box costs £6·50?

(d) Harriett can save money by buying a pre-payment certificate for one year.

How much will she save if the certificate costs £93·50?

Exercise 7 – Revision of Chapter 4

1 Simon earns £5·20 per hour. How much does he earn if he works 40 hours each week?

2 Jimmy earns £12 220 in a year. He is given a 5% pay rise. Calculate:

(a) his new annual salary

(b) his new weekly wage.

3 Joseph sells encyclopaedias. His basic monthly salary is £775. He is also paid 3% commission on all his sales. Calculate his total earnings for a month in which he sells £12 500 worth of encyclopaedias.

4 Tracy is comparing the prices of tubes of toothpaste. She sees offers for the type she wants in two shops. In the first shop it costs £1·65 per tube. In the second shop it costs £1·85 per tube, and one free tube is given with every five bought.

Where should Tracy buy toothpaste if she wants six tubes at the lowest price? Explain your answer.

5 Asfa earns £7·80 per hour. How much will she earn in a week in which she works 40 hours at the basic rate and 6 hours overtime at time and a half?

6 Jessica buys a washing machine on hire purchase. She pays a deposit of £15 and then 24 instalments of £15·75. Find the extra cost of buying the washing machine on hire purchase if the cash price is £350.

7 Assume that £1 = 1·38 euros and convert:

(a) £700 to euros

(b) 897 euros to pounds sterling.

8 Alexandra is going to Russia on holiday. She changes £750 to roubles. The exchange rate is £1 = 51·88 roubles.

(a) How many roubles will Alexandra receive?

(b) She spends 37 500 roubles in Russia. How many roubles has she left?

(c) She then changes the remaining roubles back to pounds sterling at the same exchange rate. How much will she receive? (Answer to the nearest penny.)

Exercise 7 – Revision of Chapter 4 continued

9 Tom wishes to insure his buildings which are worth £87 500. The insurance premium is £2·24 per £1000. How much will it cost Tom to insure his buildings?

10 Dughall is buying revision guides to study for his exams. He is interested in the following subjects:

Maths £7·99 *English £8·99* *Art £7·49* *French £6·99* *Biology £8·49*

He can afford to spend £24 and wishes to buy **three** revision guides.

The entry in the following table shows one way of doing this.

Maths (£7·99)	English (£8·99)	Art (£7·49)	French (£6·99)	Biology (£8·49)	TOTAL COST
✓		✓	✓		£22·47

Copy and complete the table to show all the possible ways of buying the three revision guides.

Summary

(The calculations in this Chapter are important ones and you should be sure you understand each of the following points.)

1 People in work usually receive a ***pay increase*** each year.

If a salary of £13 800 is increased by 2%, then 2% of £13 800 = 0·02 × 13 800 = £276.

Hence new salary = £(13 800 + 276) = £14 076.

Summary continued

2 Some sales people earn ***commission*** for their sales.

Suppose Daniel earns a basic salary of £6500 and receives 4% commission on his sales.

If his sales are £52 000, he earns £6 500 + 4% of £52 000 = 6500 + 0·04 × 52 000

= £(6500 + 2080) = £8580.

3 ***Overtime*** is normally paid at time and a half (1·5 × basic wage) or double time (2 × basic wage).

Suppose Marion, whose basic wage is £5·76 per hour, works 4 hours overtime at time and a half.

Then she earns 4 × 1·5 × 5·76 (=£34·56) for her overtime.

4 Certain goods may be bought on ***hire purchase***.

Suppose a television set can be bought on HP by paying a deposit of £19 followed by 24 instalments of £12 each. It would cost £19 + 24 × 12 = £19 + 288 = £307.

5 ***Insurance*** protects people against unfortunate events.

Suppose it costs £2·80 per £1000 to insure an amount of £74 500. Then:

premium $= \dfrac{74\,500}{1000} \times 2{\cdot}80 =$ £208·60.

6 ***Exchange rates*** are used when we change pounds sterling to *foreign currency* and vice versa.

Suppose £1 = 1·42 euros, then £500 = £500 × 1·42 = 710 euros.

(Also 710 euros = 710 ÷ 1·42 = £500.)

End Of Unit Tests

Test One (Non-Calculator)

1 Work out:

(a) 6·23 – 3·5

(b) 0·512 ÷ 8

(c) 6·35 × 400.

2 Express $\frac{13}{1000}$ as a decimal.

3 Work out:

(a) 32 × 12

(b) 200 + 50 × 3

(c) 340 – 70 × 2.

4 Work out:

(a) $\frac{3}{5}$ of 85 metres

(b) $\frac{5}{8}$ of £184.

5 Work out:

(a) 30% of £80

(b) 75% of 108 km

(c) 5% of £180.

6 A rectangle is 25 centimetres long and 13 centimetres broad. Calculate its area.

7 Calculate the volume of each cuboid shown in the following diagram.

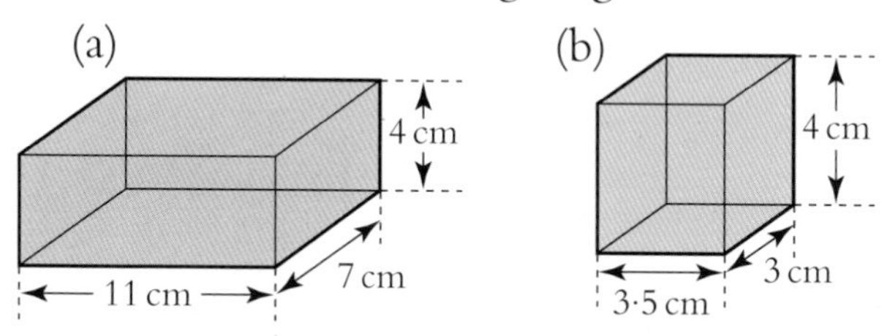

8 The 'machine chain' shown in the diagram is used by a shopkeeper to work out the cost of hiring a lawnmower to customers.

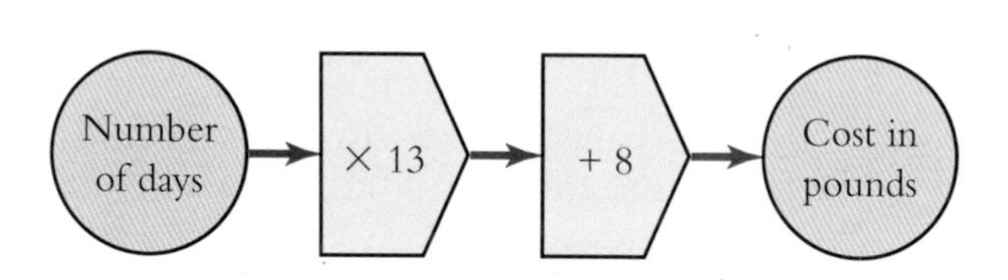

Calculate the cost of hiring out the lawnmower for 1 week.

9 Malcolm is a glazier. He fits panes of glass into windows. He has a formula for working out how long a job takes. This is:

Number of minutes = 20 + (Number of panes × 15).

How long will it take Malcolm to fit 11 panes? (Give your answer in hours and minutes.)

10 The rule used to calculate the cost of gas in pounds is:

Cost = 23 + (number of units × 0·08).

Calculate the cost of 500 units of gas.

11 The cost of 7 copies of the 'Inquirer' newspaper is £4·55. How much will 5 copies of the 'Inquirer' cost?

12 Annette changes £400 to euros. If the exchange rate is £1 = 1·41 euros, how many euros will she receive?

13 Mrs Stewart buys a television set on hire purchase by paying a deposit of £25 plus 12 monthly instalments of £16.

If the television set costs £199, calculate the extra cost of using hire purchase.

14 Molly pays an annual fee of £600 to join a sports club. The following year the fee is increased by $\frac{1}{10}$ but there is a reduction of 5% for paying by cash.

Calculate Molly's fee if she pays cash.

Test Two (Calculator)

1 How much interest would be earned on £750 invested for 1 year at an interest rate of 4·6% per annum?

2 The cost of a bill is £75·58 + VAT. If VAT is charged at 17·5%, find the total cost of the bill. (Give your answer to the nearest penny.)

3 The exchange rate for Mexican pesos is £1 = 19·15 pesos.

(a) Change £260 to pesos.

(b) Change 6950 pesos to pounds sterling. (Give your answer to the nearest penny.)

4 There are 540 pupils on a school roll and 297 of them are girls.

Express the number of girls as a percentage of the school roll.

5 A cuboid is 45 cm long, 30 cm broad and 20 cm high.

Calculate its volume in litres.

6 A circle has a radius of 17 centimetres.

Calculate (a) its circumference (b) its area.

7 The shape of a room in a museum can be thought of as a rectangle and a right-angled triangle. Its sizes are shown in the diagram.

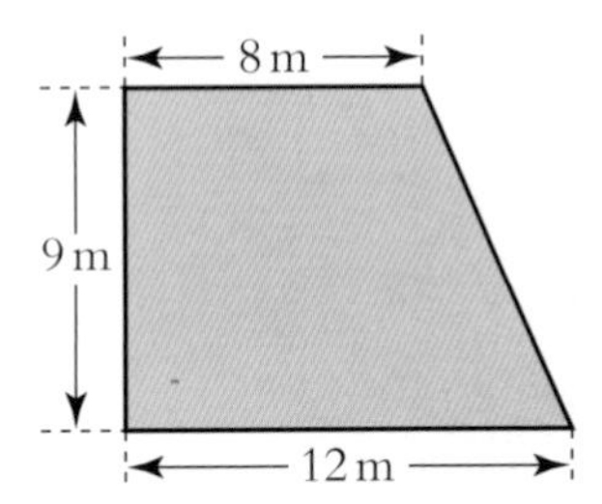

Calculate the area of the room.

8 The floor of a lift is in the shape of a rectangle and a semi-circle. Its sizes are shown in the diagram.

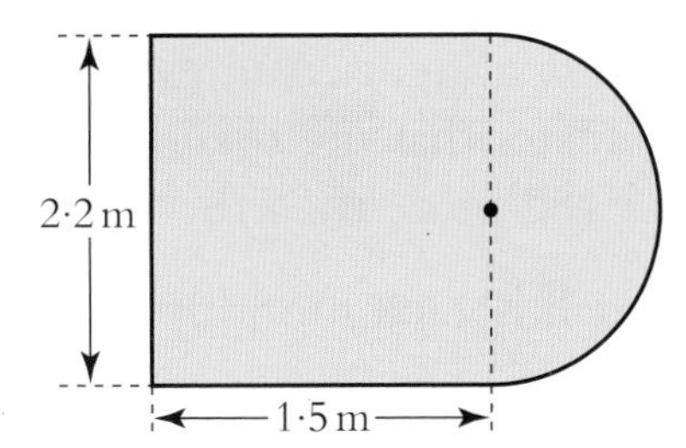

Calculate the area of the floor. (Give your answer to one decimal place.)

9 The formula $D = 4\sqrt{h}$ is used to calculate the distance, D kilometres, of the horizon from the top of a cliff of height, h metres. Find D when $h = 81$.

10 Calculate the value of $3a^2$ when $a = 7$.

11 Janice earns a basic salary of £720 per month plus 2·5% commission on all her sales. Calculate her total salary in June if her sales are £75 000.

12 Bill wishes to insure his computer for £1800. The premium is £1·35 per £100 of the value. How much will the premium be?

13 Andrew buys a bicycle on hire purchase. The bicycle costs £350 new. The hire purchase terms are: deposit of 20% of the cash price, plus 12 monthly instalments of £24·99. Find the *extra* cost of buying the bicycle on hire purchase.

14 Benedykta goes to the bank to change £400 to Polish zloty. The exchange rate is £1 = 5·5 Polish zloty. In addition, the bank has a 2% commission charge. How many zloty will Benedykta receive?

Test Three (A/B Content)

1 A survey was carried out to find how many children in a group were left-handed.

Out of 75 children asked, 9 were left-handed. What percentage is this?

2 Calculate the interest earned on a sum of £1600 invested for 9 months at an interest rate of 4·4% per annum.

3 The cross-section details of a concrete moulding used as a water runway are shown in the following diagram. The channel is semi-circular in shape.

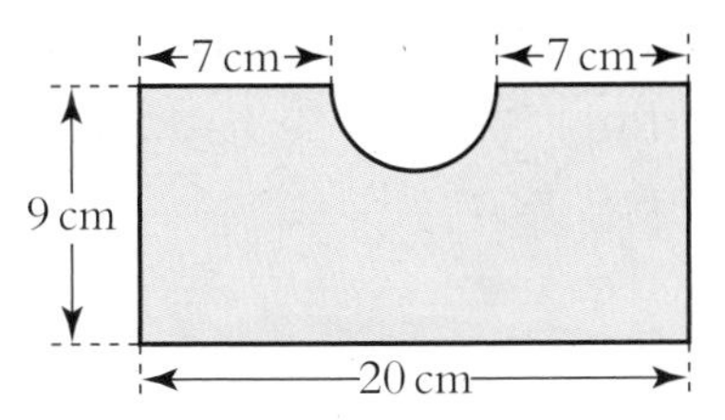

Calculate the cross-sectional area of the moulding correct to one decimal place.

4 The currency in Bulgaria is the lev.

If the exchange rate is £1 = 2·91 lev, convert 695 lev to pounds.

(Give your answer to the nearest penny.)

5 The prices of a series of books on antiques are as follows:

Furniture £7, Porcelain £8, Toys £8·50, Weapons £9, Clocks £9·50.

Eugene wishes to buy 3 different books. He cannot spend more than £25.

The entry in the following table shows one way of doing this.

Furniture (£7)	Porcelain (£8)	Toys (£8·50)	Weapons (£9)	Clocks (£9·50)	TOTAL
✓		✓	✓		£24·50

Copy and complete the table to show all the possible ways of buying the books.

6 The cash price of a second hand car is £3500. It can be bought on hire purchase by paying a deposit of 20% of the cash price followed by 24 equal monthly instalments. Audrey buys the car and calculates that the total hire purchase cost is £182 more than the cash price.

How much is each monthly instalment?

7 The perimeter of a rectangle can be found using the formula $P = 2l + 2b$.

Calculate b when $P = 24 \cdot 6$ and $l = 7 \cdot 5$.

UNIT 2

5 Integers

Mathematicians have long been interested in *number systems*, and have arranged numbers in various sets. For example:

the set of prime numbers : {2, 3, 5, 7, 11, 13, 17, ...}

the set of natural numbers: {1, 2, 3, 4, ...}

the set of whole numbers: {0, 1, 2, 3, 4, ...}

In this chapter we are interested in the set of integers {... −3, −2, −1, 0, 1, 2, 3, ...}.

Integers are positive or negative whole numbers, and they occur in a great many problem situations. In this chapter we begin by plotting points defined by integers on a coordinate diagram. Then we move on to adding, subtracting, multiplying, and dividing integers.

In this section, you should be able to plot and read coordinates in all four quadrants.

Coordinate Integers

Remember

Any point P on a two-dimensional surface can be defined by two coordinates relative to axes at right angles to each other. These axes are called x and y. A point P with coordinates (2, 3) is plotted by travelling 2 units in the x direction, followed by 3 units in the y direction.

Example Write down the coordinates of the points P, Q, R, S, T, and U, in the diagram.

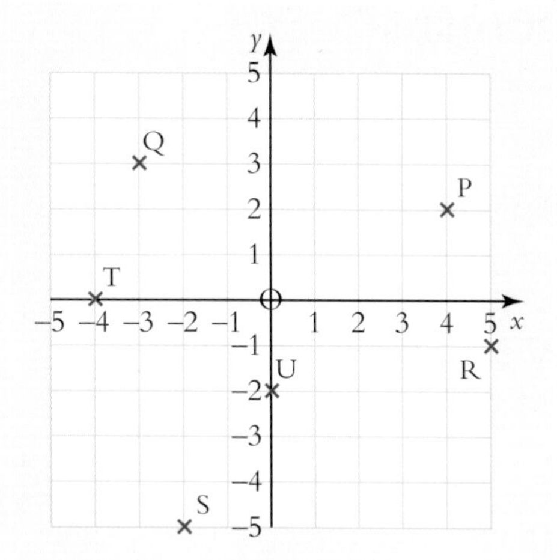

Solution P (4, 2), Q (–3, 3), R (5, –1), S (–2, –5), T (–4, 0), U (0, –2).

Exercise 1

1 Write down the coordinates of A, B, C, D, E, F, G, and H, in the diagram.

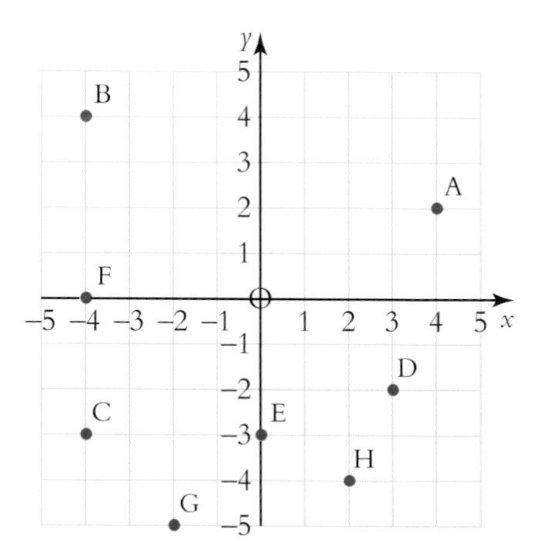

2 Using squared paper, draw a coordinate diagram and plot the points K (3, 5), L (6, –1), M (0, 4), N (–5, 5), P (–3, –4) and Q (–2, 0). (Number the axes from –7 to 7.)

Exercise 1 continued

3 The diagram shows a rectangle ABCD, a kite EFGH, a rhombus IJKL and a parallelogram MNPQ.

Write down the coordinates of the corners of the shapes shown in the diagram.

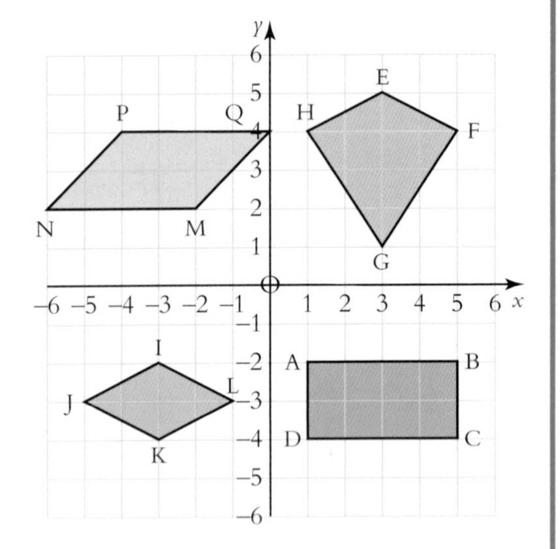

4 (a) Using squared paper, plot the points R (3, 5), S (3, –1), T (–5, –1) and U (–5, 5) on a coordinate diagram.

(b) Join R→S→T→U→R.

(c) What shape is RSTU?

(d) Join R→T and S→U and write down the coordinates of the point where they intersect.

5 The diagram shows some landmarks in a village. The church is at position (–4, 3).

(a) Write down the coordinates of the farm.

(b) Write down the coordinates of the windmill.

(c) David walks from the church to the hotel. In what direction does he walk?

(d) What landmark is at position (5, –3)?

(e) There is an oak tree midway between the hotel and the windmill.

Write down its coordinates.

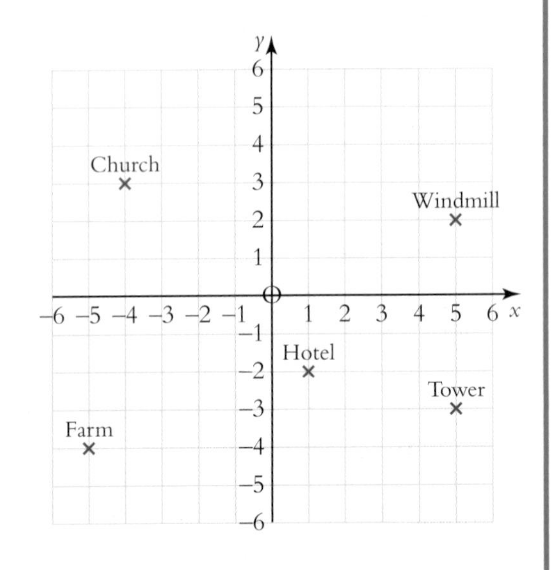

6 (a) Plot the points A (2, 7), B (6, 5) and C (2, –2) on a coordinate diagram.

(b) Plot a fourth point D so that ABCD is a kite.

(c) Write down the coordinates of D.

Exercise 1 continued

7 Some students have made up coordinate codes for their names.

SAM is (–3, 3), (4, 1), (2, –2).

Find the names of the four students whose codes are:

(a) (2, 1), (4, 1), (2, 4) (0, –2) (4, 1)

(b) (–2, –2), (–3, –3), (4, 1), (–4, –4)

(c) (–3, –1), (–2, –3), (2, 1), (1, 3), (0, 0), (2, 4)

(d) (3, –2), (2, 4), (4, 1), (–4, –4), (0, 2).

(e) Write the name KEN in code.

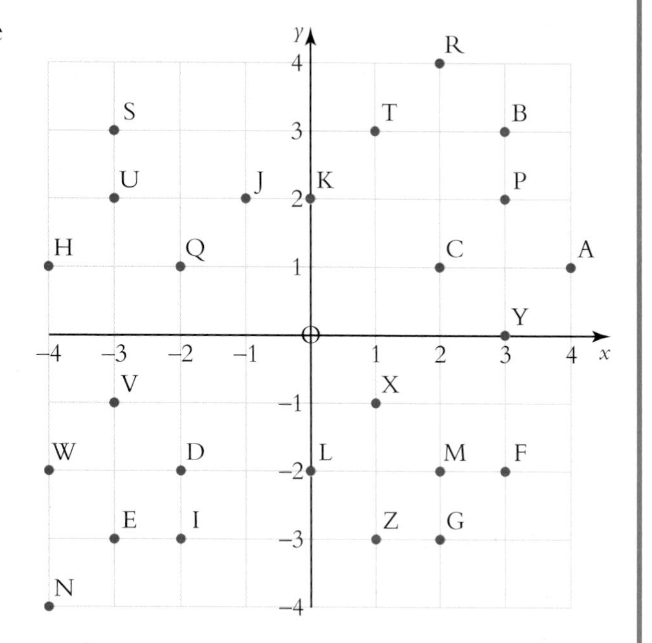

8 Three corners of a parallelogram PQRS have coordinates P (–5, 5), Q (1, 3), and R (3, –7).

(a) Find the coordinates of S.

(b) Find the coordinates of the point of intersection of the diagonals of the parallelogram.

9 The diagram shows part of a row of identical rhombuses.

The centre, A, of the first rhombus has coordinates (–2, 2).

(a) Write down the coordinates of B, the centre of the second rhombus.

(b) Find the coordinates of C, the centre of the third rhombus.

(c) Find the coordinates of F, the centre of the sixth rhombus.

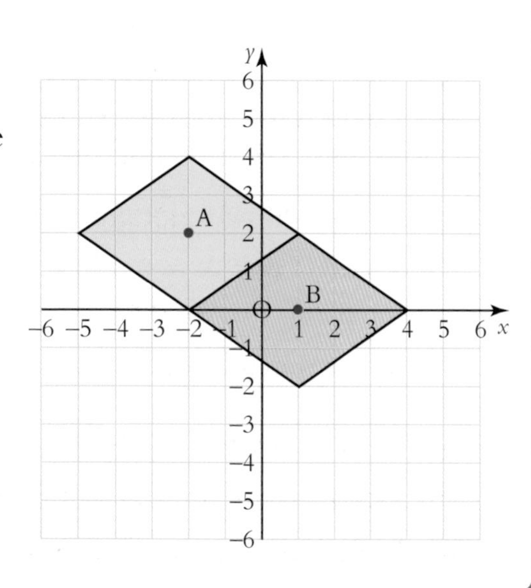

Addition of Integers

You must be able to add integers without the use of a calculator.

Using a number line

An **integer** is either a positive whole number, a negative whole number or zero, and is often shown on a **number line**. Thus:

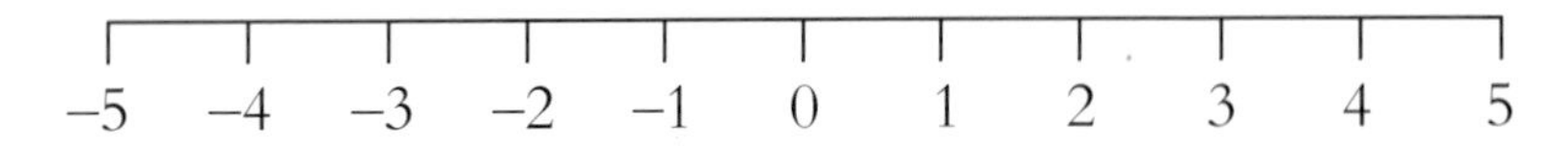

The number line can be extended in both directions, and can often be used to help you add integers.

When you are adding two integers, find the first number on the line, then:

- if you are adding a **positive** integer, count to the **right**;
- if you are adding a **negative** integer, count to the **left**.

Hence to find $-3 + 5$, we start at -3 on the number line and count 5 places to the right, finishing at 2. Therefore $-3 + 5 = 2$.

Also to find $4 + (-6)$, we start at 4 on the number line and count 6 places to the left, finishing at -2. Therefore $4 + (-6) = -2$.

Example The lowest temperature during a night in winter was –5° C. By mid-morning it had risen by 8° C. What was the mid-morning temperature?

Solution $-5 + 8 = 3$. The mid-morning temperature was 3° C.

Exercise 2

Do not use a calculator

Use a number line to help you.

1 Work out:

(a) $-5 + 2$ (b) $-3 + 7$ (c) $-1 + 4$ (d) $-4 + 4$
(e) $-2 + 7$ (f) $2 + 4$ (g) $-7 + 5$ (h) $-9 + 12$
(i) $0 + 7$ (j) $-10 + 1$.

2 Work out:

(a) $4 + (-3)$ (b) $7 + (-7)$ (c) $0 + (-2)$ (d) $-2 + (-3)$
(e) $-1 + (-1)$ (f) $8 + (-7)$ (g) $-4 + (-5)$ (h) $12 + (-5)$
(i) $-9 + (-1)$ (j) $-8 + (-5)$.

3 Work out:

(a) $-7 + 10$ (b) $7 + (-14)$ (c) $-20 + 5$
(d) $-30 + 30$ (e) $20 + (-7)$.

4 Work out:

(a) $-7 + 5 + (-3)$ (b) $8 + (-5) + 2$
(c) $-2 + (-7) + 10$ (d) $3 + (-2) + (-6)$.

5 The temperature late one evening was –7°C. By late morning the following day, it had risen by 15°C. What was the temperature after this rise?

6 Work out:

(a) $-30 + 5$ (b) $-40 + 12$ (c) $60 + (-25)$
(d) $-50 + 15$ (e) $-90 + 30$.

7 Jim is overdrawn at his bank. His bank statement shows his account stands at –£15·00. What will his bank statement show if he deposits £100·00 in his account?

Exercise 2 continued

Do not use a calculator 

8 Wendy, Hui-Lan and Patricia play three rounds of a game. The scores for each round are integers. The scores are added and the girl with the highest total wins.

The scores are as follows:

	Round 1	Round 2	Round 3	Total
Wendy	6	–8	3	
Hui-Lan	–4	7	–5	
Patricia	0	–4	–3	

(a) Find the total score for each girl.

(b) Who won the game?

9 A **magic square** is a square where the three rows, the three columns and the two diagonals all add up to the same **magic total**.

8	1	6
3	5	7
4	9	2

In the example shown, the magic total is 15.

Copy and complete the following three magic squares.

(a)

		1
	–2	
–5		–3

(b)

5	–2	3
		–1

(c)

	–5	
	–1	
–2	3	

Subtraction of Integers

You must be able to subtract integers without the use of a calculator.

Again, you can make use of a number line.

When you are subtracting two integers, find the first number on the number line, then:

- if you are subtracting a **positive** integer, count to the **left**;
- if you are subtracting a **negative** integer, count to the **right**.

Hence to find 4 – 7, we start at 4 on the number line and count 7 places to the left, finishing at –3. Therefore 4 – 7 = –3.

Also to find –2 –3, we start at –2 on the number line and count 3 places to the left, finishing at –5. Therefore –2 – 3 = –5.

To find 1 – (–4), we start at 1 on the number line and count 4 places to the right, finishing at 5. Therefore 1 – (–4) = 5.

(Note that subtracting (or taking away) a negative number is just like adding the number. The two negatives cancel each other. So 1 – (–4) = 1 + 4 = 5.)

Finally to find –3 – (–6), we start at –3 on the number line and count 6 places to the right, finishing at 3. Therefore –3 – (–6) = –3 + 6 = 3.

Daily Temperature Chart

	Minimum	Maximum
Edinburgh	-3	8
London	-1	12
Paris	0	12
Madrid	9	22

Example The temperature at 5 pm in the afternoon is 5° C. During the night it drops by 8°C. What will the temperature be then?

Solution 5 – 8 = –3. The temperature will be –3° C.

Exercise 3

Do not use a calculator

Use a number line to help you.

1 Work out

(a) 5 – 7 (b) 3 – 4 (c) –2 – 6 (d) 0 – 3

(e) –5 – 2 (f) 4 – 10 (g) 8 – 11 (h) 2 – 9

(i) 5 – 12 (j) –4 – 6.

Exercise 3 continued

Do not use a calculator 

2 Work out:

(a) $2 - (-2)$ (b) $1 - (-3)$ (c) $0 - (-5)$ (d) $-1 - (-3)$
(e) $-5 - (-2)$ (f) $-6 - (-4)$ (g) $5 - (-2)$ (h) $-1 - (-8)$
(i) $-7 - (-1)$ (j) $-10 - (-8)$.

3 Work out:

(a) $20 - (-5)$ (b) $-15 - (-10)$ (c) $-30 - 50$
(d) $15 - 20$ (e) $-25 - (-35)$.

4 The temperature in Moscow one afternoon in winter was a steady −5° C. During the night however the temperature fell by 12° C. What was the temperature after this fall?

5 In a game, Marjorie scored 11 points and Anne scored −4 points. How many points more did Marjorie score than Anne?

6 The diagram shows the control panel on a lift.

The numbers on the buttons show the floors at which the lift can stop.

The number 0 indicates the ground floor.

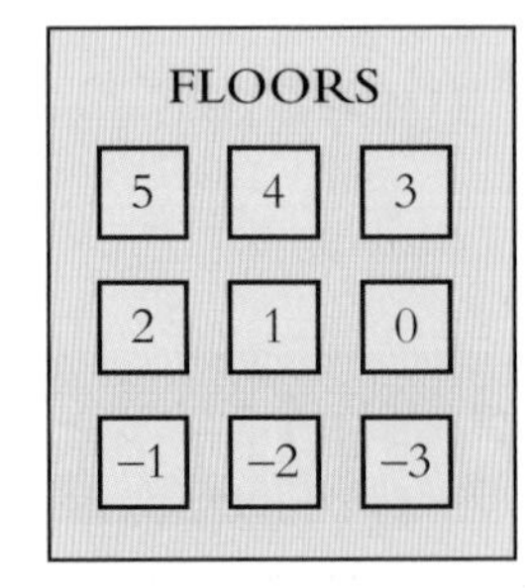

(a) What do you think floor −3 means?

(b) Brian enters the lift at floor 2 and goes down 4 floors. At which floor does he get off?

(c) Alison enters the lift at floor 5 and leaves at floor −2. How many floors did she go down?

7 The temperatures at noon on 1 February in four Scottish cities were:

Glasgow 3° C, Edinburgh −1° C, Dundee −2° C, Aberdeen −4° C.

(a) Find the difference in temperature between Glasgow and Edinburgh.

(b) Find the difference in temperature between Dundee and Aberdeen.

(c) Which city had the highest temperature?

(d) Which city had the lowest temperature?

(e) What was the difference in temperature between the warmest and coldest cities?

(f) Later that day the temperature in Aberdeen had dropped by 7° C. What was the temperature after this drop?

Multiplication of Integers

Once again you must be able to deal with integers without the use of a calculator.

The multiplication rules for integers are summarised below.

Multiplying two integers

Main points

- When we multiply two **positive** integers, the answer is **positive** ($2 \times 4 = 8$).
- When we multiply a **positive** integer by a **negative** integer, the answer is **negative** ($2 \times (-4) = -8$).
- When we multiply a **negative** integer by a **positive** integer, the answer is **negative** ($-2 \times 4 = -8$).
- When we multiply two **negative** integers, the answer is **positive**. The two negatives cancel each other ($-2 \times (-4) = 8$).
- Remember too that when you are multiplying any two integers, the order of multiplication is unimportant, and $3 \times 5 = 5 \times 3 = 15$
- If you multiply any integer by zero, the answer is always zero ($-4 \times 0 = 0, 0 \times 3 = 0$).

$+ \times + = +$

$+ \times - = -$

$- \times + = -$

$- \times - = +$

Multiplying three integers

If you have to multiply three integers, multiply two of them together first, then multiply the answer by the third integer.

Thus to find $-2 \times 5 \times (-6)$, we multiply -2×5 ($= -10$), then multiply $-10 \times (-6)$($= 60$). Hence $-2 \times 5 \times (-6) = 60$. (Remember to multiply integers in the easiest order. Thus to find $-50 \times 9 \times 2$, it is easiest to multiply -50×2 ($= -100$) first, then to multiply -100×9 ($= -900$).

Exercise 4

Do not use a calculator 

1 Work out:

(a) -3×7 (b) $5 \times (-3)$ (c) 4×6 (d) $-2 \times (-3)$

(e) $0 \times (-7)$ (f) -1×12 (g) $6 \times (-7)$ (h) $-3 \times (-3)$

(i) -19×0 (j) $-4 \times (-10)$.

2 Work out:

(a) -25×4 (b) $-7 \times (-7)$ (c) $8 \times (-6)$ (d) -9×3

(e) 11×23 (f) $-50 \times (-3)$ (g) $0 \times (-23)$ (h) -7×12

(i) $-12 \times (-11)$ (j) $(-5)^2$.

3 Chris plays a game with 4 rounds. He scores -8 points in every round. What is his total score?

4 Work out:

(a) $-2 \times 3 \times 4$ (b) $5 \times (-3) \times 2$ (c) $-9 \times (-8) \times 0$

(d) $-4 \times (-4) \times (-1)$ (e) $-25 \times 7 \times (-4)$ (f) $8 \times 4 \times (-5)$.

5 If $x = -3$, $y = 5$ and $z = -2$, find the value of

(a) xy (b) xz (c) yz

(d) $4x + 9$ (e) $6y - 35$ (f) $7z - x$.

Exercise 4 continued

Do not use a calculator 

6 The following diagram shows a **multiplication cross**.

6	×	5	=	30
×		×		×
2	×	4	=	8
=		=		=
12	×	20	=	240

Study this multiplication cross and then copy and complete the two which follow.

(a)

2	×	4	=	
×		×		×
–3	×	–5	=	
=		=		=
	×		=	

(b)

–3	×	10	=	
×		×		×
–6	×	–2	=	
=		=		=
	×		=	

Division of Integers

When you are dividing integers, you should do so without a calculator.

Division is the reverse process of multiplication and the rules for division are the same as those for multiplication. Since $5 \times (-3) = -15$, then $-15 \div 5 = -3$. This may be summed up as follows

$$+ \div + = +$$
$$+ \div - = -$$
$$- \div + = -$$
$$- \div - = +$$

Example Work out: (a) $-20 \div 5$ (b) $-30 \div (-6)$ (c) $\frac{40}{-5}$.

Solution (a) -4 (b) 5 (c) -8.

Exercise 5

Do not use a calculator 

1 Work out:

(a) $27 \div 9$ (b) $14 \div (-2)$ (c) $-32 \div 8$ (d) $-21 \div (-3)$

(e) $-80 \div 10$ (f) $-18 \div (-9)$ (g) $35 \div (-7)$ (h) $0 \div (-1)$

(i) $-36 \div (-6)$ (j) $-11 \div (-1)$ (k) $50 \div (-2)$ (l) $-48 \div 3$.

2 Work out:

(a) $\frac{-51}{3}$ (b) $\frac{75}{5}$ (c) $\frac{80}{-4}$ (d) $\frac{-98}{7}$ (e) $\frac{64}{-4}$ (f) $\frac{-140}{-5}$.

Exercise 5 continued

Do not use a calculator

3 (a) Barry has £76. He divides it equally amongst his four grandchildren.

How much does each grandchild receive?

(b) Walter's bank account shows –£96, (he is £96 overdrawn). He has to cancel this balance in two equal instalments. What will his bank account show after the first instalment?

4 Phil's final score in the Masters Golf Championship was –20 (20 under par).

If there were 4 rounds in the Championship, what was his average score per round?

5 In a seaside town, the temperature is taken every day at 9 am during a week in winter.

The results are as shown:

6° C, –8° C, 2° C, –3° C, –2° C, –6° C, –3° C.

Find the mean (average) temperature for the week.

6 Repeat Question 5 for a week when the temperatures were:

–6° C, –5° C, 1° C, –4° C, –7° C, 2° C, –9° C.

7 Copy and complete the following two machine chains.

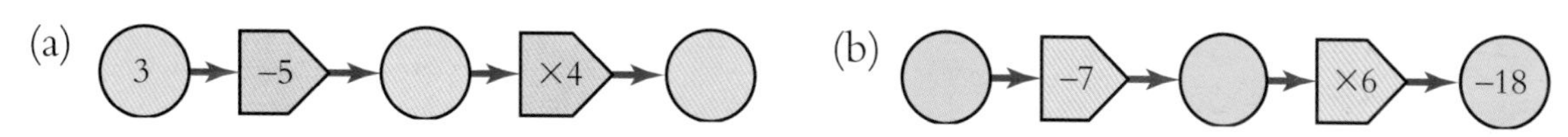

8 Copy and complete the following three multiplication crosses.

(a)

–5	×	–2	=	
×		×		×
4	×	–3	=	
=		=		=
	×		=	

(b)

–2	×		=	
×		×		×
	×		=	–10
=		=		=
	×	–8	=	–80

(c)

3	×		=	–6
×		×		×
	×	10	=	
=		=		=
	×		=	240

Exercise 6 – Revision of Chapter 5

Do not use a calculator 

1 Write down the coordinates of points W, X, Y and Z in the diagram.

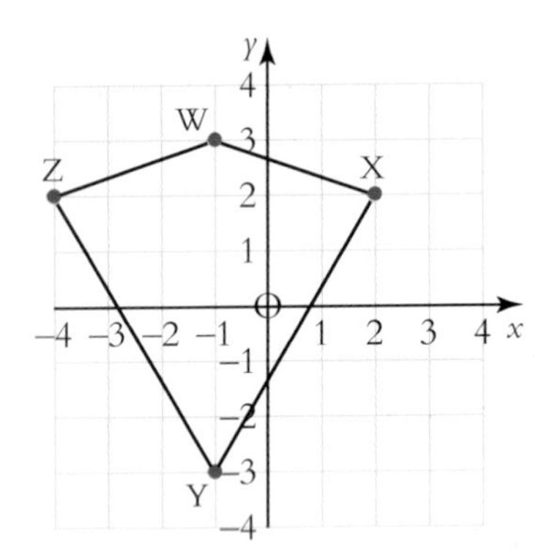

2 The points A (–5, 2), B (–1, 5) and C (3, 2) are three corners of a rhombus ABCD, Find the coordinates of the fourth corner D.

3 The lowest temperature one night was –7° C. By late morning it had risen by 12° C. What was the temperature in the late morning?

4 Work out:

(a) –3 + 8 (b) 12 + (–12) (c) –80 + 15 (d) 75 + (–45).

5 Work out:

(a) 25 – (–15) (b) 7 – 13 (c) –8 – (–20) (d) –12 – (–11).

6 The scale below can be used to convert temperatures from degrees Fahrenheit to degrees Celsius and vice versa.

Degrees Fahrenheit (° F)	–13	–4	5	14	23	32	41	50	59
Degrees Celsius (° C)	–25	–20	–15	–10	–5	0	5	10	15

(a) Use the scale to change –4° F to degrees Celsius.

(b) Use the scale to change –10° C to degrees Fahrenheit.

(c) The temperature falls from 5° C to –10° C. By how many degrees Celsius has the temperature fallen?

(d) By how many degrees Fahrenheit has the temperature fallen?

Exercise 6 – Revision of Chapter 5 continued

Do not use a calculator 

7 Copy and complete this magic square so that the three rows, the three columns and the two diagonals all add up to the same total.

−1	0	−5
	−2	

8 Work out:

(a) -15×4 (b) $6 \times (-9)$ (c) -19×0 (d) $-14 \times (-3)$.

9 Work out:

(a) $-5 \times (-6) \times 3$ (b) $-4 \times (-3) \times (-25)$.

10 If $a = 3$ and $b = 4$, find the value of

(a) $a - b$ (b) $ab - 20$.

11 Work out:

(a) $-110 \div (-10)$ (b) $154 \div (-7)$ (c) $\frac{-57}{3}$.

12 Find the mean of the integers −6, −2, 3, 2 and −2.

Summary

(Whenever a question on Integers appears, you should be able to tackle it using one or more of the following techiques.)

1 ***Coordinates and Integers***

You should be able to plot points and read coordinates.

The first coordinate, x, is 'along' from the origin (to the left or right).

The second coordinate, y, is 'up' or 'down' from the origin.

The point P (−2, 4) is 2 to the left of the origin and then 4 'up'.

Summary continued

2 ***Addition of Integers***

A number line can help you to add integers.

For example, $7 + (-5) = 2$, $-5 + 8 = 3$.

3 ***Subtraction of Integers***

Again, the number line will help you.

For example, $6 - 8 = -2$, $5 - (-3) = 8$, $-6 - (-5) = -1$.

4 ***Multiplication of Integers***

Recall: $+ \times + = +$

$+ \times - = -$

$- \times + = -$

$- \times - = +$

For example, $6 \times (-3) = -18$, $-5 \times 4 = -20$, $-7 \times (-2) = 14$.

Also, $-5 \times 3 \times (-2) = -15 \times (-2) = 30$.

5 ***Division of Integers***

We use the same rules as for multiplication.

For example, $-10 \div 2 = -5$, $12 \div (-3) = -4$, $-20 \div (-5) = 4$.

6 Speed, Distance, and Time

Did you know that although light travels at a fantastic 186 000 miles per second, it still takes about 8 minutes for light from the Sun to reach Earth? That tells us something about the distance between the Earth and the Sun!

Calculations involving speed, distance, and time are some of the most frequent calculations we perform – especially for those who drive for a living, or spend much of their working days travelling.

Relationships involving speed, distance, and time can also be dealt with by a graphical approach, and that is how we begin this chapter. This introduction to graphs also provides a useful preparation for Chapter 8, which is concerned with graphs, charts, and tables.

Distance–Time Graphs

The graphical approach to a problem often gives visual clues and so makes the problem easier to understand and to deal with.

In this section, you should be able to

- interpret distance–time graphs
- recognise the significance of the point of intersection of two lines on the same graph.

Example John goes to watch the first football match of the season.

The graph shows his journey from leaving home, to the football ground, and home again.

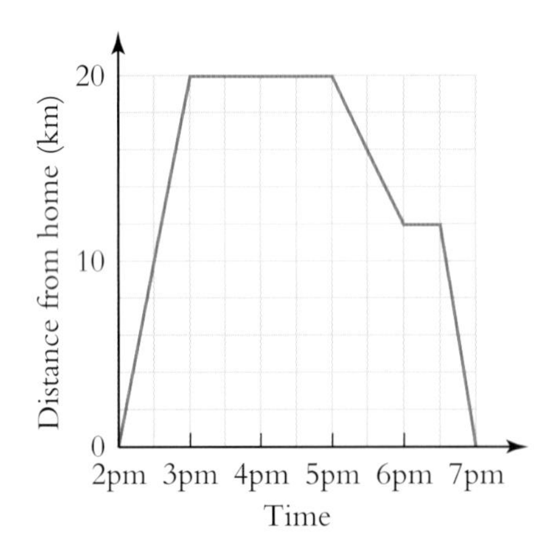

(a) How far is it from John's home to the football ground?

(b) How long did he spend in the football ground?

(c) On the way home he visits a pub. How far is the pub from the football ground?

(d) How long did he spend in the pub?

Solution

(a) 20 km (b) 2 hours

(c) 8 km (d) 30 min.

Main points

1 In the graph in the last Example: Horizontal lines represent times when John is not travelling. For example, when he is in the football ground.

2 Sloping lines represent times when John is travelling.

3 The steeper the slope of a line, the faster he is travelling.

Example Ben and May are next-door neighbours. Their journeys to school on a particular morning are shown in the following graph.

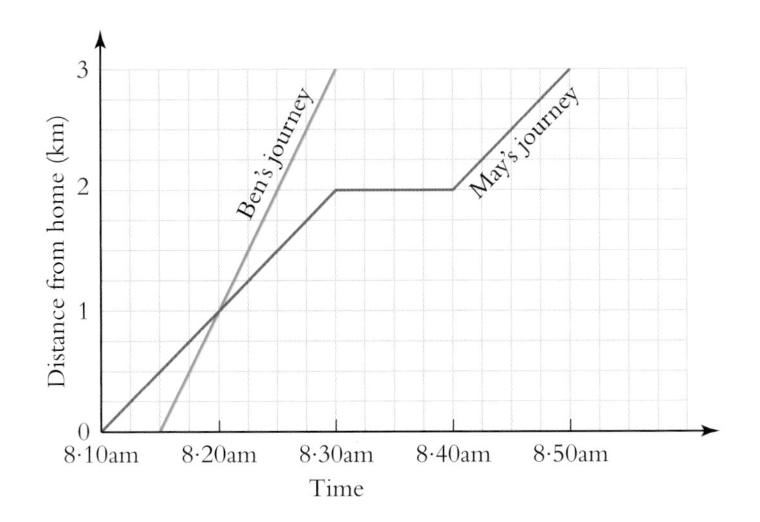

(a) How long did Ben take to travel to school?

(b) How much longer was May's journey?

(c) May stops off at a friend's house on the way. How long does she spend at her friend's house?

(d) One of them travelled by bicycle, the other walked.
Who travelled by bicycle?

(e) At what time did Ben pass May?

Solution

(a) 15 minutes (b) 25 minutes

(c) 10 minutes (d) Ben (e) 8.20 am.

(Notice that Ben passes May where the two graphs intersect. Although May left earlier than Ben (at 8.10 am), he caught up with her on his bicycle and passed her at 8.20 am. Notice too that Ben's graph is steeper than May's indicating that he travelled faster.)

Exercise 1

1 Mrs. Jones takes a coach trip from Glasgow to Perth. The coach stops for a tea break on the way. The journey is represented in the distance–time graph.

(a) How far is it from Glasgow to Perth?

(b) How far was the coach from Perth when it stopped for a tea break?

(c) How long did the tea break last?

(d) How can you tell that the speed for each part of the journey was the same?

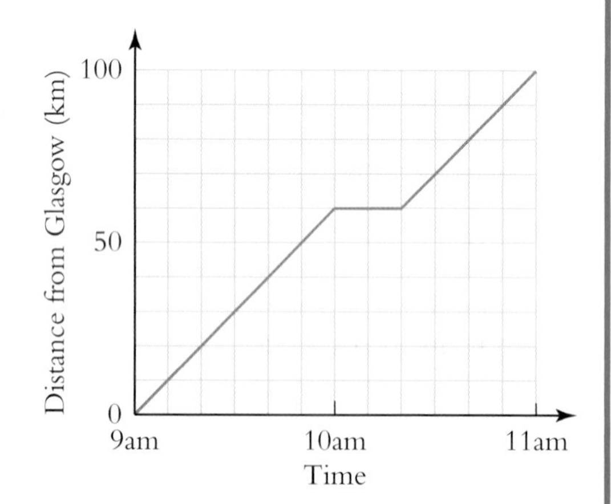

2 A group of students sets out from college by bus to visit the zoo. On the way back from the zoo, the bus has a puncture. The whole outing is represented in the distance–time graph.

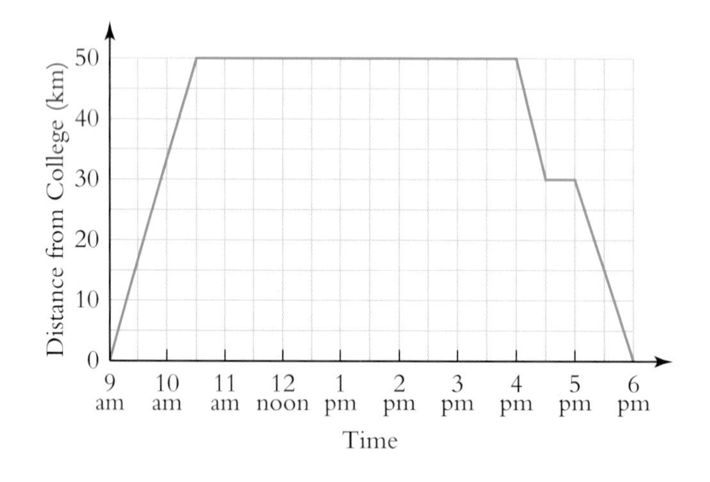

(a) How far is it from the college to the zoo?

(b) How long did the group spend at the zoo?

(c) How far were they from home when the bus had a puncture?

(d) How long did it take to repair the puncture?

(e) On the return journey, did the bus travel faster *before* or *after* the puncture?

Give a reason for your answer.

Exercise 1 continued

3 Shaheen sets out by train from Glasgow to Edinburgh at the same time as Sarah sets out by train from Edinburgh to Glasgow. Their journeys are represented in the distance–time graph.

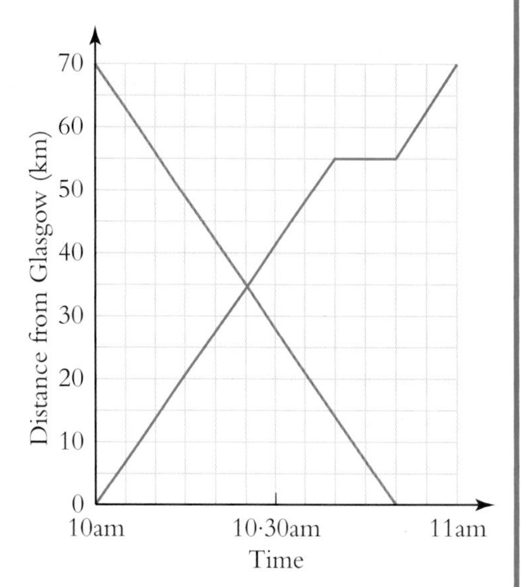

(a) How far is it from Glasgow to Edinburgh?

(b) At what time did the two trains pass each other?

(c) Whose train suffered a delay?

(d) How long did the delay last?

(e) Who arrived at her destination first?

4 Willie is travelling by car from Aberdeen to Glasgow. He stops to meet friends on the way. Jimmy also travels from Aberdeen to Glasgow. He travels without a break. They both arrive in Glasgow at the same time.
Their journeys are represented in the distance–time graph.

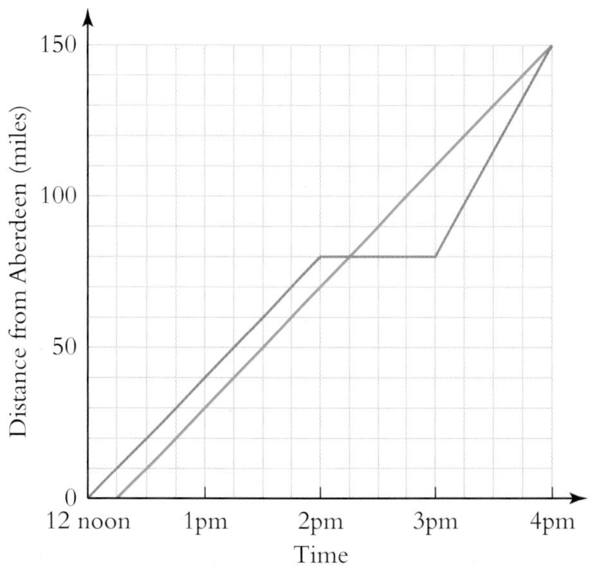

(a) At what time did Jimmy leave Aberdeen?

(b) At what time did Jimmy pass Willie?

(c) How long did Willie spend with his friends?

(d) Did Willie travel faster before or after he met his friends? Explain your answer.

(e) What distance was Jimmy from Glasgow when he passed Willie?

5 Don has to travel 100 miles from his home to Dundee for a meeting. He sets out at 10 am and travels 60 miles in an hour in his car. He then stops for half an hour for lunch. It takes him another hour to drive the remaining distance to Dundee.

(a) Draw a distance–time graph to show Don's journey.

(b) When does Don arrive in Dundee?

(c) Did he travel faster before or after his lunch? Explain your answer.

Calculating Time Intervals

It is important that you should be able to calculate time intervals, including those over midnight or midday, both on the 12-hour clock and the 24-hour clock. You should study the following examples carefully.

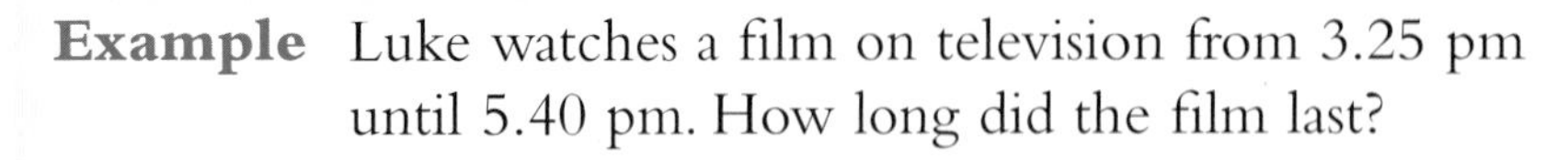

Example Luke watches a film on television from 3.25 pm until 5.40 pm. How long did the film last?

(The safest method is to count in 1 hour intervals from the start time until you are **within** 1 hour of the finish time. Then add on the extra minutes. Never use a calculator to work out a time interval.)

Solution

3.25 → 4.25 → 5.25 = 2 hours

5.25 → 5.40 → = (40 – 25) minutes = 15 minutes

The film lasted 2 hours 15 minutes.

Example Tom's train departs at 22 40. It arrives at its destination at 03 10 the following morning. How long did the train journey last?

Solution

22 40 → 23 40 → 00 40 → 01 40 → 02 40 = 4 hours

02 40 → 03 00 = (60 – 40) = 20 minutes

03 00 → 03 10 = 10 minutes

The journey time was 4 hours + (20 +10) minutes. That is, 4 hours 30 minutes.

Exercise 2

1 Jonathon watches a film on television. It starts at 9.25 pm and finishes at 11.50 pm. How long did the film last?

2 Archie watches the big game on TV. The programme starts at 7.30 pm and finishes at 10.15 pm. How long does the programme last?

Exercise 2 continued

3 The 10 30 train from Glasgow arrived in Edinburgh at 11 18. How long did the journey last?

4 Justin goes out jogging from 7.30 pm until 9.00 pm.

(a) For how long was he jogging?

(b) If his speed was a steady 8 miles per hour, what distance did he jog?

5 Derek has to travel each day as part of his job. One week in December, he had the following journey times:

Monday	Depart	07 30	Arrive	11 55
Tuesday	Depart	12 30	Arrive	15 25
Wednesday	Depart	08 50	Arrive	12 20
Thursday	Depart	15 20	Arrive	17 05
Friday	Depart	08 55	Arrive	15 20

Find his journey time each day.

6 Sean takes the overnight train from London to Dundee. The train leaves London at 21 30 hours and arrives in Dundee at 06 15 hours the following morning. How long does the train journey last?

7 The following programmes were shown on a television channel one evening:

6.00 pm	News Report
6.30 pm	Local News
6.45 pm	Scenes from Soaps
7.30 pm	European Football Live
10.00 pm	Film – *The Curse of Pi*
11.30 pm	Evening News
11.45 pm	Wee Brother

(a) How long does 'Scenes from Soaps' last?

(b) Brian plans to watch 'Scenes from Soaps', European Football, and also the film. How long does he plan to spend watching television?

(c) Due to extra time in the football, all programmes are delayed by 50 minutes. When will 'Wee Brother' start now?

Exercise 2 continued

8 Bert is working on night shift. He starts work at 10.40 pm and finishes at 7.25 am the next day. How long does he work?

9 Part of the Amsterdam to Paris railway timetable is as follows:

Amsterdam	18 57	19 57	22 25
Rotterdam	19 54	20 54	23 24
Antwerp	20 56	21 56	00 47
Brussels	21 40	22 40	01 30
Paris	23 05	00 10	02 55

(a) How long does the 18 57 train take from Amsterdam to Paris?

(b) Pierre misses the 19 57 train from Amsterdam to Paris by one minute. How long will he have to wait until the next train?

(c) Paul takes the 22 40 train from Brussels to Paris. How long does his journey take?

(d) How long does the 22 25 train take to travel from Amsterdam to Paris?

10 Martin has a DVD which lasts for 135 minutes. If he starts running it at 8.50 pm, when will it finish?

11 Amy has to check in at the airport at 3.20 pm. If it takes her 4 hours 30 minutes to travel from home to the airport, what is the latest time she can leave home?

12 Shona has an interview at 10.15 am. She has allowed 2 hours 30 minutes to travel from her home to the venue for the interview. What is the latest time she can leave home to be in time for the interview?

Calculating Speed

Key words and definitions

Speed is the term we use to describe how quickly something or someone is moving.

Speed may be measured in miles per hour (mph), kilometres per hour (km/h) or metres per second (m/s).

To find the *average speed* for a journey, we divide the distance travelled by the time taken for the journey.

$$\textbf{Speed} = \frac{\textbf{Distance}}{\textbf{Time}} \quad \textbf{or} \quad S = \frac{D}{T}$$

Example A lorry travels 144 kilometres in 3 hours. Find its average speed for the journey.

Solution

$S = \frac{D}{T} = \frac{144}{3} = 48$ Hence the average speed is 48 km/h.

Example Thomas leaves home at 9.50 am in his car and drives 135 miles to his destination, arriving at 12.05 pm. What was his average speed for the journey?

Solution

9.50 → 10.50 → 11.50 = 2 hours

11.50 → 12.05 = 15 minutes

so time taken = 2 hours 15 minutes = $2\frac{1}{4}$ hours = 2·25 hours

$S = \frac{D}{T} = \frac{135}{2{\cdot}25} = 60$ Hence average speed = 60 mph.

(Remember that 15 minutes = $\frac{1}{4}$ hour = 0·25 hour;

30 minutes = $\frac{1}{2}$ hour = 0·5 hour;

45 minutes = $\frac{3}{4}$ hour = 0·75 hour.

For any other number of minutes, we divide the number of minutes by 60 to find the decimal fraction of an hour. For example, 36 minutes = $\frac{36}{60}$ = 0·6 hour.)

Example Dave cycles 30 kilometres in 1 hour 12 minutes. Find his average speed.

Solution

Time = 1 hour 12 minutes = $\left(1 + \frac{12}{60}\right)$ hours = 1·2 hours.

$S = \frac{D}{T} = \frac{30}{1{\cdot}2} = 25$ Hence average speed = 25 km/h.

Exercise 3

1 Find the average speed for each of these journeys:

(a) A jogger runs 18 miles in 3 hours.

(b) A train travels 520 kilometres in 4 hours.

(c) A car covers 280 kilometres in 5 hours.

(d) Cheryl walks 15 kilometres in 5 hours.

(e) An aeroplane flies 2703 kilometres in 3 hours.

(f) A space shuttle flies 210 960 miles in 12 hours.

2 An athlete is declared the 'fastest man in the world' after running 100 metres in 9·77 seconds. Find his average speed in metres per second. (Give your answer correct to 2 decimal places.)

3 The cheetah is the fastest land animal in the world. One cheetah was timed running 760 metres in 24 seconds. Find its average speed in metres per second, correct to 2 decimal places.

4 Find the average speed for each of these journeys:

(a) Jensen drives 200 miles in 2 hours 30 minutes.

(b) Lance cycles 135 miles in 3 hours 45 minutes.

(c) Paula runs 40·5 kilometres in 2 hours 15 minutes.

5 Noelle drives to work each morning. The journey is 36 kilometres and it takes her 45 minutes to get there. Find her average speed.

6 Find the average speed for the following journeys:

(a) A bus travels 48 miles in 2 hours 24 minutes.

(b) A car travels 124 kilometres in 3 hours 6 minutes.

7 Edith leaves home in her car at 7.45 am. She drives 90 miles to a meeting. If she arrives at 9.30 am, what was her average speed for the journey? (Give your answer to the nearest kilometre per hour.)

8 Lorenzo takes the 07 30 train from Rome to Milan. The distance is 630 kilometres. If the train arrives in Milan at 11 00 hours, find its average speed.

Exercise 3 continued

9 Donalda flies from Stornoway to Bristol. She has to change planes at Edinburgh. Her complete journey is represented in the following distance–time graph.

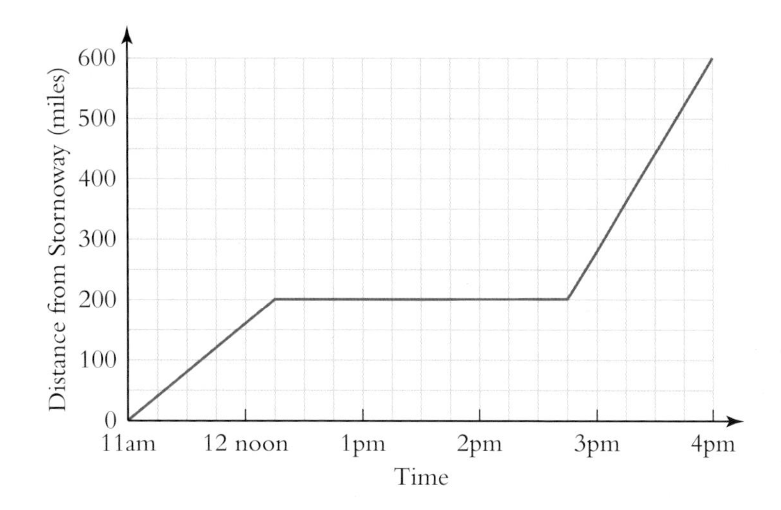

(a) What distance is the flight from Stornoway to Edinburgh?

(b) Find the average speed of the flight from Stornoway to Edinburgh.

(c) How long did Donalda have to wait at Edinburgh?

(d) What distance is the flight from Edinburgh to Bristol?

(e) Find the average speed of the flight from Edinburgh to Bristol.

10 Dr Cameron leaves his surgery to visit one of his patients at his home. He then returns to the surgery. The distance–time graph for his visit is as shown.

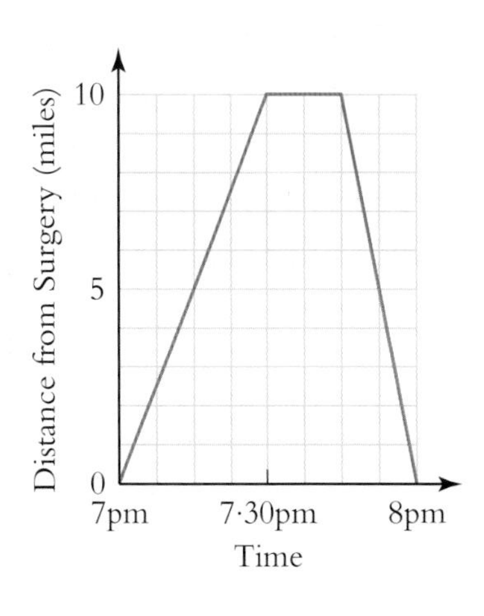

(a) How far did Dr Cameron drive altogether?

(b) How long did he spend with the patient?

(c) What was his average speed on the return journey?

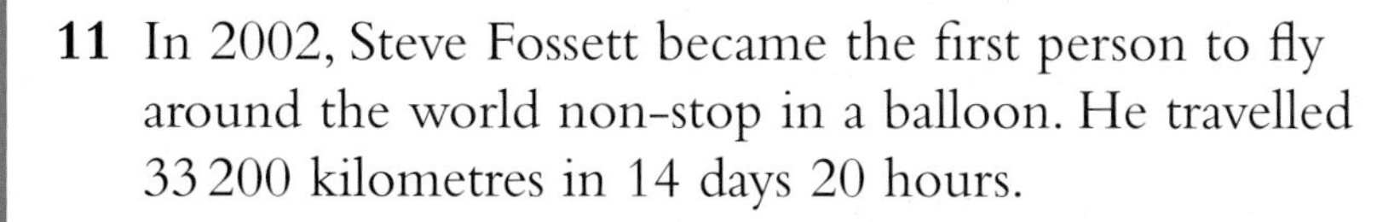

11 In 2002, Steve Fossett became the first person to fly around the world non-stop in a balloon. He travelled 33 200 kilometres in 14 days 20 hours.

Find his average speed to the nearest kilometre per hour.

Calculating Distance

Now that you have tackled the questions in Exercise 3, it is perhaps obvious to you how we calculate distance instead of average speed. (Usually in a car journey however we can *read* the distance travelled from the trip counter, and then use it to calculate our average speed.)

To calculate the distance travelled in a journey, we can multiply the average speed by the time taken.

Distance = Speed × Time or $D = S \times T$

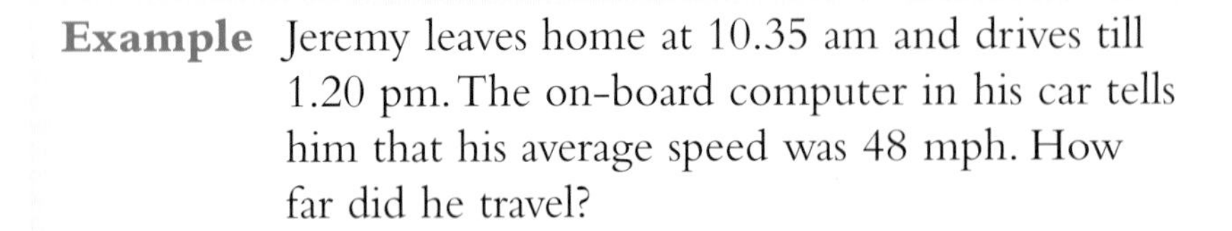

Example Jeremy leaves home at 10.35 am and drives till 1.20 pm. The on-board computer in his car tells him that his average speed was 48 mph. How far did he travel?

Solution

10.35 → 11.25 → 12.35 = 2 hours

12.35 → 1.00 = (60 – 35) minutes = 25 minutes

1.00 → 1.20 = 20 minutes

The journey time was 2 hours + (25 + 20) minutes = 2 hours 45 minutes.

Also, 2 hours 45 minutes = $2\frac{3}{4}$ hours = 2·75 hours.

Hence $D = S \times T = 48 \times 2{\cdot}75 = 132$.
Hence distance travelled = 132 miles.

Exercise 4

1 Find the distance travelled in:

(a) 3 hours at an average speed of 55 km/h

(b) 5 hours at an average speed of 44 mph

(c) 30 seconds at an average speed of 8 m/s

(d) 7·5 hours at an average speed of 80 km/h

(e) 12 hours at an average speed of 30 mph.

Exercise 4 continued

2 Joseph drives from 5 pm until 8 pm at an average speed of 50 mph. How far did he drive?

3 Find the distance travelled in:

(a) 2 hours 30 minutes at an average speed of 70 km/h

(b) 3 hours 15 minutes at an average speed of 52 km/h

(c) 6 hours 45 minutes at an average speed of 88 mph.

4 Mohammed drives from 6.50 pm until 9.05 pm and at an average speed of 64 km/h.

What distance does he travel?

5 Janet drives for 45 minutes at an average speed of 40 mph. Calculate the distance she travels.

6 Benjamin travels overnight by train. The departure time is 21 50 hours and the train travels at an average speed of 108 km/h, arriving at its destination at 03 20 hours.

What distance did Benjamin travel?

7 Zander drives from Aberdeen to Inverness at an average speed of 64 km/h. He leaves Aberdeen at 9.50 am and arrives in Inverness at 12.35 pm.

How far do these figures suggest it is from Aberdeen to Inverness?

8 Find the distance travelled in:

(a) 3 hours 42 minutes at an average speed of 50 km/h

(b) 1 hour 24 minutes at an average speed of 75 km/h

(c) 39 minutes at an average speed of 60 km/h

(d) 1 hour 51 minutes at an average speed of 40 mph

(e) 2 hours 18 minutes at an average speed of 30 mph

(f) 2 hours 20 minutes at an average speed of 30 mph.

9 Isla drives for 3 hours 30 minutes at an average speed of 70 mph on the motorway. Then she drives for 2 hours 15 minutes at an average speed of 40 mph on minor roads.

How far did she travel altogether?

Calculating Time

Now that you are able to calculate average speed (using distance and time) and distance (using average speed and time), you should have little difficulty in calculating time, using distance and average speed.

To calculate the time taken for a journey, we divide the distance travelled by the average speed.

$$\textbf{Time} = \frac{\textbf{Distance}}{\textbf{Speed}} \quad \textbf{or} \quad T = \frac{D}{S}$$

Example Roberta drives 63 kilometres at an average speed of 28 kilometres per hour. How long did her journey take?

Solution $T = \frac{D}{S} = \frac{63}{28} = 2{\cdot}25.$

Since 2·25 hours = $2\frac{1}{4}$ hours we can say that Roberta's journey took 2 hours 15 minutes.

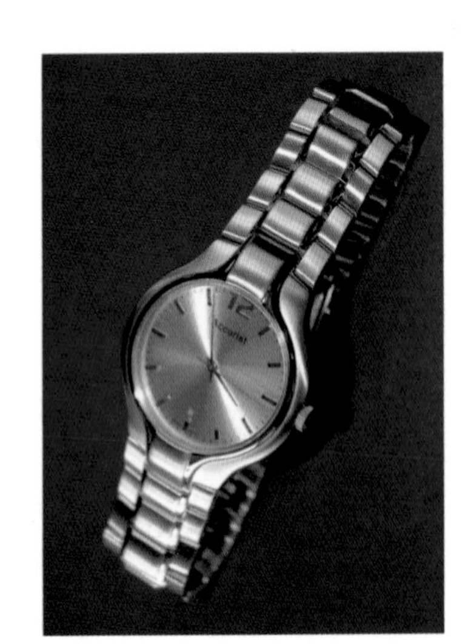

Example An ambulance transporting a patient, leaves Oban at 8.30 am. It travels 92 miles to Glasgow at an average speed of 40 mph. When will the ambulance arrive in Glasgow?

Solution

$T = \frac{D}{S} = \frac{92}{40} = 2{\cdot}3.$

Now 2·3 hours = 2 hours + 0·3 × 60 minutes = 2 hours 18 minutes.

The ambulance arrives 2 hours 18 minutes after 8.30 am. That is, at 10.48 am.

Take care

When you are changing a *decimal fraction of an hour to minutes*, multiply by 60, e.g. 0·9 hours = 0·9 × 60 = 54 minutes.

Exercise 5

1 Calculate the time taken for each of the following journeys:

(a) 210 km at an average speed of 70 km/h

(b) 160 m at an average speed of 8 m/s

(c) 175 miles at an average speed of 35 mph

(d) 78 cm at an average speed of 3 cm/s

(e) 1360 km at an average speed of 160 km/h.

(Give your answers to the following questions in hours and minutes.)

2 Calculate the time taken for each of the following journeys:

(a) 150 miles at an average speed of 60 mph

(b) 44 km at an average speed of 16 km/h

(c) 350 km at an average speed of 56 km/h.

3 Sasha drives 180 miles at an average speed of 40 mph. How long will her journey take her?

4 How long did it take for a speedboat to go from Hastings (on the south coast of England) to Cherbourg (on the north coast of France), given that the two towns are 195 km apart and that the speedboat recorded an average speed of 60 km/h?

5 Jan set out for work at 7.50 am and drove the 18 miles to her office at an average speed of 24 mph. When did she arrive at the office?

6 Steve takes the train from Birmingham to London, a distance of 180 kilometres. If the train sets out at 09 04 hours and travels at an average speed of 72 kilometres per hour, when does Steve arrive in London?

7 Calculate the time taken for each of the following journeys:

(a) 228 km at an average speed of 60 km/h

(b) 99 miles at an average speed of 30 mph

(c) 94 km at an average speed of 40 km/h

(d) 120 km at an average speed of 90 km/h.

Exercise 5 continued

8 The distance from Inverness to London is 552 miles. Richard takes the 20 20 train from Inverness to London. This train travels at an average speed of 60 miles per hour.

When will Richard arrive in London?

9 Mary drives 24 miles from Glasgow to Gourock. Her average speed is 48 mph. She then waits 15 minutes before taking the ferry from Gourock to Dunoon. The crossing is exactly 3 miles and the average speed of the ferry is 10 mph.

How long did it take Mary to get from Glasgow to Dunoon?

10 Tony *must* reach the airport by 15 10 hours in order to book in for a flight. He plans to travel the 35 miles from his home to the airport by minibus. The minibus can only do this journey at an average speed of 28 mph. What is the latest time that Tony can leave his home?

11 A train in a theme park travels round a circular track of radius 100 metres.

(a) Calculate the circumference of the track. (Take $\pi = 3{\cdot}14$.)

(b) If the train has an average speed of 7 m/s, how long will it take to complete one circuit of the track? (Answer to the nearest second.)

Speed, Distance, Time – Calculating any one given the other two

We now have three formulae for calculating average speed, distance, and time. These are:

$$S = \frac{D}{T}, \quad D = S \times T, \quad T = \frac{D}{S}.$$

To help us memorise and apply these formulae, we draw the 'distance, speed, time' triangle. Thus:

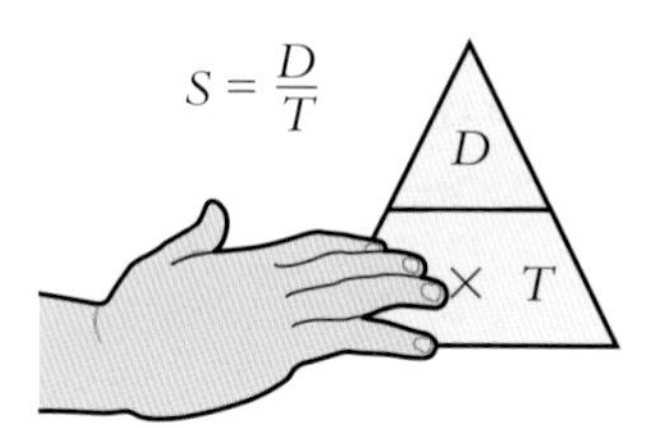

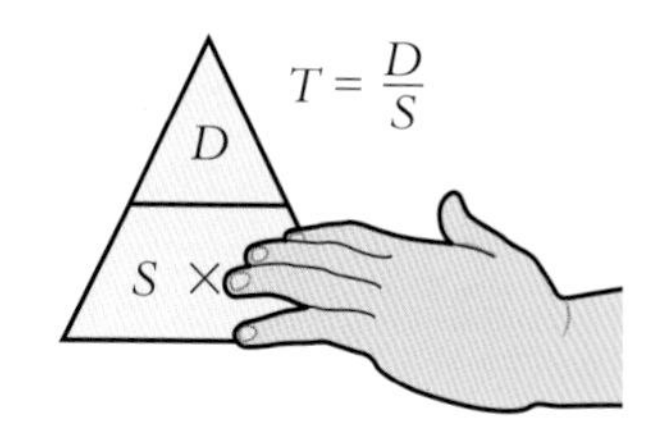

To use the triangle to calculate any of the distance, speed, time, cover up with your finger the quantity you want to find – distance, speed, or time. What you can still see shows you the correct formula to use to find the quantity. So if you want to find speed, cover up *S*, and you will see $\frac{D}{T}$!

If you want to find distance, cover up *D* and you will see $S \times T$!

If you want to find time, cover up *T*, and you will see $\frac{D}{S}$!

Exercise 6

(Draw a '*D S T*' triangle and practice using it in the following questions.)

1 Find the average speed for each of these journeys:

(a) Shaun drives 245 miles in 7 hours

(b) An aircraft flies 1125 kilometres in 1 hour 15 minutes

(c) Fatima drives 171 kilometres in 4 hours 45 minutes.

2 Calculate the time taken for each of the following journeys:

(a) 132 km at an average speed of 12 km/h

(b) 351 miles at an average speed of 52 mph

(c) 95 km at an average speed of 38 km/h.

3 Calculate the distance travelled in journeys whose times and average speeds are:

(a) 4 hours, 69 mph

(b) 3 hours 15 minutes, 64 km/h

(c) 5 hours 30 minutes, 92 km/h.

Exercise 6 continued

4 Camilla leaves London at 8.55 am to drive to Truro in Cornwall, a distance of 450 kilometres. If she averages a speed of 60 kilometres per hour, when will she arrive in Truro?

5 A high speed train travels from Hamburg to Munich in Germany. The train leaves at 22 15 hours and travels at an average speed of 140 km/h. If the train arrives in Hamburg at 04 00 hours the following morning, find the distance it travels.

6 William drives coaches. One morning he drove from 9.50 am until 11.35 am without stopping. When he checked his trip counter, he discovered that he had travelled 112 miles. Calculate his average speed for the drive.

Exercise 7 – Revision of Chapter 6

1 Part of the Glasgow to Stranraer train timetable is as shown:

Glasgow	d.	22 05
Kilmarnock	a.	22 41
Kilmarnock	d.	22 47
Ayr	a.	23 17
Maybole	a.	23 28
Girvan	a.	23 43
Barrhill	a.	00 07
Stranraer	a.	00 44

(a) Ivan wishes to catch the 22 05 train from Glasgow. He must set out from his home in Fife $2\frac{1}{2}$ hours before the train departs. What is the latest time he can leave home?

(b) What do 'd' and 'a' stand for?

(c) How long does the train journey from Glasgow to Stranraer take?

(d) How long does the train wait in Kilmarnock?

Exercise 7 – Revision of Chapter 6 continued

(e) Which two stations are closest together (assuming the train maintains a constant speed throughout)?

(f) Ayr is 25 kilometres from Kilmarnock? Calculate the average speed of the train between Kilmarnock and Ayr in kilometres per hour.

(g) After Ivan arrives in Stranraer, he waits 2 hours 16 minutes for the Stranraer to Belfast ferry. The ferry journey then takes 2 hours 50 minutes. At what time will Ivan arrive in Belfast? (Give your answer in 24-hour time.)

2 (a) Jamie drives 100 kilometres in 2 hours 30 minutes. Find his average speed.

(b) Fiona drives for 5 hours at an average speed of 85 kilometres per hour. How far does she drive?

(c) Mark drives 225 miles at an average speed of 60 mph. How long does his journey take?

3 A teacher takes a class of students to see the ballet *Sleeping Beauty*.

They set out from school at 6.00 pm for the theatre.

(a) How far is it from the school to the theatre?

(b) How long did the students spend at the theatre?

(c) Calculate the average speed of the return journey from the theatre back to school.

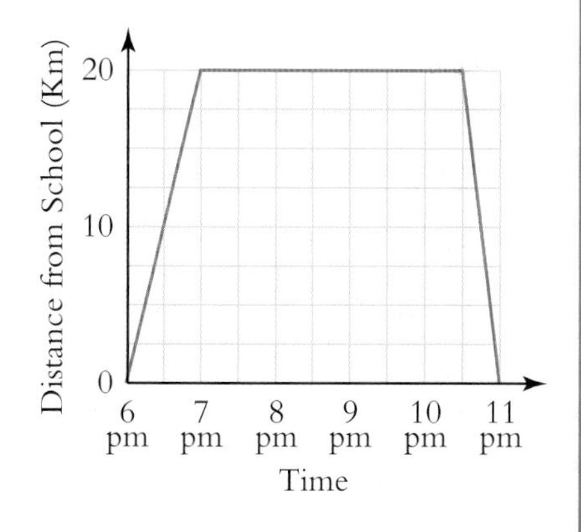

4 Marco drives at an average speed of 60 mph from 10.50 am till 1.35 pm. What distance does he travel in that time?

5 As part of her marathon training, Kayleigh sets out to run to her mother's house which is 31.5 km away. If she leaves at 5.45 pm and runs at an average speed of 14 kilometres per hour, when will she meet her mother?

6 Stan has to make a motorway journey of 434 miles. He knows that he can keep up an average speed of 56 miles per hour. At what time should he set off if he is to arrive at his destination at 05 30 hours?

Summary

(This Chapter contains three main ideas and these are explained again briefly in this short summary)

1 *Distance–Time Graphs*

Where two lines on the graph cross, this indicates a meeting point.

Horizontal lines indicate a pause in a journey.

The steeper the slope of the line, the greater the speed in that portion of the journey.

Here, Brian drives 100 miles from his home, stops at 7 pm for 30 minutes, then drives another 100 miles. His speed is increased after his stop.

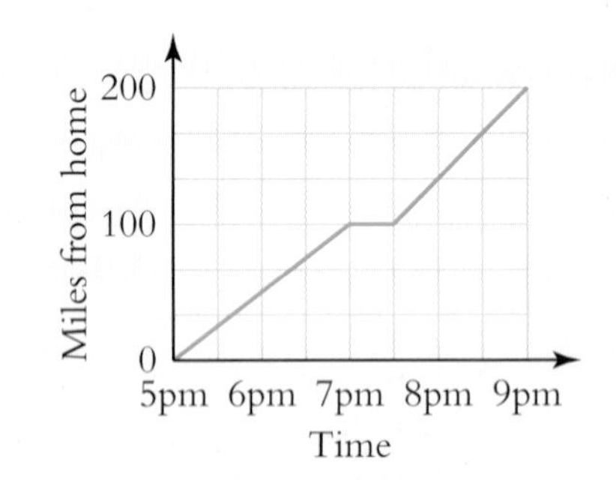

2 *Time Intervals*

We count in 1 hour intervals from the *start* until *within an hour* of the finish.

We then add on any minutes. For example from 2235 to 0405:

(2235 = 10.35 pm, 0405 = 4.05 am)

10.35 → 11.35 → 12.35 → 1.35 → 2.35 → 3.35 = 5 hours.

3.35 → 4.05 = 30 minutes.

Hence time interval = 5 hours 30 minutes.

3 *Calculating Speed, Distance and Time*

$S = \frac{D}{T}$, $D = S \times T$, $T = \frac{D}{S}$. Remember to use the triangle!

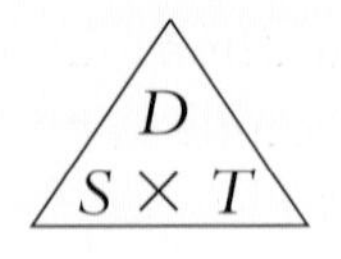

4 *Time conversions*

- 30 minutes = 0·5 hours, 15 minutes = 0·25 hours, 45 minutes = 0·75 hours
- to change minutes to hours, divide by 60 (12 minutes = $\frac{12}{60}$ = 0·2 hours)
- to change hours to minutes, multiply by 60 (0·6 hours = 0·6 × 60 = 36 minutes).

7 The Theorem of Pythagoras

The name of Pythagoras is perhaps the most famous name in Mathematics. Almost everyone has heard of it, and almost everyone who has left school, including those who left *many* years ago, will probably be able to tell you that 'Pythagoras' has something to do with 'the square on the hypotenuse'.

Pythagoras was a Greek mathematician who lived in the 6th Century BC.

He is remembered for his work in mathematics and music, and his ideas about the human soul. He was also among the first philosophers to believe that the Earth was a sphere – most people thought that it was flat!

He is given credit for the theorem which we study in this chapter. This theorem helped to lay the foundations of a great deal of geometry.

A theorem is a mathematical statement, which may be proved to be true by a chain of reasoning.

The Theorem of Pythagoras concerns right-angled triangles. The side opposite the right angle in such triangles is called the **hypotenuse**, and so the hypotenuse is the longest side in any right-angled triangle.

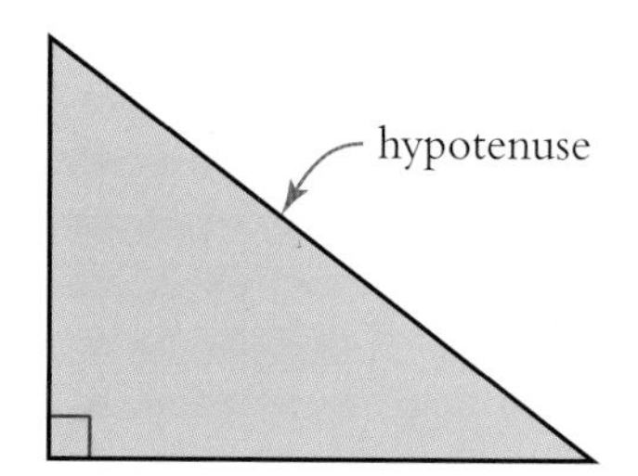

A statue of Pythagoras in Greece

Pythagoras showed that the square on the hypotenuse equals the sum of the squares on the other two sides. This means that the areas of the two smaller squares in the following diagram add up to the area of the larger square on the hypotenuse. This is true for every right-angled triangle.

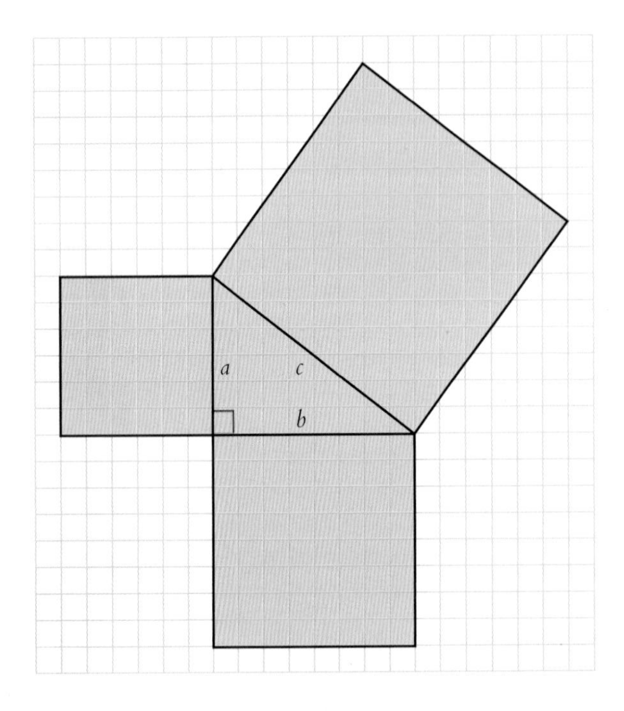

This can be written as a formula: $c^2 = a^2 + b^2$

The following diagram shows how the squares on sides a and b can be incorporated to complete the square on side c, the hypotenuse. This has involved dividing the square on side b into four equal sections.

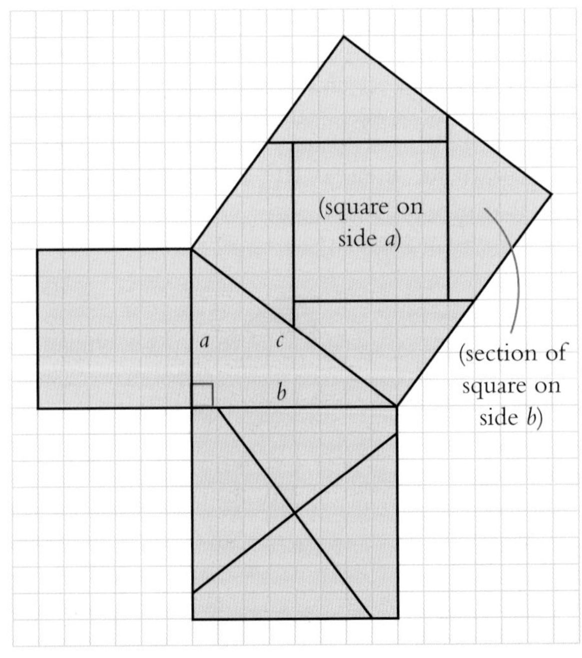

The theorem of Pythagoras can always be used to calculate the length of one side in a right-angled triangle when the lengths of the other two sides are known.

Calculating the length of the hypotenuse

Here you will see how to calculate the length of the hypotenuse in a right-angled triangle, when you are given the lengths of the two shorter sides.

Example Calculate the length of the hypotenuse, c, in the triangle.

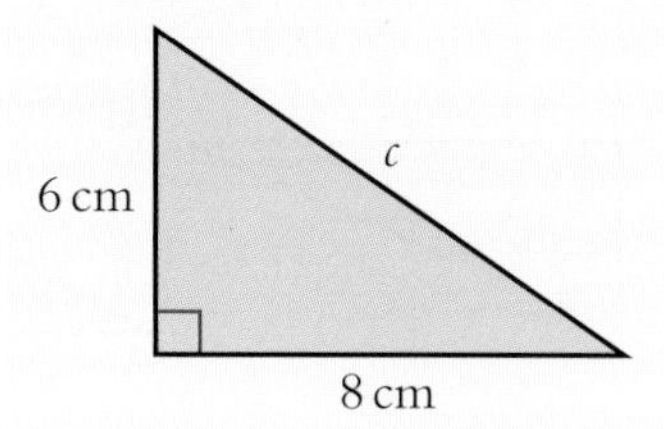

Solution

Using the Theorem of Pythagoras:
$$\begin{aligned} c^2 &= a^2 + b^2 \\ &= 6^2 + 8^2 \\ &= 36 + 64 \\ &= 100. \end{aligned}$$
$$\begin{aligned} \text{Hence } c &= \sqrt{100} \\ &= 10. \end{aligned}$$

The hypotenuse is 10 cm long.

(To find the square of any number, such as 8^2, you can either work out 8×8 or use the x^2 button if you have a scientific calculator.)

Example Calculate y in the following diagram. (Give your answer correct to one decimal place.)

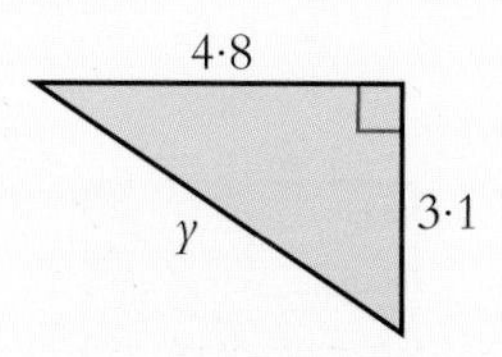

Solution

$$\begin{aligned} y^2 &= 4{\cdot}8^2 + 3{\cdot}1^2 \\ &= 23{\cdot}04 + 9{\cdot}61 \\ &= 32{\cdot}65. \end{aligned}$$

$$\begin{aligned} \text{Hence } y &= \sqrt{32{\cdot}65} \\ &= 5{\cdot}7 \text{ (correct to one decimal place).} \end{aligned}$$

Exercise 1

1 Calculate the length of the hypotenuse in each of the following four triangles.

(a)
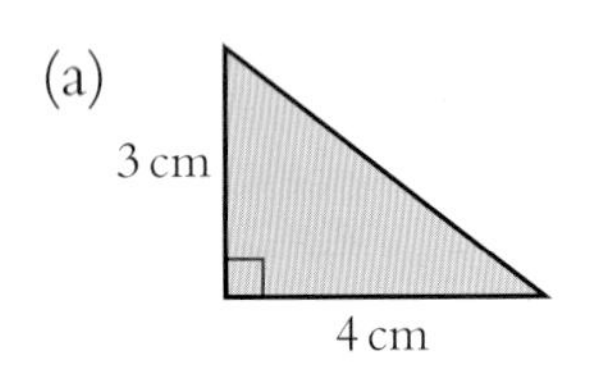

(b)
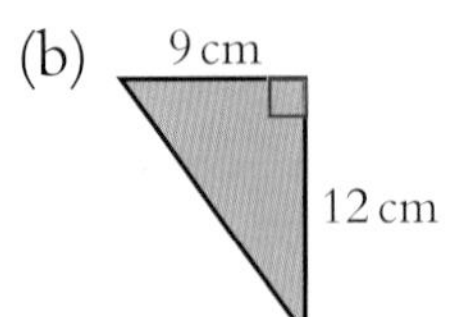

(c)
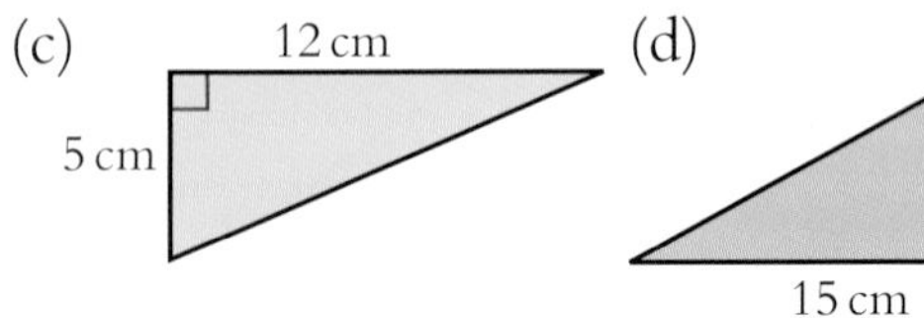

(d)
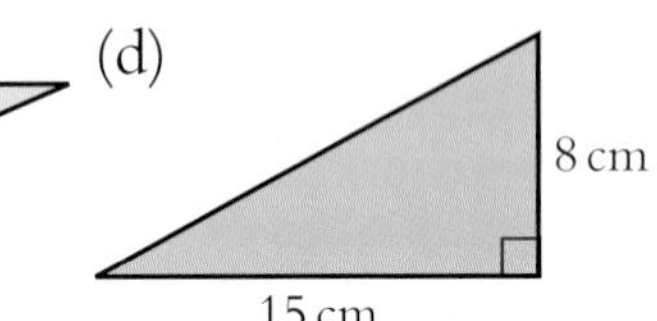

2 Calculate the size of side x in each of the following four triangles.

(a) (b) (c)
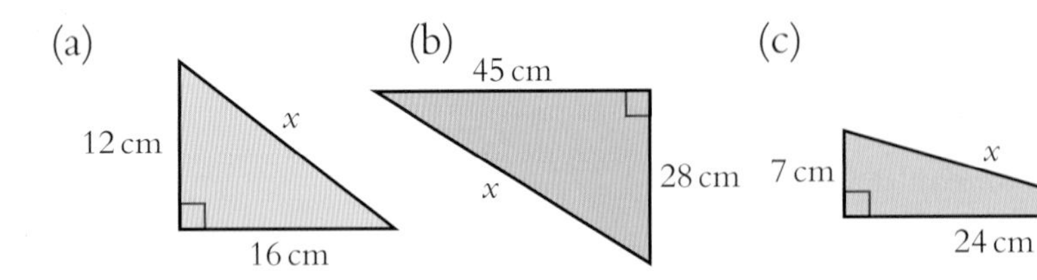

(d)
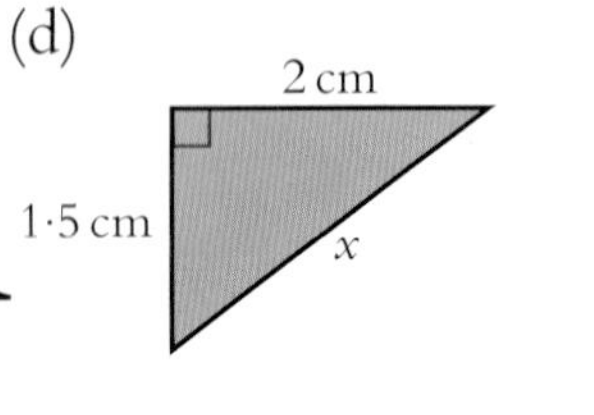

3 Calculate the size of side c in each of the following four triangles. (Give your answer correct to one decimal place.)

(a)
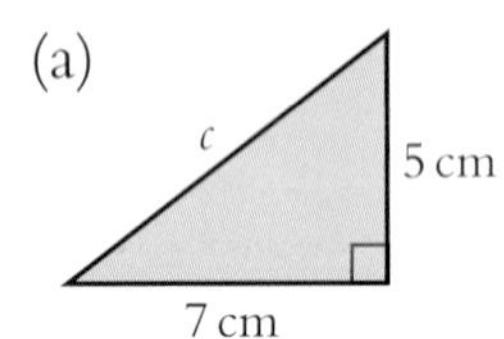

(b)
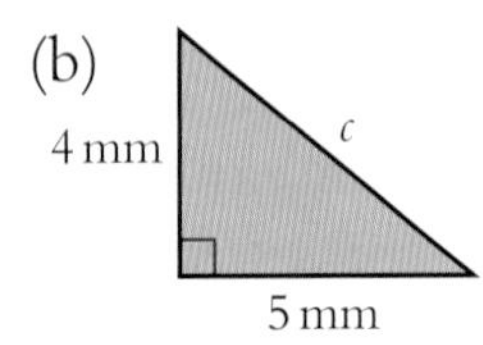

(c)
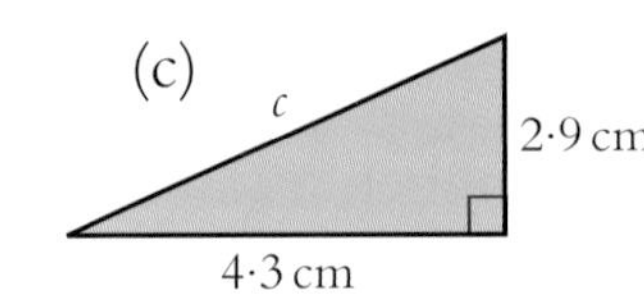

(d)
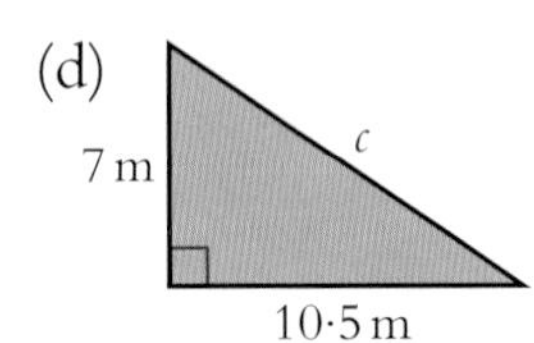

4 Percy has an allotment in the shape of a right-angled triangle.

Calculate the length of the hypotenuse in his allotment.

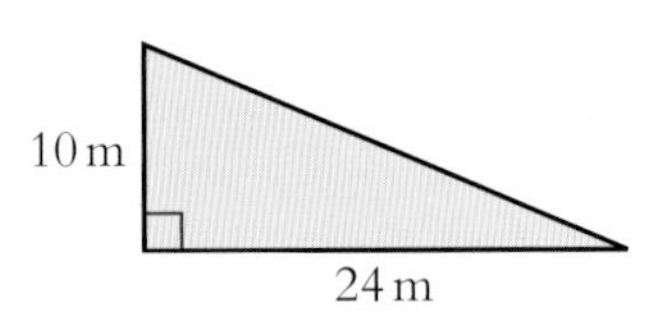

5 Calculate the length of the diagonal in the square.

(Give your answer correct to two decimal places.)

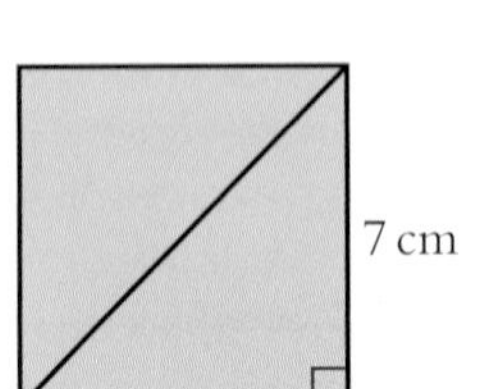

6 Calculate the length of the diagonal in the rectangle. (Give your answer correct to one decimal place.)

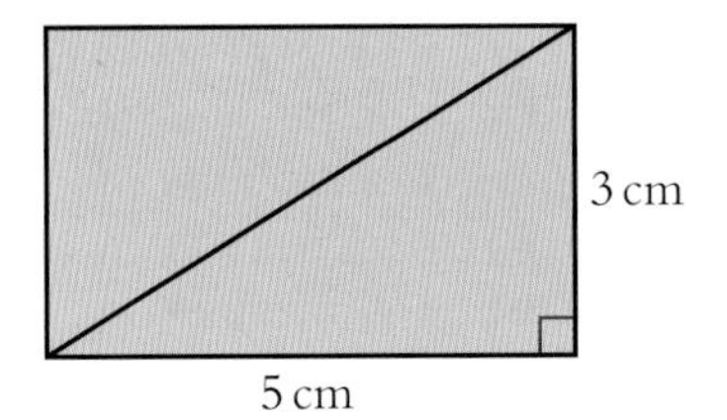

Exercise 1 continued

7 A ladder rests against a vertical wall. The foot of the ladder is 1·5 metres from the foot of the wall and the top of the ladder reaches 8 metres up the wall as shown in the diagram.

Calculate the length of the ladder to one decimal place.

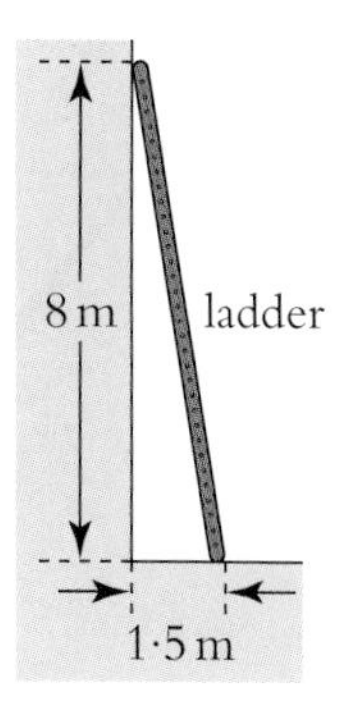

8 The diagram shows an isosceles triangle PQR.

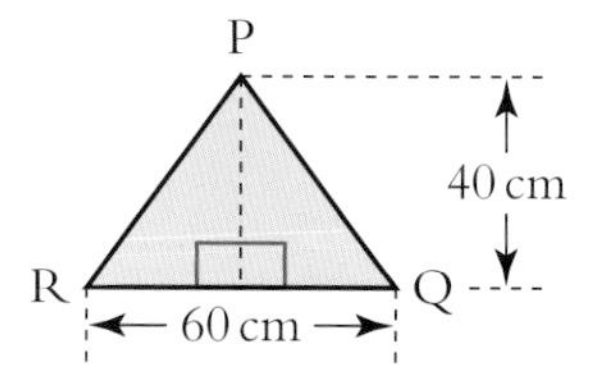

Calculate the length of side PQ.

Calculating the length of a shorter side

Here you will see how to calculate the length of a shorter side in a right-angled triangle, given the lengths of the hypotenuse and the other shorter side.

Technique

When you are finding the length of the hypotenuse (the longest side) in a right-angled triangle, you first **add** the squares of the other two sides. When you are finding the length of one of the shorter sides however, you must first subtract the squares of the other two sides.

Triangular images are more common than you might realise – take a closer look

Example Find the length of side a in the following diagram.

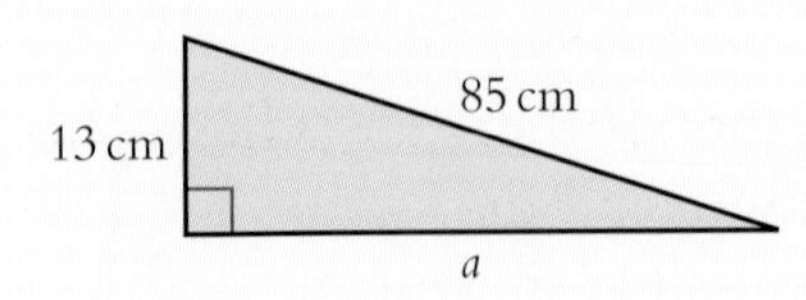

Solution Here

$$85^2 = a^2 + 13^2$$
$$\Rightarrow a^2 = 85^2 - 13^2$$
$$= 7225 - 169$$
$$= 7056$$
$$\Rightarrow a = \sqrt{7056}$$
$$= 84.$$

The length of side a is thus 84 cm.

Exercise 2

1 Calculate the size of side x in each of the following four triangles.

(a)
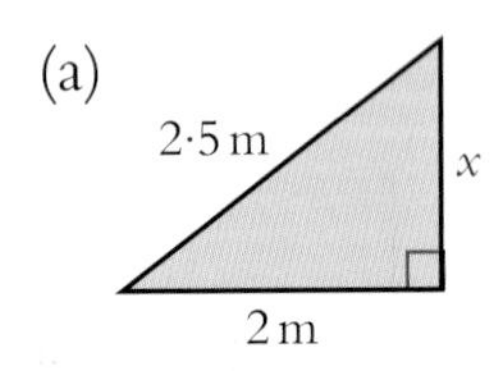

(b)
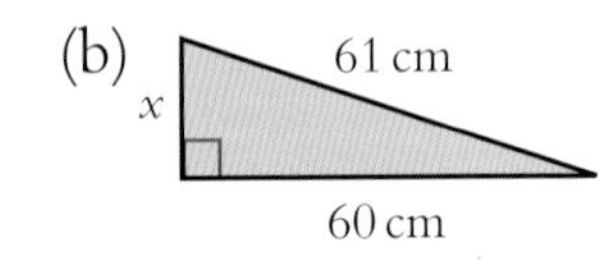

(c)
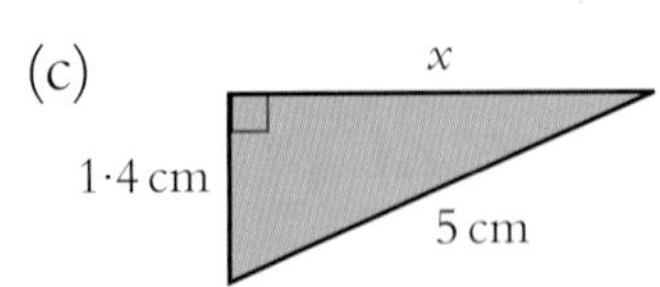

(d)
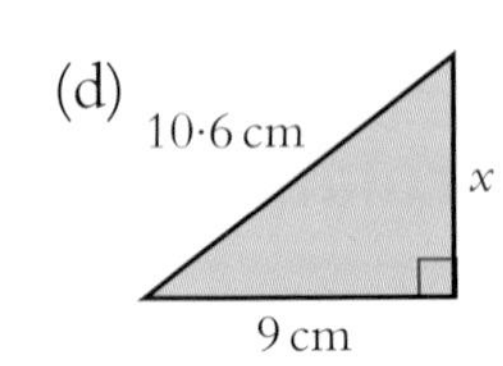

2 Calculate the sizes of p, q, r, and s in the following triangles. (Give your answers correct to one decimal place.)

(a)
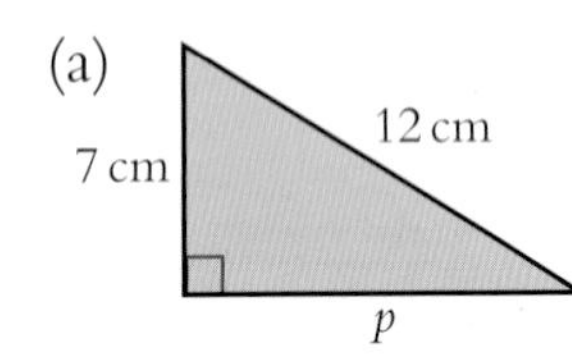

(b)
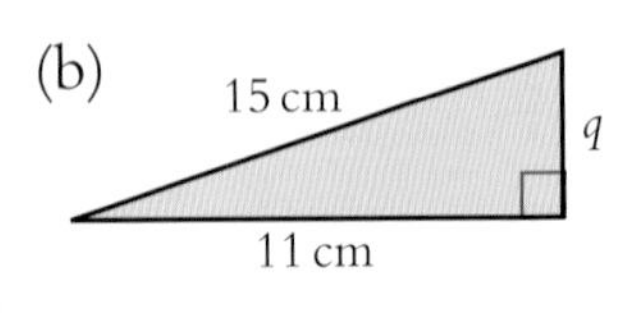

(c)
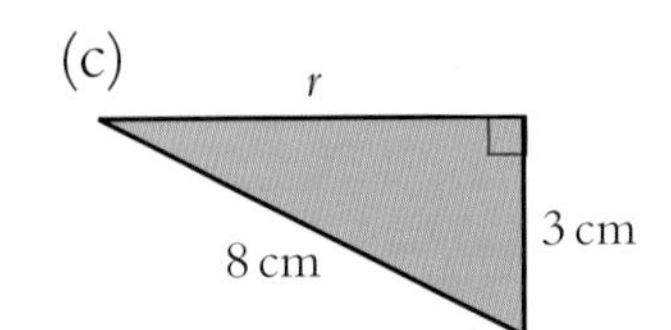

(d)
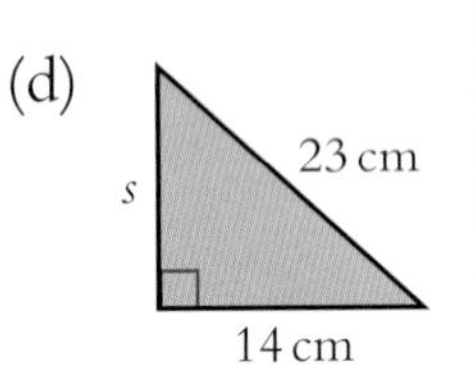

3 Find the sizes of e and f in the following triangles.
(Give your answers to two decimal places.)

(a)
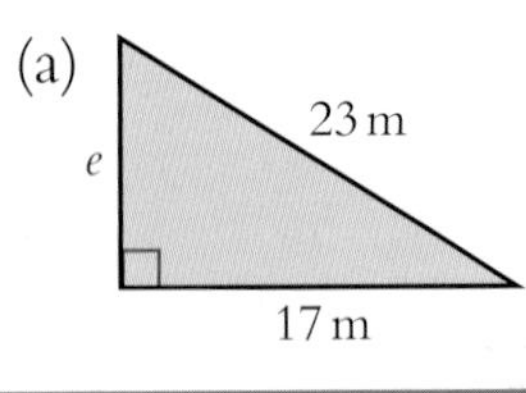

(b)
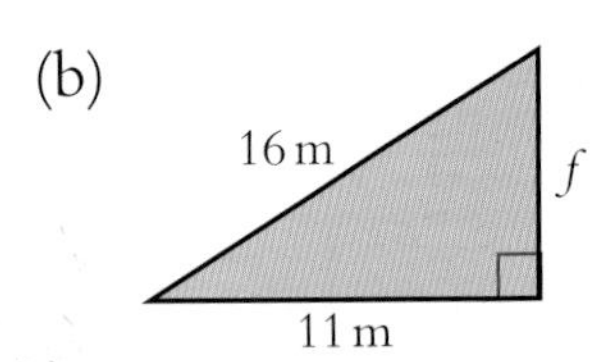

Exercise 2 continued

4 The hypotenuse in a right-angled triangle is 32·5 cm long. One of the other sides is 12·5 cm long. Find the length of the remaining side.

5 A ladder is 3 metres long. A window cleaner rests it against a wall so that the foot of the ladder is 0·8 metres from the foot of the wall. How far up the wall does the ladder reach. (Give your answer to 2 decimal places.) (Hint: Draw a sketch.)

6 The diagonal in a rectangle is 13·6 cm long. The length of the rectangle is 12 cm.

Calculate the breadth of the rectangle.

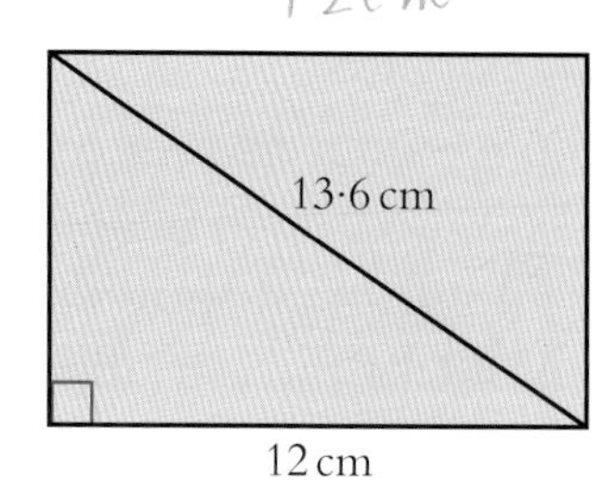

7 Find the sizes of x and y in the following triangles. (Give your answers correct to one decimal place.)

(a)

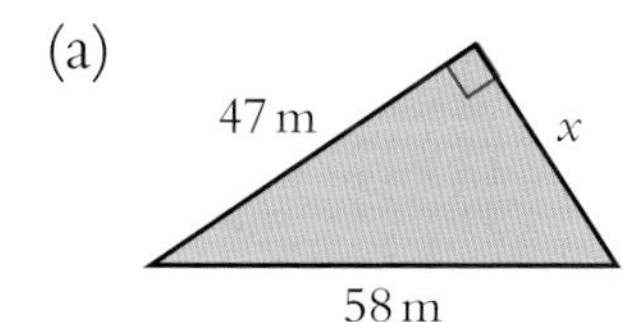

(b)

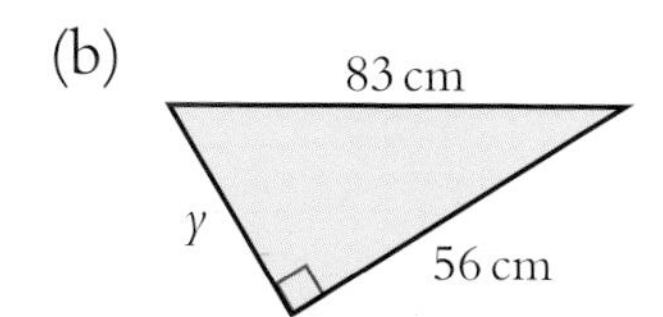

8 A country park is bounded by three villages, which make up a right-angled triangle as shown in the diagram.

How far is it from Vorey to Weston to the nearest mile?

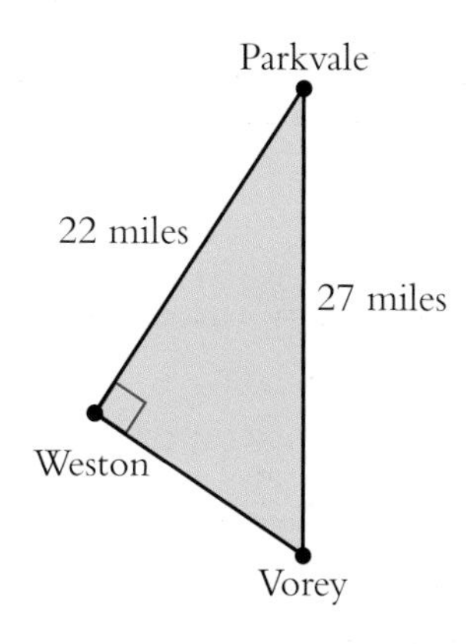

Exercise 2 continued

9 A flagpole is supported by two wires as shown in the diagram.

The wires meet at a point P on the flagpole which is 20 metres above the ground.

Each wire is 21 m long.

How far apart are the points Q, R, where the wires are attached to the ground?

(Give your answer to the nearest metre.)

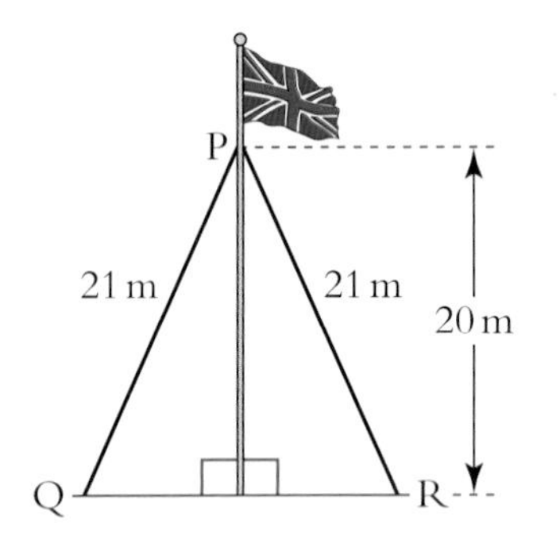

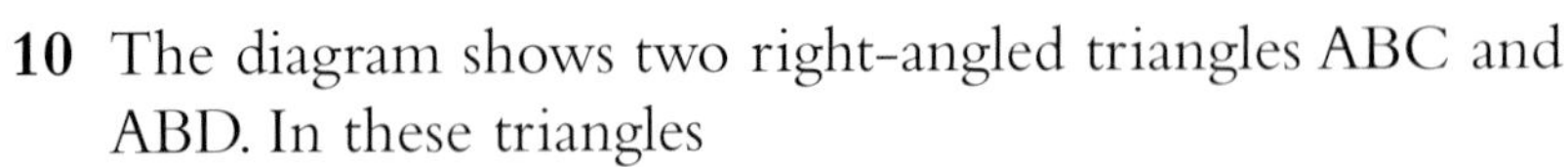

10 The diagram shows two right-angled triangles ABC and ABD. In these triangles AC = 15 cm, BC = 9 cm and AD = 18 cm.

(a) Use the Theorem of Pythagoras to find the length of AB.

(b) Hence, find the length of BD, giving your answer correct to one decimal place.

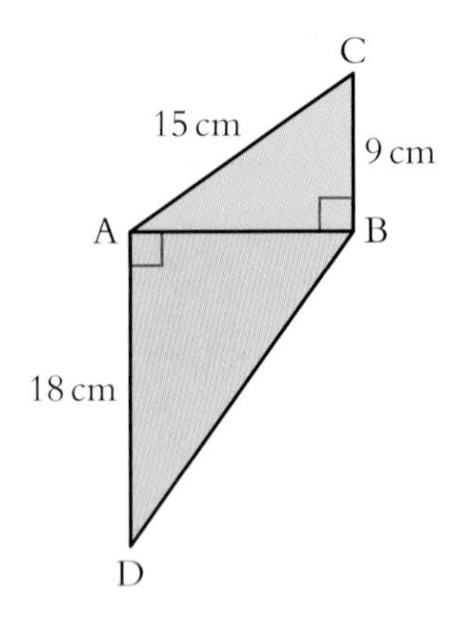

Problem solving using the Theorem of Pythagoras

In this section we tackle more demanding problems by applying the Theorem of Pythagoras to right-angled triangles contained within more complex geometric shapes.

Example A trapezium is a quadrilateral with one pair of parallel sides.

The diagram shows a sketch of trapezium ABCD.

In this trapezium:

AD = 4 cm, DC = 8 cm, and BC = 7 cm.

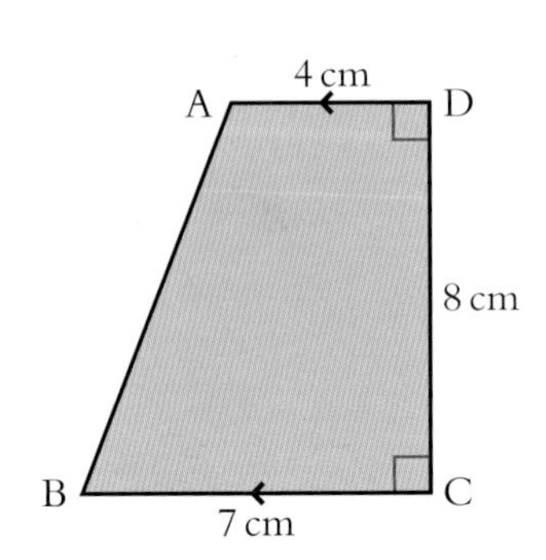

Angles ADC and BCD are both 90°.

AD is parallel to BC

Calculate the length of side AB correct to one decimal place.

Solution We divide trapezium ABCD into a rectangle ADCE and a right-angled triangle ABE as shown:

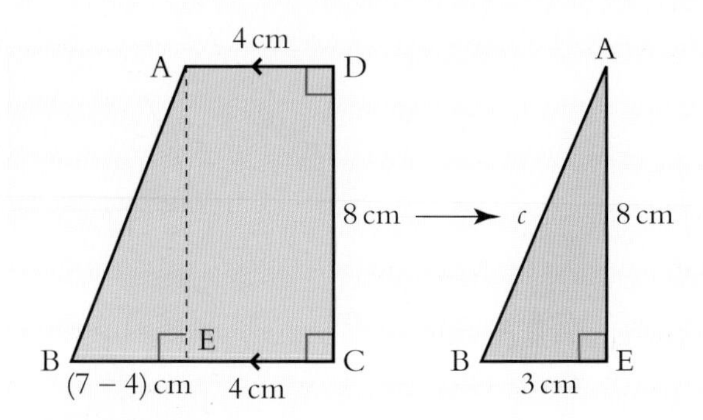

Using the Theorem of Pythagoras in triangle ABE:

$$c^2 = a^2 + b^2$$
$$= 8^2 + 3^2$$
$$= 64 + 9$$
$$= 73$$
$$\Rightarrow c = \sqrt{73}$$
$$= 8{\cdot}5.$$

Hence AB = 8·5 cm (correct to 1 decimal place)

Example Calculate the distance between two points A (3, 2) and B (7, 5).

Solution We plot both points on a coordinate diagram, and make AB the hypotenuse of a right-angled triangle. We then use the Theorem of Pythagoras.

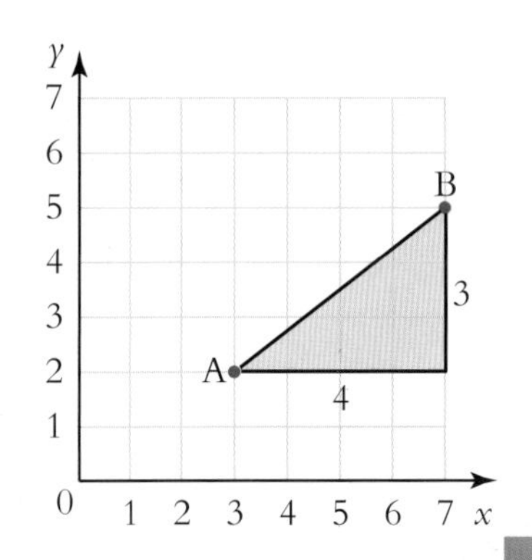

Hence $(AB)^2 = 4^2 + 3^2$
$= 16 + 9$
$= 25$

$\Rightarrow$ AB $= \sqrt{25} = 5$. Hence AB = 5 units.

Exercise 3

1 Find the sizes of x, y, z, w, in the following triangles.

(a) x 4·5 cm 6 cm

(b)

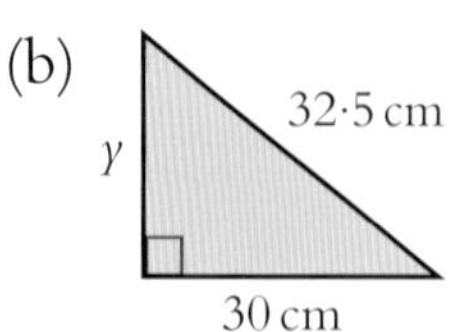

(c)

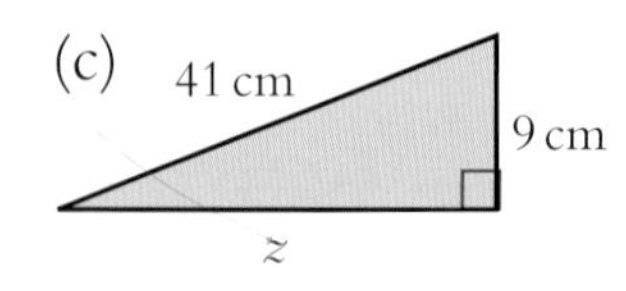

(d)

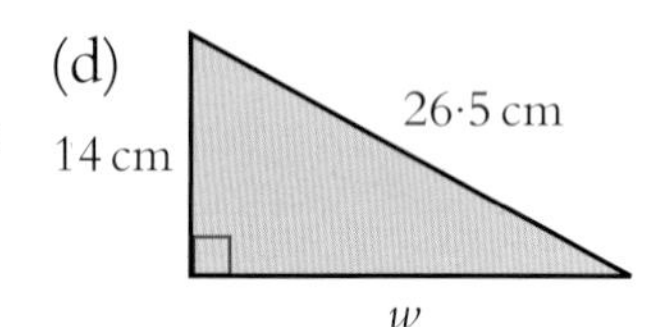

2 EFGH is a trapezium whose sizes are shown in the diagram.

Calculate the length of side EH.

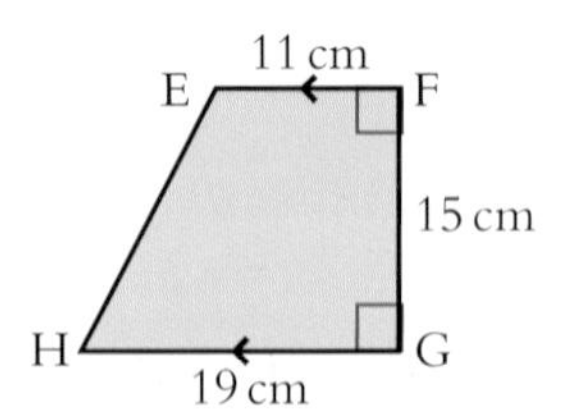

3 Charlie has a garden shed with a sloping roof. A side view of the shed is shown in the diagram, with the sizes shown.

Calculate the length of the sloping roof.
(Give your answer correct to one decimal place.)

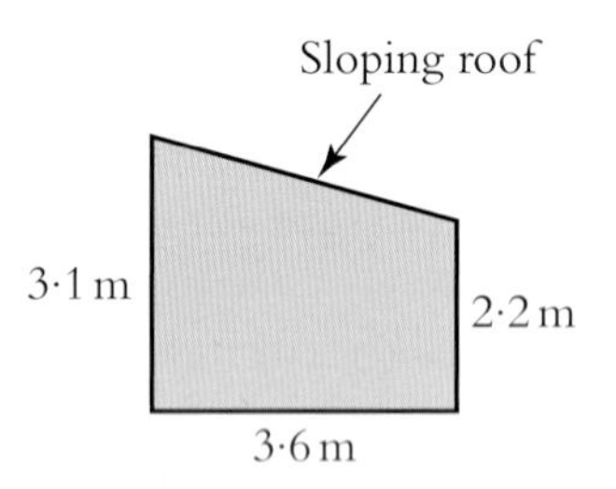

4 PQRS is a rhombus, in which

PR = 24 cm and QS = 32 cm.

Find the length of QR.

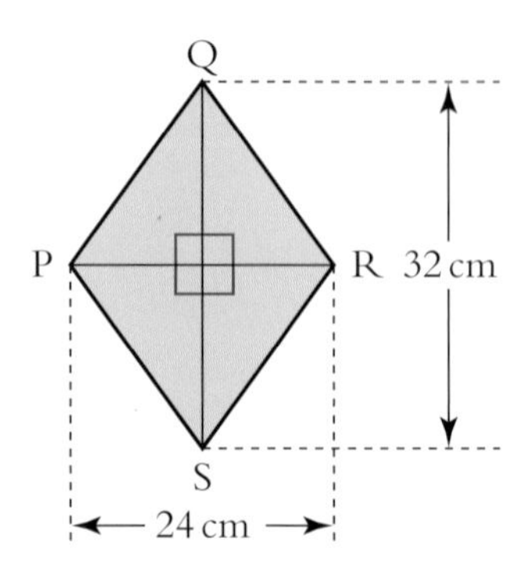

5 VWXY is another rhombus, in which

VY = 19·5 cm and VX = 15 cm.

Find the length of WY.

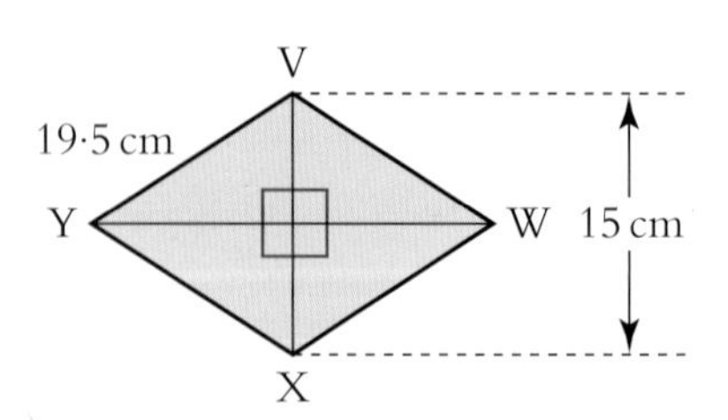

6 Calculate the distance between the points:

(a) C (4, 1) and D (8, 4)
(b) E (1, 1) and F (7, 9)
(c) G (–5, 2) and H (7, 7).

Exercise 3 continued

7 A side view of the shape of a swimming pool is shown in the following diagram.

The base of the pool slopes down from the shallow end to the deep end.

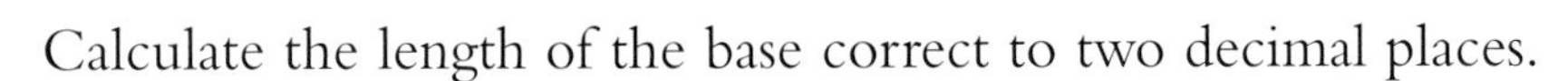

Calculate the length of the base correct to two decimal places.

8 ABCD is a kite in which

AD = 4 cm, DC = 12 cm, BD = 6 cm.

Find the length of AC.

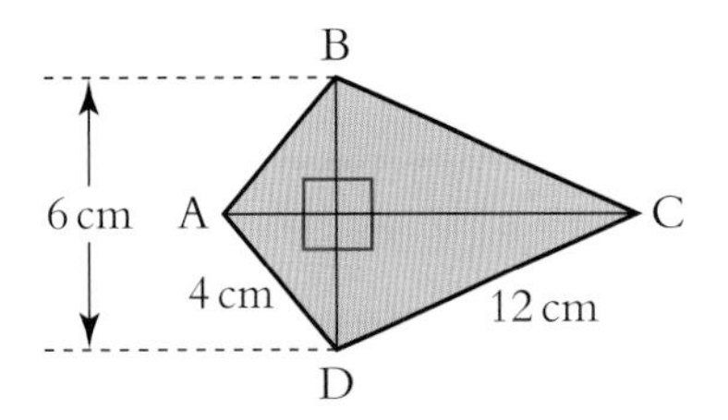

9 A jeweller displays bracelets on a square piece of card as shown in the diagram.

Calculate the length of the longest bracelet shown.

(Give your answer to the nearest centimetre.)

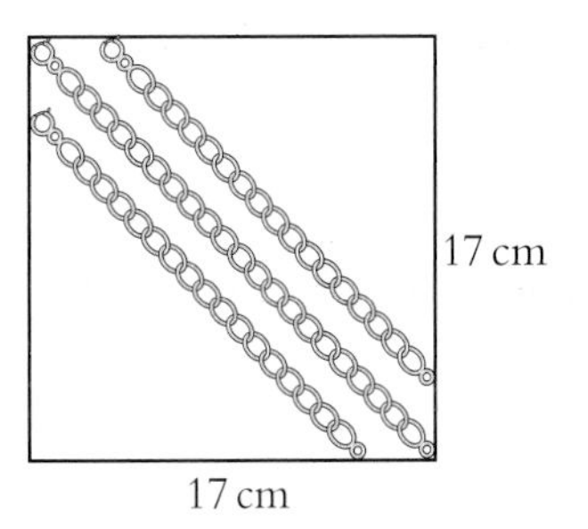

10 The 14th hole on Loch Bonnie Golf Course has a 'dog-leg' fairway as shown.

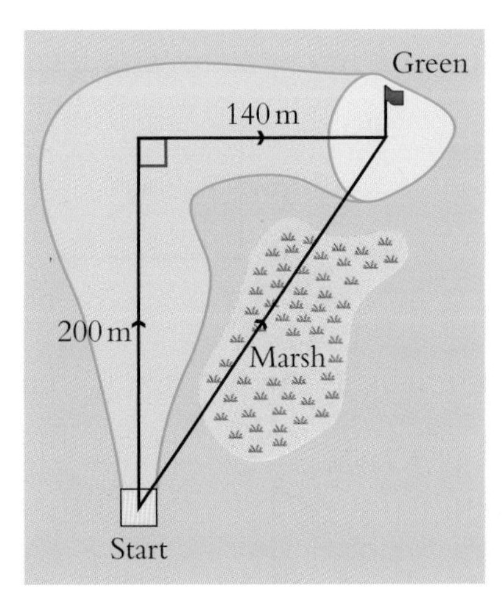

Laura plays her first shot 200 metres down the fairway, then plays her second shot (at right angles) 140 metres on to the green. Michelle plays her shot over the marsh, directly on to the green. How much further did Laura play to get on to the green than Michelle?

Exercise 3 continued

11 The diagram shows a circle with its centre at the Origin. The circle passes through the point (8, 6).

Calculate the area of the circle. **Take π = 3·14.** (**Do not use a calculator**)

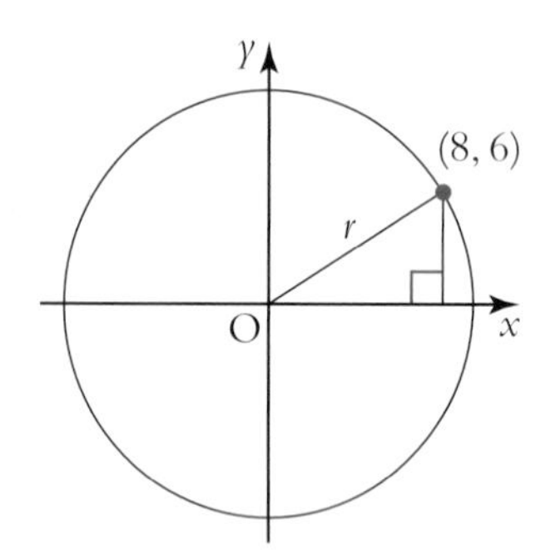

Exercise 4 – Revision of Chapter 7

1 Calculate the length of the hypotenuse in each of these triangles.

(a) 13·5 cm, 18 cm

(b)

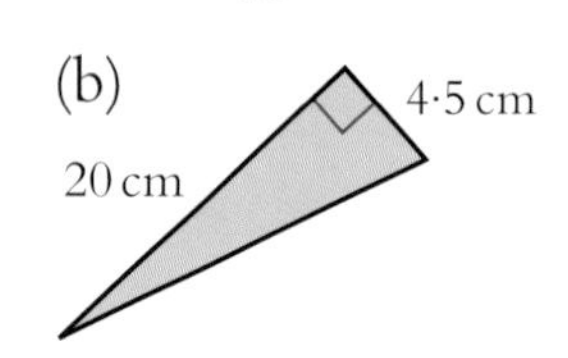

2 Calculate the sizes of a and b in the following triangles.

(a) 113 m, 15 m, a

(b)

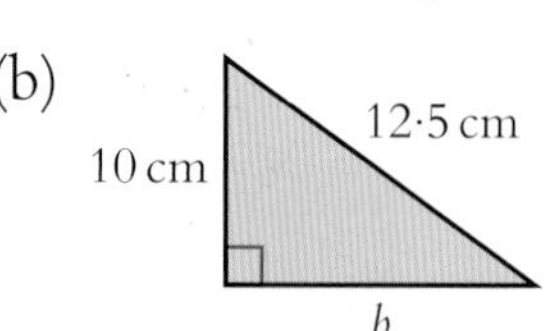

3 Calculate the sizes of s, t, u, v, in the following triangles. (Give your answers correct to one decimal place.)

(a) 7 cm, 11 cm, s

(b)

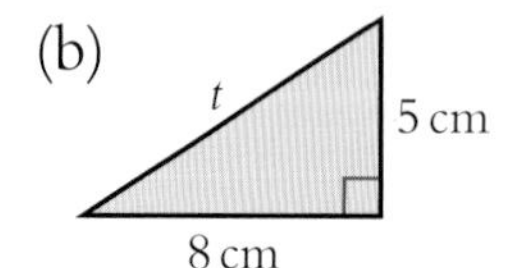

(c)

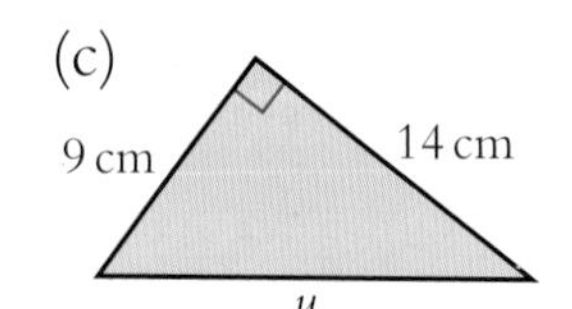

(d)

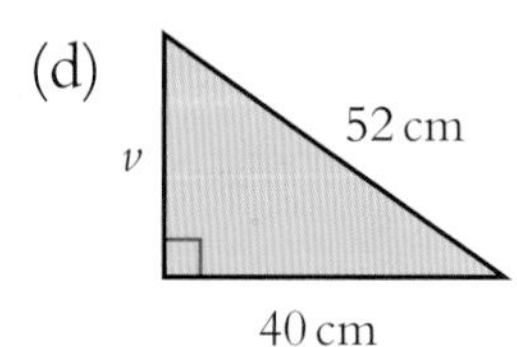

4 Find the length of the diagonal of a rectangular field which measures 80 metres by 60 metres.

Exercise 4 – Revision of Chapter 7 continued

5 Jackie's garage has a sloping roof as illustrated in the diagram.

Calculate the length of this sloping roof. (Give your answer to 2 decimal places.)

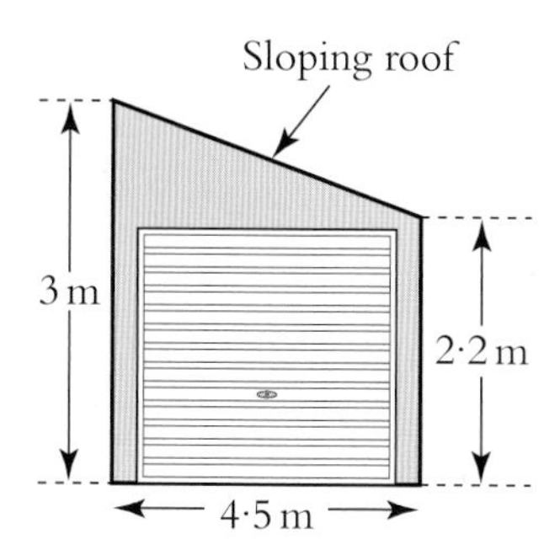

6 Find the distance between the points M (1, 2) and N (6, 5). (Give your answer correct to 1 decimal place.)

7 The diagram shows an end view of a house.

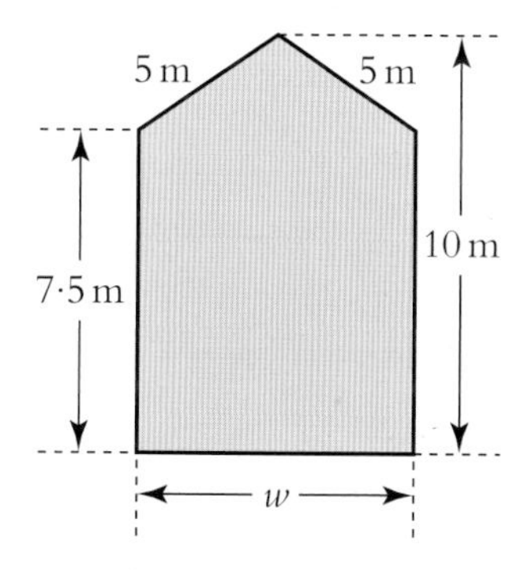

Calculate the width of the house w, to the nearest centimetre.

8 The diagram shows a circle with its centre at the Origin. The circle passes through the point (8, 6).

Calculate the circumference of the circle. **Take π = 3·14. (Do not use a calculator.)**

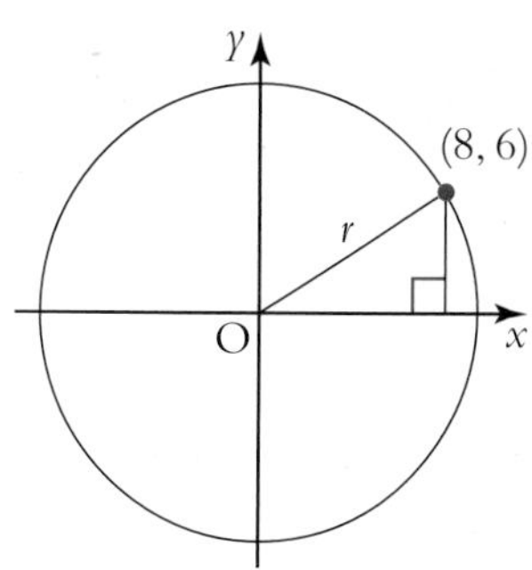

Summary

(This is perhaps the most famous theorem in Geometry, and you should be able to apply it, bearing in mind the following points.)

1 The Theorem of Pythagoras can be used to find the length of the third side in a right-angled triangle when you know the lengths of the other two sides.

$a^2 + b^2 = c^2$

2 (a) When finding the length of the hypotenuse (the longest side opposite the right angle), we square and **add** the other two sides.

(b) When finding the length of one of the two shorter sides, we square and **subtract** the other two sides.

(a)

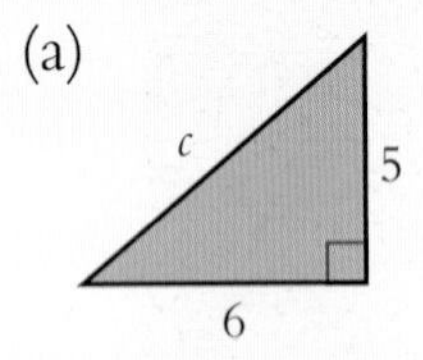

(b)

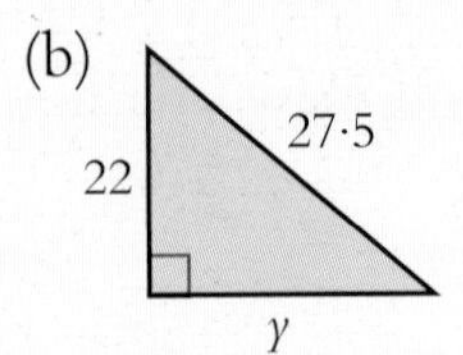

(a)

$$c^2 = 5^2 + 6^2$$
$$= 25 + 36$$
$$= 61.$$
$$\text{So } c = \sqrt{61}$$
$$= 7{\cdot}8 \quad \text{(to 1 d.p.)}.$$

(b)

$$27{\cdot}5^2 = y^2 + 22^2$$
$$\Rightarrow y^2 = 27{\cdot}5^2 - 22^2$$
$$= 756{\cdot}25 - 484$$
$$= 272{\cdot}25.$$
$$\text{So } y = \sqrt{272{\cdot}25}$$
$$= 16{\cdot}5.$$

3 Many more complex problems can be solved using the Theorem of Pythagoras. If we are dealing with a Trapezium for example, we divide the shape into a rectangle together with a right-angled triangle. The theorem is then applied to the right-angled triangle.

Summary continued

4 To find the distance between two points, given their coordinates, we plot the points, construct a right-angled triangle, then use the Theorem of Pythagoras. Thus:

Given A(3, 1) and B(5, 4), then

$AB^2 = 3^2 + 2^2 = 9 + 4 = 13 \Rightarrow AB = \sqrt{13} = 3{\cdot}6$ (to one decimal place).

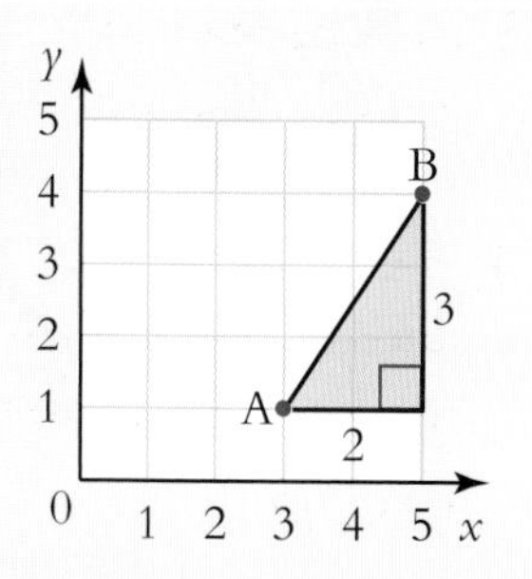

8 Graphs, Charts, and Tables

In this chapter, and in the next, we will be looking at some Statistics. *Statistics* is the branch of Mathematics concerned with trends and patterns in data. The approach in this chapter is very much 'visual statistics', and we use graphs, charts, and diagrams to present data in a way which makes a *visual* impact on the reader. This can often simplify our understanding and interpretation of data in a way that say rows and rows of numbers cannot.

Bar Graphs and Line Graphs

In this section, you should be able to:

- extract and interpret data from a bar graph
- construct a bar graph
- extract and interpret data from a line graph
- construct a line graph.

Example The bar graph shows the takings (in hundreds of pounds) at a new restaurant each day over a period of one week.

(a) On which day did the restaurant take most money?
(b) On which day were the takings £450?
(c) How much did the restaurant take on Tuesday?
(d) Find the total takings for the week.

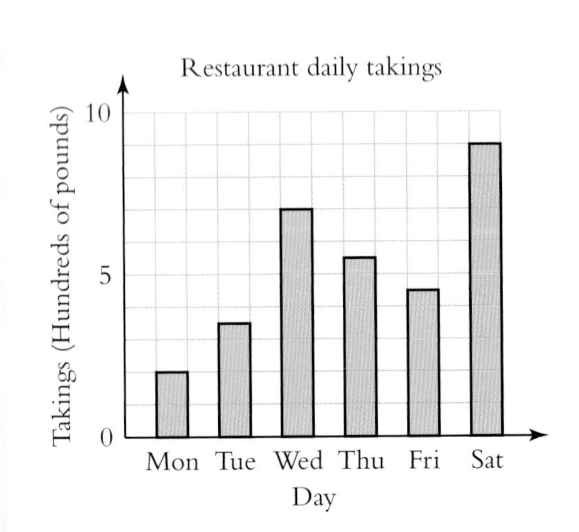

Solution (a) Saturday (b) Friday (c) £350

(d) £(200 + 350 + 700 + 550 + 450 + 900) = £3150.

Technique

When drawing a line graph (or a bar graph), you must:

- give the graph a title
- clearly label each axis
- choose a suitable scale for each axis
- plot time on the horizontal axis.

Example Linda Kildare has been admitted to hospital with fever. Her temperature is being checked every three hours.

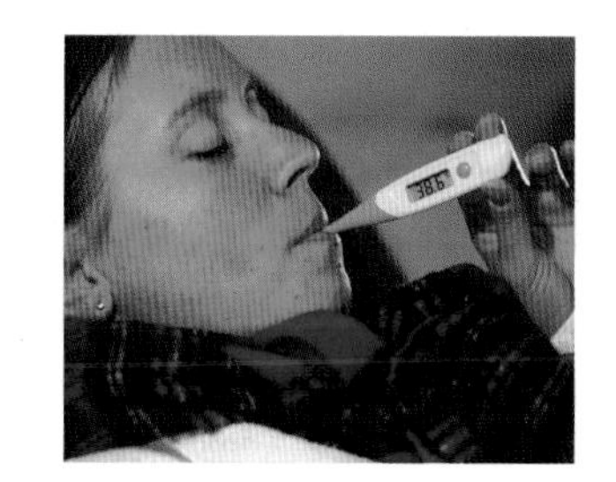

Time	4 am	7 am	10 am	1 pm	4 pm	7 pm
Temp (°C)	40·3	40·5	39·4	39·0	38·2	37·5

Illustrate this data on a line graph

Solution
The *trend* of this graph is downward since Linda's temperature is decreasing.

(The 'zigzag' below 37 on the vertical scale indicates that the numbers between 0 and 37 are not shown.)

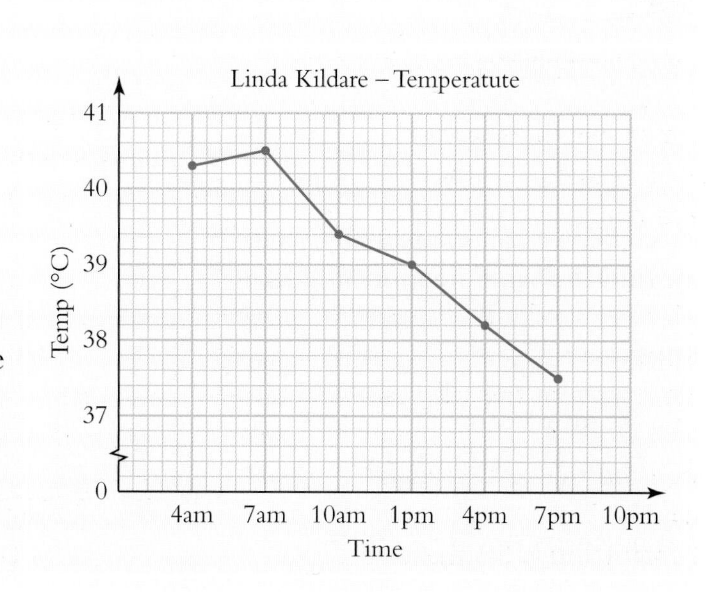

The Pie Chart (or diagram)

The Pie Chart (or diagram) is another method of presenting information in an attractive and effective way. Sectors of a circle are used to represent different items of data.

Example The pie chart shows the nationalities of the first team squad of players in a Premier League football team.

There are 24 players in the first team squad.

(a) How many of them are Scottish?

(b) How many Welsh players are in the squad?

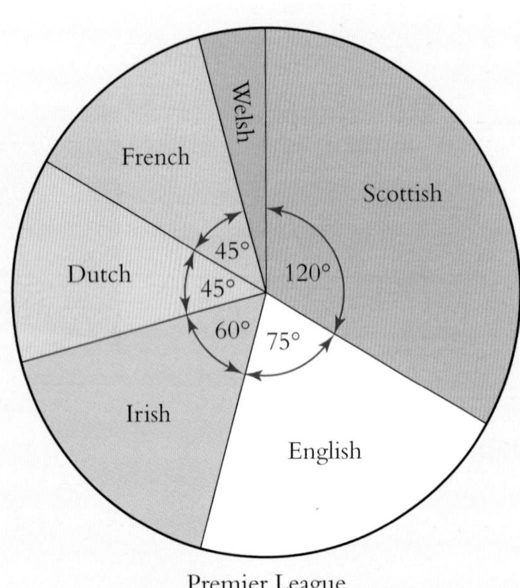

Premier League
First team squad nationality

Solution

(Since there are 360° in a circle, we can express any sector as a fraction of 360°, and then find the same fraction of 24.)

(a) Number of Scottish players $= \frac{120}{360} \times 24 = 8$.

(b) The angle representing Welsh players
$= 360° - (120 + 75 + 60 + 45 + 45)° = 15°$.

Hence the number of Welsh players $= \frac{15}{360} \times 24 = 1$.

Exercise 1

1 Chris Low has drawn a bar graph to show how many pets he has in his pet shop.

The graph is as shown.

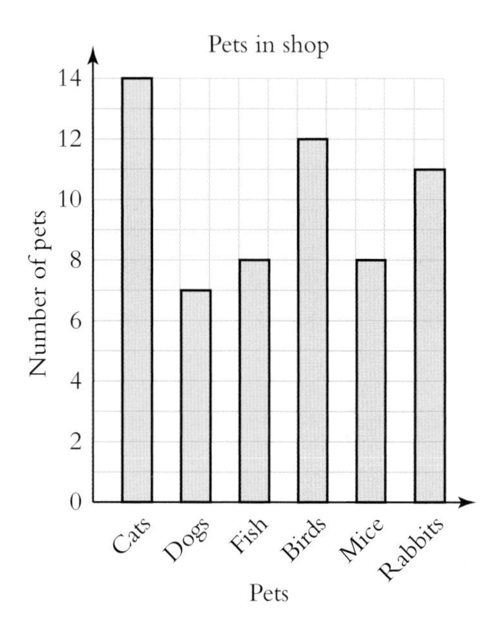

(a) What is the most numerous pet in the shop?

(b) How many rabbits are there in the shop?

(c) Chris buys some gerbils. After buying them, the **total** number of pets in the shop is 68. How many gerbils did he buy?

2 A survey was carried out to find how a group of pupils travelled to school. The results were as follows:

Bus	*Car*	*Bicycle*	*Train*	*Walk*
10	7	2	5	9

Illustrate this data on a bar graph.

3 The grades of 800 students who sat a mathematics exam are as shown.

Grade	*A*	*B*	*C*	*D*
No. of Candidates	120	240	350	90

Illustrate this data on a bar graph.

Exercise 1 continued

4 The temperature at Strathcowan was recorded every hour on the first morning of March.

The line graph shows the results.

(a) What was the temperature at 12 midnight?

(b) At what time was the temperature 2° Celsius?

(c) What is the general trend of the graph?

(d) By how many degrees did the temperature rise between 3 am and 11 am?

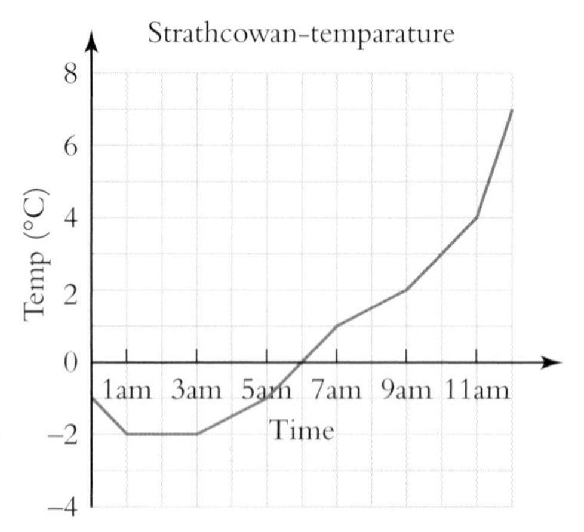

5 Some water was heated over a candle flame. The temperature of the water was recorded every two minutes, with the following results.

Time in minutes	0	2	4	6	8	10	12
Temperature (°C)	15	30	44	57	67	74	77

(a) Illustrate this data on a line graph.

(b) Describe the trend of the graph.

6 Two hundred and forty people were asked about their favourite hot drink.

The pie chart shows the results.

How many people said their favourite hot drink was

(a) tea

(b) coffee?

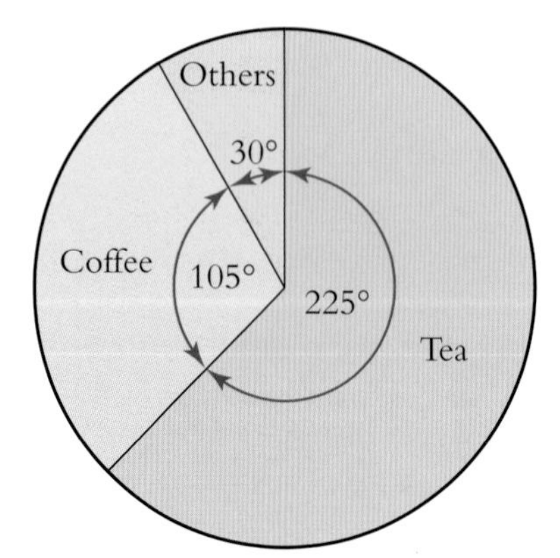

Favourite hot drink

Exercise 1 continued

7 The following pie chart describes the people who attended a concert.

If 1080 people attended the concert, how many were girls?

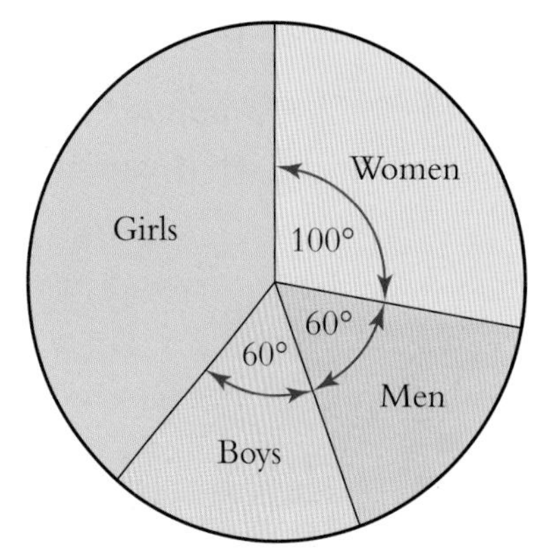

Concert Attendance

8 The pie chart shows how the population of the United Kingdom is made up.

(a) What is the population of Scotland to the nearest thousand?

(b) 50 years ago it was said that 1 in 10 people in the United Kingdom was Scottish. Is this still true? (Comment.)

UK Population – Total 60·209 million

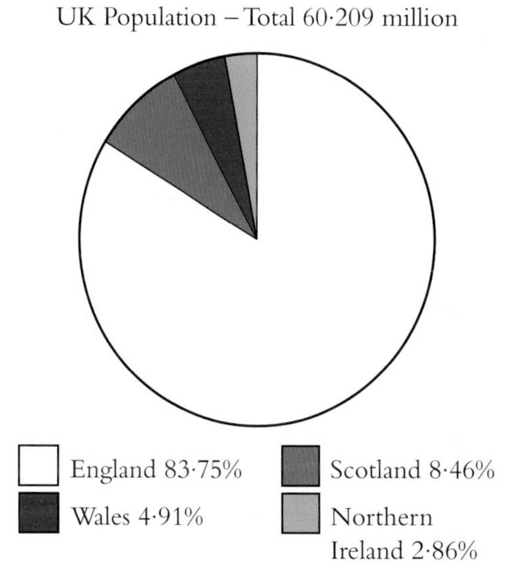

9 The graph compares the interest rates offered by the Regent Bank and the Caledonia Bank on their current accounts.

(a) How much interest would you receive in a year if you invested:

(i) £5000 with the Caledonia Bank

(ii) £10 000 with the Regent Bank?

(b) Jessica is about to invest money in one of these banks.
What advice would you give her?

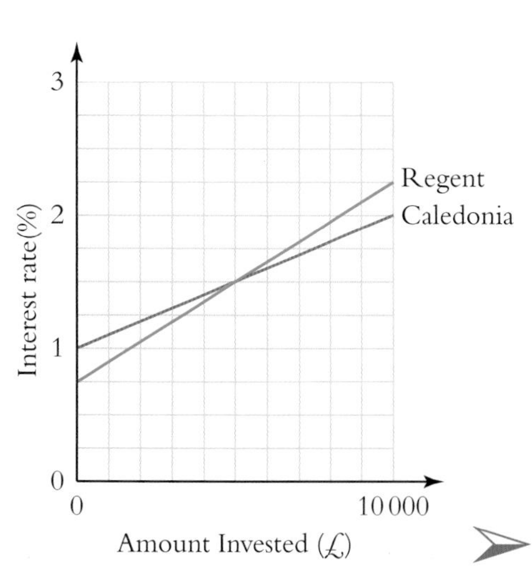

Exercise 1 continued

10 Sometimes graphs can be deliberately misleading!

Draw this graph again with a correct vertical scale, and explain in what way it is misleading.

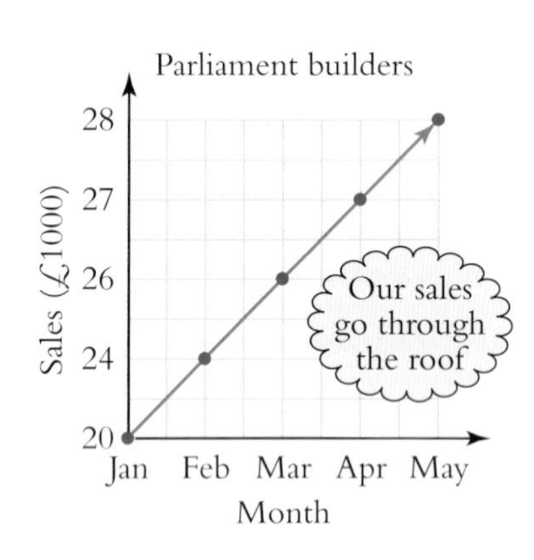

11 The town of Cartlee has two newspapers, the 'Echo' and the 'Advertiser.'

The following compound bar graph shows their sales over a 6-month period.

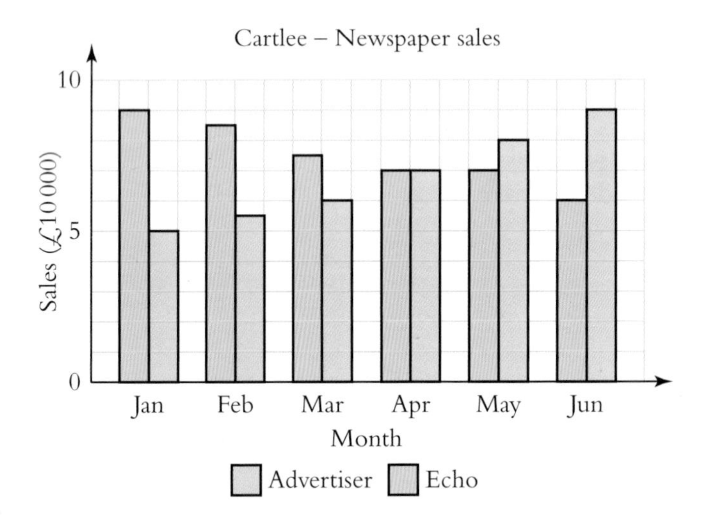

(a) What was the value of sales for the 'Echo' in January?

(b) What was the value of sales for the 'Advertiser' in February?

(c) Describe the trend in sales for the 'Advertiser'.

(d) In which month did the sales of the 'Advertiser' first overtake the sales of the 'Echo'?

Stem and Leaf Diagrams

The stem and leaf diagram presents numerical data in a very structured way and so allows us to see an order as well as any trend in the data.

In this section, you should be able to:

- extract and interpret data from a stem and leaf diagram
- construct a stem and leaf diagram

Example The numbers of goals scored by the twenty teams in the Premiership in season 2005-06 was as follows:

72 72 57 68 53 51 47 49 52 45

34 48 41 48 43 42 37 28 31 26.

Show this data in a stem and leaf diagram.

Solution Premiership Goals 2005-6

Stem	Leaves
2	6 8
3	1 4 7
4	1 2 3 5 7 8 8 9
5	1 2 3 7
6	8
7	2 2

(n = 20) (2|6 means 26 goals)

Remember

(When constructing a stem and leaf diagram, remember the following points.)

- The diagram should have a **title** at the top.
- The **stem** is the column of figures to the left of the vertical line.
- The **leaves** are the figures to the right of the line (increasing in value outward from this stem.)

- Each row is called a **level**.
- The number of items of data should appear at the bottom left of the diagram.
- A **key** should appear at the bottom right of the diagram to explain the meaning of individual entries.

Exercise 2

1 The following stem and leaf diagram shows the marks of class 5A in a class exam.

Class 5A Maths Exam Marks

Stem	Leaf
2	3 9
3	4 7 9
4	0 3 3 7
5	1 1 2 5 7 8 9
6	3 6 8
7	5

$n = 20$ 2 | 3 means 23

(a) How many students are in Class 5A?

(b) Level 2 can be read as 23 and 29. Write down the marks in level 3.

(c) Albert was first in the class with a mark of 75. In what position was Marilyn whose mark was 57?

2 The following stem and leaf diagram shows the heights of students in class 5A.

Heights of Students in Class 5A:

Stem	Leaf
14	8
15	3 5 9 9
16	3 3 4 5 6 8
17	2 5 6 7 7 7 9
18	1 3

($n = 20$) (14 | 3 means 143 centimetres)

(a) What height was the tallest person in the class?

(b) What height was the smallest person in the class?

(c) How many students were over 170 cm tall?

Exercise 2 continued

3 The marks of class 5B in the maths exam are listed as follows:

46 53 67 72 53 40 35 28 76 60
42 55 71 65 39 24 46 51 59 47.

Illustrate this data in a stem and leaf diagram.

4 The number of copies of a school magazine which sold each day was counted over 3 weeks. The results are as follows:

49 67 58 46 50
56 75 43 48 51
58 42 40 39 31.

Illustrate this data in a stem and leaf diagram.

5 The weights of a number of 3-month old babies were recorded in kilograms as follows:

4·8 5·3 6·7 5·6 5·8 6·0 4·9 5·2 5·1
4·9 5·8 5·6 6·5 6·1 5·3 5·3 5·8 6·9.

(a) How many babies were weighed?

(b) Illustrate the data in a stem and leaf diagram.

(c) Find the difference in weight between the heaviest and lightest babies.

6 Jo works as a hairdresser. She records in minutes how long it takes her to do 24 haircuts. The results are as follows:

24 35 38 21 15 41 32 7 34 11 21 30
8 30 27 42 39 20 16 19 21 31 36 15.

Illustrate this data in a stem and leaf diagram.

Technique

To compare **two** sets of data, a **back-to-back stem and leaf diagram** can be constructed. For example, suppose we wish to compare the marks of the boys and girls in class 5A in their exam.

Exercise 2 continued

Exam marks for class 5A

Girls		Boys
9	2	3
9	3	4 7
7	4	0 3 3
9 8 7 5 1	5	1 2
8	6	3 6
5	7	

$(n = 10)$ $(n = 10)$

(| 2 | 3 means 23)
(9 | 3 | means 39)

The back-to-back stem and leaf diagram shows that a girl had the highest mark (75) and a boy had the lowest mark (23). It also suggests that the girls seemed to do better than the boys.

7 The exam results for class 5B are shown in the following back-to-back stem and leaf diagram.

Exam marks for class 5B

Girls		Boys
8	2	4
9	3	5
7 6 0	4	2 6
9 5	5	1 3 3
7	6	0 5
6 2	7	1

$(n = 10)$ $(n = 10)$

(| 2 | 4 means 24)
(9 | 3 | means 39)

Compare and comment upon the performance of the girls and boys.

8 Two rugby teams are about to play each other. The ages of the players in the teams are as shown.

Border Storm	23	25	23	19	21	20	27	25	20	18	17	35	21	22	28
Highland XV	25	29	35	32	41	19	40	34	31	23	27	30	26	27	29

(a) Illustrate this data in a back-to-back stem and leaf diagram.

(b) Compare the ages of the two teams.

Frequency Tables

The data collected from tests or experiments may have little obvious meaning until they have been arranged and classified in some logical way. Larger sets of data are organised into *frequency tables*, and a frequency table is something you should be able to construct.

Example Nathan rolls a standard die (faces numbered from 1 to 6) thirty times and records the results as follows:

6 6 1 3 4 4 2 5 1 5
4 5 5 3 1 6 2 6 4 2
1 3 2 6 5 4 3 6 2 4

(a) Using tally marks, construct a frequency table to illustrate these results.

(b) Do the results suggest the die is unbiased or not?

Solution

(a)

Score	Tally	Frequency
1	||||	4
2	卌	5
3	||||	4
4	卌 |	6
5	卌	5
6	卌 |	6

(b) Since each number on the die appears with roughly the same frequency, this suggests that the die is unbiased. It is not 'loaded' so that for example, a '6'comes up much more often than any other number.

Exercise 3

1 A teacher gave her class of 33 pupils a test which was marked out of 10. The results were as follows:

5 6 7 8 7 6 3 2 9 6 8
4 9 8 3 2 5 5 6 7 4 6
6 5 3 7 8 3 2 6 5 8 10

(a) Use tally marks and hence construct a frequency table for these results.

(b) What was the most common score?

2 The label on a matchbox states 'average contents 48'. To check the accuracy of this statement Harriett counts the contents of 24 matchboxes, with the following results:

48 47 49 48 50 46 45 48 48 48 48 50
46 47 48 47 49 48 50 48 48 47 48 51

(a) Construct a frequency table to illustrate these results.

(b) What was the most frequent result?

(c) Do you think it is fair to state 'average contents 48'? Explain your answer.

3 Quentin conducts a survey to find out which fruit is most popular among his classmates.

He uses this code:
Orange (O), Apple (A), Pear (P), Grape (G), Strawberry (S), Melon (M).

The results he obtained were as follows:

O A O M O A P S S O A O A M
A A G P S O O A A A S O A P

(a) Express this data in a frequency table

(b) Draw a bar graph to show the results of his survey.

4 A travel agency manager constructs a frequency table to show which countries were chosen to be visited by his customers during the first week in May. The results are shown in the following table.

Country	*USA*	*France*	*Spain*	*Greece*	*UK*	*Malta*	*Italy*
Frequency	4	6	10	3	2	3	5

Illustrate the results of this frequency table in a bar graph.

Exercise 3 continued

5 Steven rolls a die thirty times. His results are as follows:

6 6 4 3 6 5 2 6 3 6 4 3 2 6 6

5 6 1 5 2 2 3 4 6 6 4 6 5 6 6

(a) Construct a frequency table to show these results.

(b) Do you think this die is unbiased? Explain.

Scattergraphs

Two sets of data which are related in some way are often presented in a *scattergraph*. The scattergraph is a collection of points on a coordinate diagram with a line of *best fit* drawn through them. You should be able to construct a scattergraph and draw a line of best fit. You should also be able to interpret the information presented in a scattergraph.

Example A group of students has completed tests in Unit 1 and Unit 2 of the Intermediate 1 Maths course. The results are shown in the following table.

Student	A	B	C	D	E	F	G	H	I	J	K	L
Unit 1	58	45	69	79	50	42	75	87	50	57	64	78
Unit 2	51	43	60	72	49	40	65	81	40	54	64	70

(a) Illustrate these results on a scattergraph.

(b) Draw a line of best fit on the scattergraph.

(c) Imran scored 90 for the Unit 1 test, but missed the Unit 2 test due to illness. Use the line of best fit to estimate a mark for Imran for Unit 2.

Solution (a) and (b)

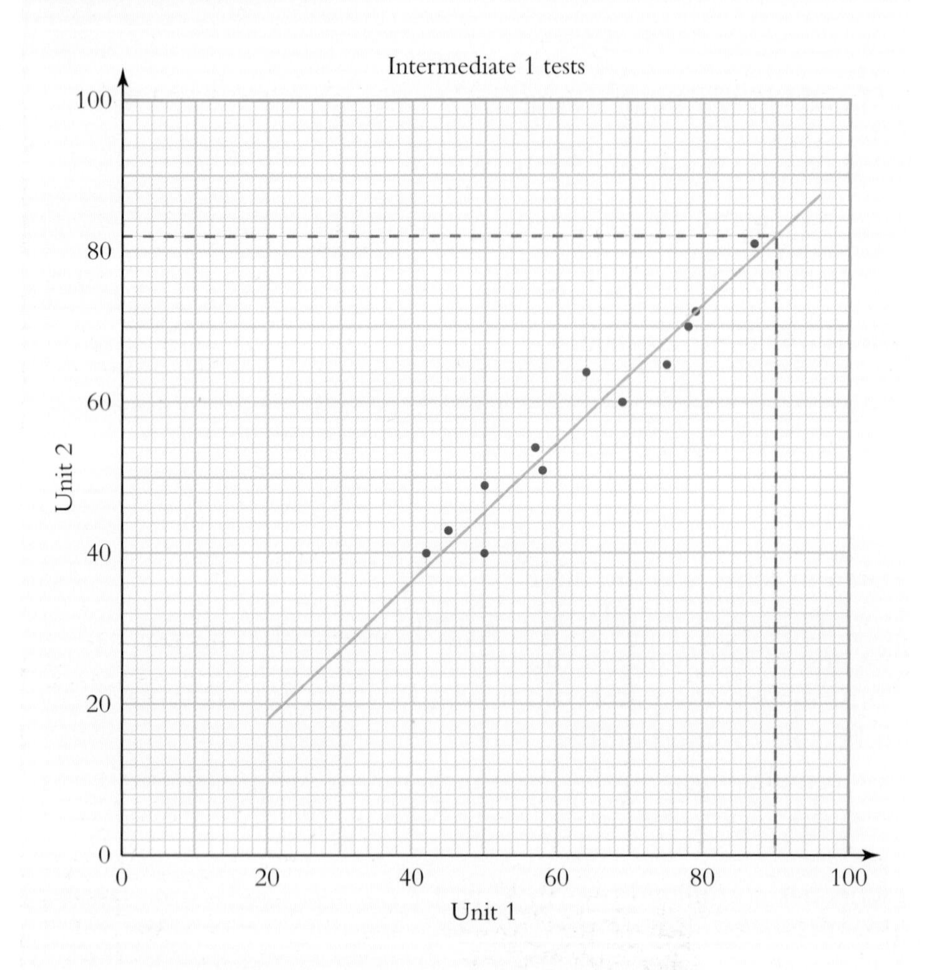

(c) Imran's estimated mark for Unit 2 is 82.

Main points

- Always put clear scales and labels on both axes when constructing your scattergraph.
- The first set of data ('Unit 1' marks in this last example) should always be plotted on the horizontal axis.
- The line of best fit should follow the general path of the dots, with as many dots above the line, as below.
- The line of best fit can be extended slightly in order to predict other results.

Exercise 4

Use 2 mm graph paper for questions 2–4

1 A group of students has completed two short tests in Maths and Science. The marks are shown in the following table.

Student	A	B	C	D	E	F	G	H	I	J	K
Maths mark	7	6	8	9	4	9	3	7	8	6	10
Science mark	5	4	6	6	3	5	3	4	5	5	7

(a) Illustrate these results in a scattergraph.

(b) Draw a line of best fit.

(c) Estimate the Science mark for a student who scored 5 in the Maths test.

2 Mr Jones keeps a record of how many times in the session each pupil in his class is absent. He compares this number of absences with the pupil's end-of-session test mark (out of 50). The results are as follows.

Pupil	Amy	Ben	Cleo	Dave	Emma	Fred	Gill	Herb	Ina	Jim	Kim	Leo
No. of absences	5	7	15	0	2	20	4	8	17	10	9	25
Test mark	38	36	18	48	45	12	40	32	17	29	35	4

(a) Illustrate this data in a scattergraph.

(b) Draw a line of best fit.

(c) Matt was absent on the day of the test! He already has had 12 absences. Estimate his mark.

(d) What conclusion can you draw from this graph?

Exercise 4 continued

(Use 2 mm graph paper for questions 2–4)

3 A café owner keeps a record of the temperature, in degrees Celsius, at mid-day and the sales in pounds of ice-cream. He does this during a two week period in June, with the following results.

Temp (°C)	20	22	25	18	15	21	13	10	16	23	27	24	17	8
Sales (£)	75	77	86	61	52	76	45	32	59	80	92	83	61	25

(a) Illustrate the data in a scattergraph.

(b) Draw a line of best fit.

(c) Use your line to estimate what the sales might be if the temperature is 5° Celsius.

(d) On the first day of the third week £70 worth of ice-cream was sold. Estimate the temperature at mid-day on that day.

4 Twelve students are studying chemistry and physics at college. Their first term exam marks in these subjects are shown in the following table.

Student	A	B	C	D	E	F	G	H	I	J	K	L
Chemistry	70	60	65	80	85	55	90	50	45	75	40	35
Physics	60	52	55	68	70	46	75	43	40	64	36	30

(a) Illustrate this data in a scattergraph.

(b) Draw a line of best fit.

(d) Petra scored 95 marks for Chemistry. Estimate her Physics mark.

Exercise 5 – Revision of Chapter 8

1 The following line graph shows the progress made by Scottish tennis player Andrew Murray in the World Rankings during the first 30 weeks of 2006.

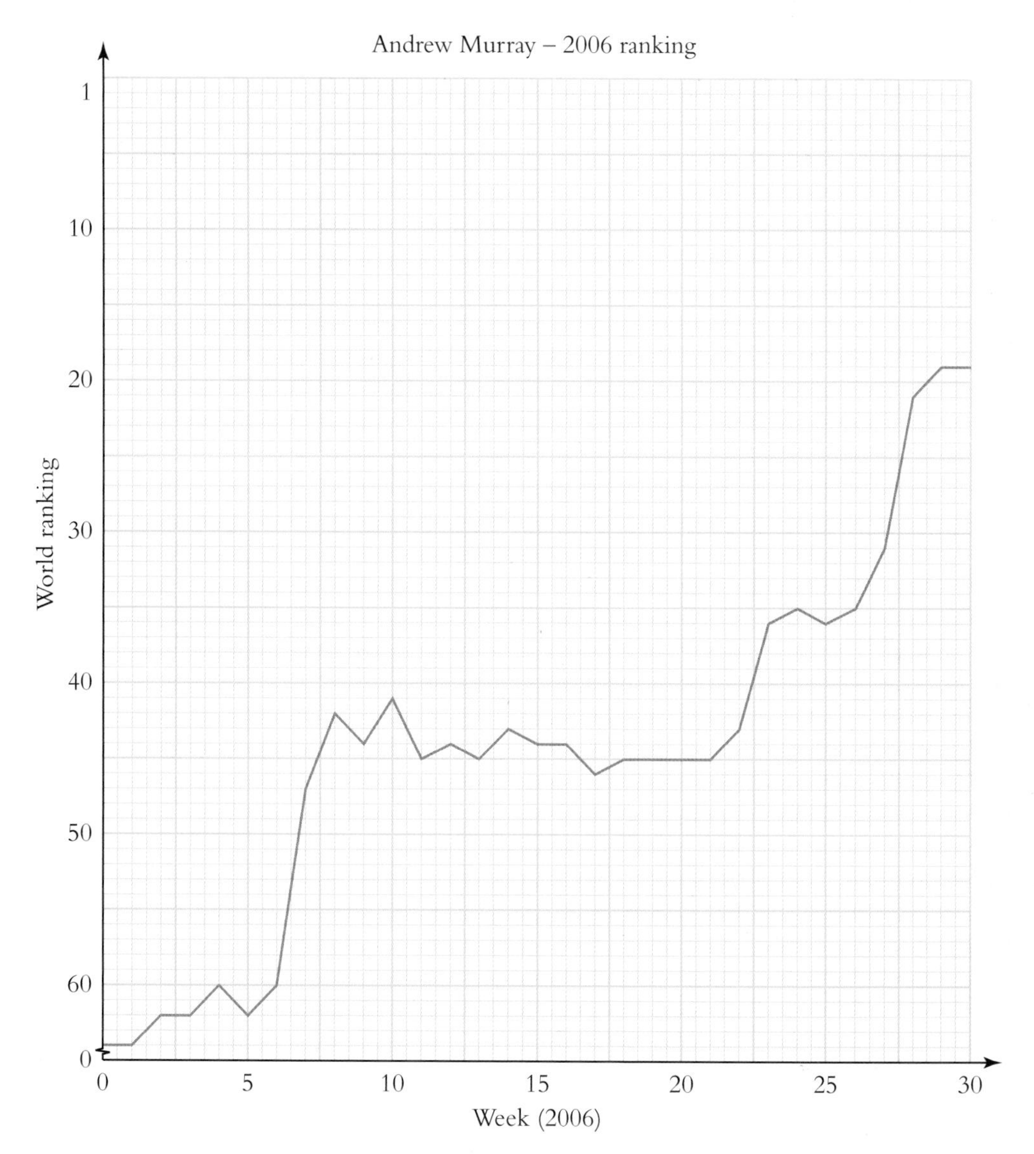

(a) What was Andrew's ranking at the start of 2006 (week 0)?

(b) In what week did Andrew first appear in the top 40?

(c) What ranking did Andrew have by week 30?

(d) By how many places did Andrew's ranking improve over the 30-week period?

(e) What is the trend of the graph?

Exercise 5 – Revision of Chapter 8 continued

2 Eric is practising throwing darts. He aims 30 darts at the '20' sector on a dartboard and keeps a record of his scores. These are as follows.

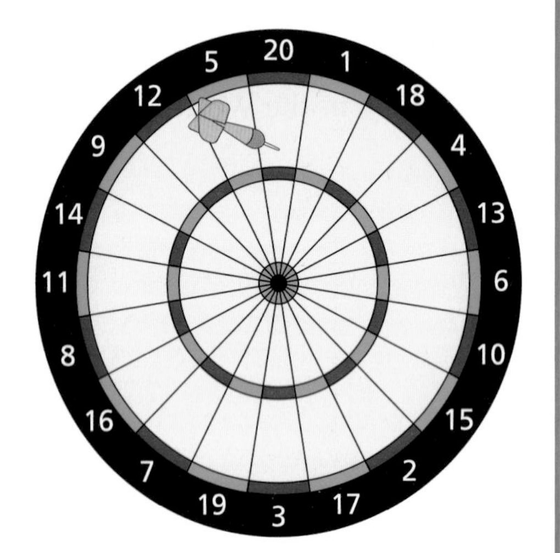

20	20	5	1	20	1	20	20	1	5
18	20	20	1	1	18	20	20	5	20
12	5	5	20	20	20	20	1	20	20

(a) Construct a frequency table to illustrate these scores.

(b) Draw a bar graph of the results.

3 The pie chart shows eye colour for all students at Spring Bay College.

(a) If there are 900 students at the college, how many have green eyes?

(b) By measuring the angle at the centre of the brown sector, calculate how many students have brown eyes.

Spring Bay College

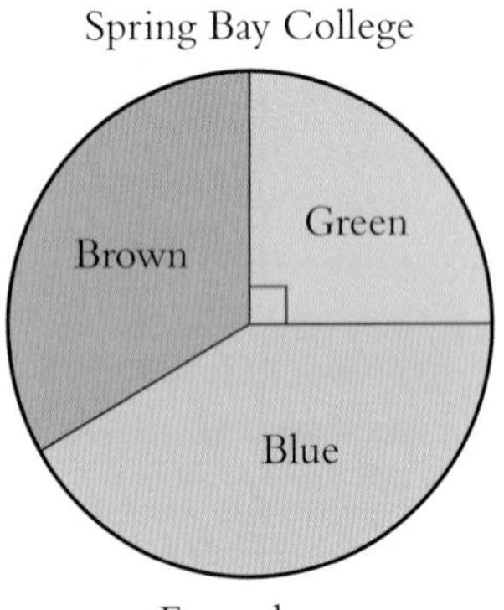

Eye colour

4 The following pie charts illustrate the grades scored by class 5C in their October and December Mathematics tests. Between the two tests, extra homework classes were organised for the students.

Class 5C – Grades in Maths

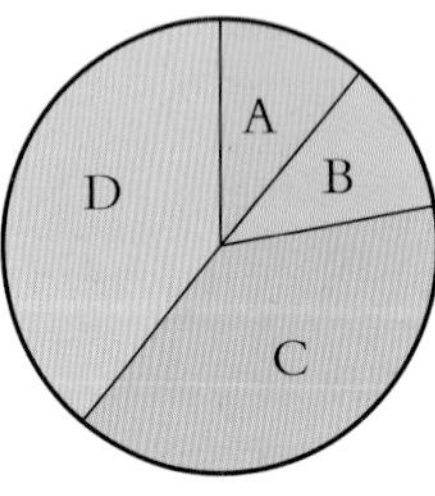

October

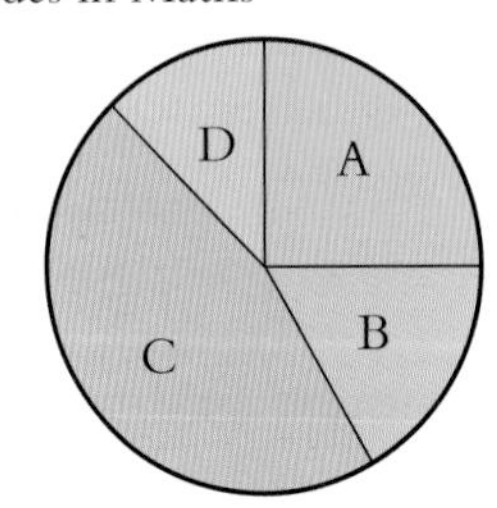

December

Describe in detail the differences in the October and December results, and comment on the effect of the extra homework classes.

Exercise 5 – Revision of Chapter 8 continued

5 John was given some data by his teacher, and asked to draw a line graph. He produced the graph which follows.

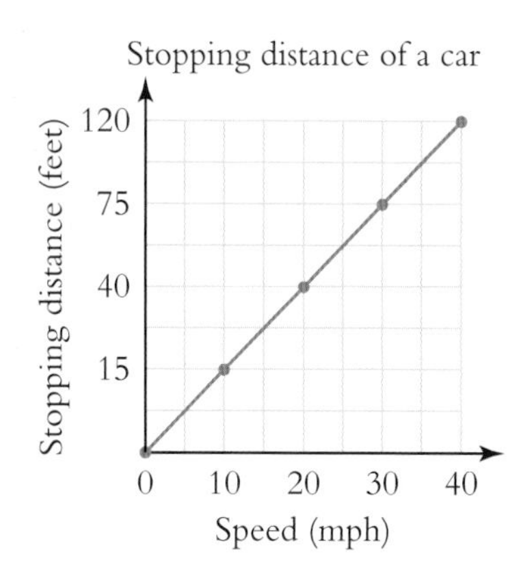

(a) What is wrong with the graph John has drawn?

(b) Redraw the graph correctly.

(c) In what way is John's graph misleading?

6 The ages of the residents of Limetree Close are recorded during a census. The ages are:

36 37 15 19 24 30 26 29 3 2 49 43

57 55 27 23 45 46 19 13 22 23 1 2.

(a) Illustrate the data in a stem and leaf diagram.

(b) Find the difference in ages between the oldest and youngest residents.

7 A doctor is conducting a survey to compare the heights and weights of a group of sportsmen. He records their heights (in inches) and weights (in pounds). The results are as follows.

Sportsman	Jon	Ken	Len	Matt	Neil	Oli	Pat	Rod	Sid	Tom	Vic	Will
Height (inches)	66	69	70	68	65	72	73	64	62	71	68	70
Weight (pounds)	154	160	170	155	133	176	190	140	125	180	152	167

(a) Illustrate the data in a scattergraph.

(b) Draw a line of best fit.

(c) Use this line to estimate the weight of another sportsman who is 75 inches tall. 

Exercise 5 – Revision of Chapter 8 continued

8 The graph shows how the rate of interest for the 'Bronze Saver' Account with the *State Bank* changed during a year.

(a) What was the interest rate in June?

(b) What is the trend of the graph?

(c) Bruce invests £900 in this account in July. How much interest will he earn in 3 months?

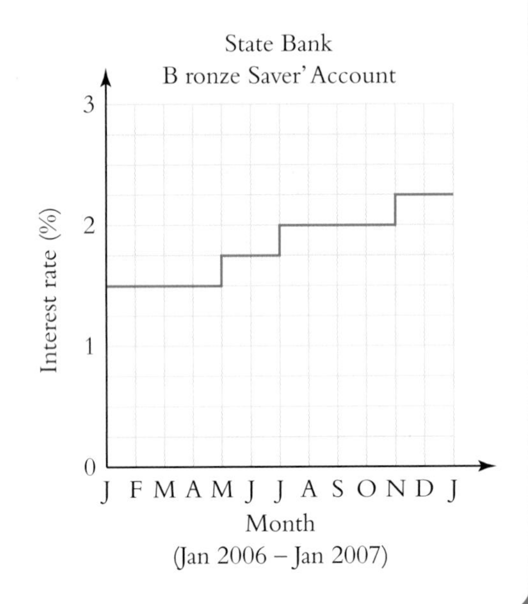

Summary

(The following points should remind you of the important approaches we use as 'visual' techniques in statistics.)

1 ***Bar Graphs and Line Graphs***

When drawing bar graphs and line graphs, remember the following points:

- give the graph a title
- label both axes clearly
- choose a suitable scale for each axis.

(The **trend** in a graph indicates whether it is going up or down.)

2 ***Pie Charts***

Each angle at the centre should be expressed as a fraction of 360°. An angle of 72° thus represents a fraction of $\frac{72}{360}$ or $\frac{1}{5}$. In a total group size of 200, a 72° sector therefore represents $\frac{1}{5}$ of 200, or 40 individuals.

Summary continued

3 ***Stem and Leaf Diagrams***

They provide a convenient way of displaying all data in a numerical ascending order. Thus:

21 32 45 23 56 45 20 26

22 30 36 39 45 51 27 56

23 45 50 34 56 46 30 21

(Test Marks)

→

Test Marks

Stem	Leaf
2	0 1 1 2 3 3 6 7
3	0 0 2 4 6 9
4	5 5 5 5 6
5	0 1 6 6 6

(n = 24) (2|0 means 20)

Back-to-back stem and leaf diagrams can be used to compare **two** sets of data.

4 ***Frequency tables*** can be constructed using tally marks. Thus:

5 6 4 5 6 8 8 9

4 7 8 6 9 5 5 9

9 7 5 6 4 7 4 5

→

Score	Tally	Frequency
4	||||	4
5	~~||||~~ |	6
6	||||	4
7	|||	3
8	|||	3
9	||||	4

5 ***Scattergraphs***

For example,

French marks	4	7	8	6	9	3	5	7	6	10
English marks	5	8	9	6	9	5	7	6	8	10

A line of best fit should follow the general path of the dots, and should have as many dots above as below, or as near as possible.

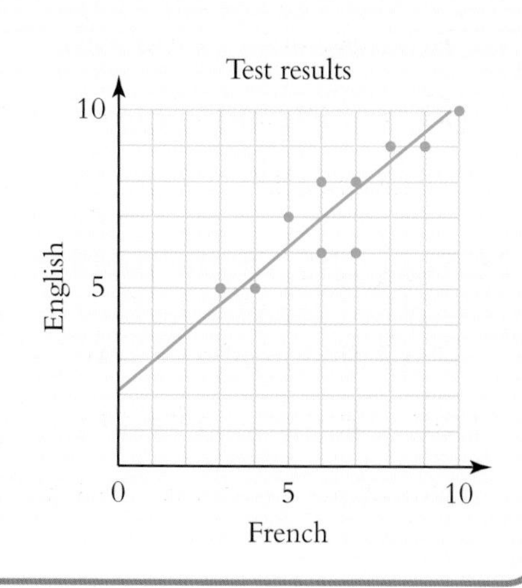

9 Using Simple Statistics

The presentation of data in graphs, charts, and tables helps us in our understanding and interpretation of the data. Statistical calculations help us too, and often provide us with a more accurate and clearer meaning of what the data represents. In this chapter we look at some simple calculated statistics and their meanings, and also at the topic of *Probability*.

Statistical 'Averages'

Most people have some idea of what is meant by an 'average'. In statistics however, it is possible and often convenient to think in terms of *three* types of average. These are called the *mean*, the *median*, and the *mode*, and you should be able to calculate any or all of them for a given set of data.

Key words and definitions

The mean, median, and mode of a data set are defined as follows:

1 For any data set the **mean** $= \dfrac{\textit{sum(total) of values}}{\textit{number of values}}$.

(For example, the mean of 6, 8, 4, 9 and 8 is $\dfrac{6+8+4+9+8}{5} = \dfrac{35}{5} = 7$.)

2 For any data set the **median** = the *middle* value in an *ordered* set. (For example, the median of 4, 6, 9, 10, 13 is 9, which is the middle number.

However, the median of 4, 6, 9, 10, 13, 14 is the mean of the two middle numbers. That is, $\dfrac{9+10}{2} = \dfrac{19}{2} = 9{\cdot}5$.)

Note that the numbers *must be in order* before you can calculate the median.

3 For any data set, the **mode** (or **modal value**) = the most *frequent* value. (For example, the mode of 2, 3, 3, 3, 4, 4, 5 is 3.)

Range of data

The *range* is an important measure of *spread* in statistics. If the range is high, the data set may be well spread out. If the range is small, the data set is more consistent and not likely to be spread out.

For any data set, the **Range** = Highest value – Lowest value. (For example, the range of 4, 5, 8, 10, 12 is 12 – 4 = 8.)

Example Calculate the mean, median, mode and Range of the following data:

7, 12, 5, 9, 14, 6, 7, 13, 2, 10.

Solution

$$\text{Mean} = \frac{7 + 12 + 5 + 9 + 14 + 6 + 7 + 13 + 2 + 10}{10}$$

$$= \frac{85}{10} = 8{\cdot}5.$$

To find the median, we first have to list the 10 numbers in order. Thus, 2, 5, 6, 7, 7, 9, 10, 12, 13, 14.

Hence median = mean of 7 and 9 = $\frac{7+9}{2} = \frac{16}{2} = 8.$

In this case the mode = 7 (the most common number)

Since Range = Highest – Lowest, then Range = 14 – 2 = 12.

Statistics are used to present information about the population

Exercise 1

Do not use a calculator in questions 1 or 2

1 Find the mean, median and mode of each of these data sets:

(a) 4, 4, 5, 8, 9

(b) 3, 4, 6, 7, 8, 8

(c) 5, 7, 9, 9, 10, 20

(d) 1, 1, 2, 6, 7, 8, 10.

2 Find the Range for each data set in question 1.

3 The ages of the players in a football team are:

22 26 28 36 19 22 27 29 21 32 24.

Find (a) the mean (b) the median (c) the mode (d) the Range.

4 Sourav plays eight innings at cricket. His scores (runs) are:

45 88 5 79 107 43 90 15.

Find (a) his mean score (b) his median score.

5 The hourly pay rates of ten workers in a grocery store are as follows:

£5·60, £5·05, £5·05, £6·50, £5·05, £5·80, £5·05, £5·80, £6·50, £8·80.

Calculate (a) the mean (b) the median (c) the modal pay (d) the Range.

6 The mean weight of a group of 4 girls is 48 kilograms. A fifth girl joins the group. She weighs 53 kilograms. Find the mean weight of the group now.

7 The heights of four boys are measured. The mean height is 157 cm. If the heights of three of the boys are 152 cm, 160 cm, and 163 cm, find the height of the fourth boy.

8 The temperature in a Scottish town was measured at mid-day every day for one week during the winter. The temperatures were as follows:

–5°C, 3°C, –4°C, –2°C, 0°C, 3°C, –2°C.

Find (a) the mean temperature (b) the median (c) the Range.

9 Colin and Paul record their scores in six medal competitions at the golf club, as follows:

Colin	74	67	70	66	69	75
Paul	73	72	71	73	70	72.

(a) Calculate the Range for Colin.

(b) Calculate the Range for Paul.

(c) Which golfer was the more consistent?

Exercise 1 continued

10 Michael times his journey from home to work each morning. After four mornings his mean time is 22 minutes. On the fifth day, he is delayed in a traffic jam. The mean increases to 25 minutes after five days.

How long did his journey take on the fifth day?

11 (a) Find the median of the numbers: 7, 10, 12, 14, 18, 22, 23.

(b) After an eighth number is added to this list, the median changes to 15. What is the number that was added to the list?

Frequency Tables

In Chapter 8 we saw how to set up a frequency table. We noted that the frequency table was useful for showing groupings within the data. Here we look again at the frequency table, and use it to help us calculate mean, median, mode and Range of the data.

Example A group of students sit a class test, and their marks are shown in the following frequency table:

(a) Write down the modal mark.

(b) Find the range of marks.

(c) Calculate the mean mark.

(d) Find the median mark.

Mark	Frequency
4	1
5	3
6	4
7	5
8	4
9	2
10	1

Solution

(a) The modal mark = 7 (the mark with the highest frequency)

(b) Range of marks = Highest – Lowest = 10 – 4 = 6.

(c)

Mark	Frequency	Mark × Frequency
4	1	$4 \times 1 = 4$
5	3	$5 \times 3 = 15$
6	4	$6 \times 4 = 24$
7	5	$7 \times 5 = 35$
8	4	$8 \times 4 = 32$
9	2	$9 \times 2 = 18$
10	1	$10 \times 1 = 10$
	(Total = 20)	(Total = 138)

The mean mark $= \dfrac{\textit{Total marks}}{\textit{Number of students}} = \dfrac{138}{20} = 6{\cdot}9.$

(d) The median mark is the middle mark. That is, between the 10th and 11th marks, since there are 20 marks in total. By counting down the frequency column, you see that the 10th and 11th marks are both 7, hence the median = 7.

Exercise 2

1 The ages of a sample of 40 pupils at a Secondary School are shown in the following table:

(a) Copy and complete the table, and hence calculate the mean age.

(b) Write down the modal age.

(c) Find the median age.

(d) What is the Range of age?

Age	Frequency	Age × Frequency
12	5	60
13	7	91
14	8	
15	9	
16	7	
17	4	
	(Total = 40)	(Total =)

Exercise 2 continued

2 Some students were asked how many times they had attended a football match during a particular month. The results are shown in the following table:

No. of games attended	Frequency	No. of games attended × Frequency
0	35	
1	24	
2	17	
3	14	
4	7	
5	3	
	(Total = 100)	(Total =)

(a) Copy and complete the table, and hence calculate the mean number of games attended.

(b) Write down the mode.

(c) Find the median number of games attended.

3 The salaries of all employees at a call centre are shown in the following frequency table:

(a) What is the range of salaries?

(b) Write down the modal salary.

(c) Copy the table and add a third column headed '*Salary × Frequency*'.

(d) Calculate the mean salary.

Salary	Frequency
£10 000	6
£12 000	4
£14 000	8
£16 000	4
£18 000	2
£20 000	1

Exercise 2 continued

4 The temperature was recorded for thirty consecutive days at Chillmarnock during January and February. The results are shown in the following frequency table:

(a) Write down the modal temperature

(b) What is the Range of temperature?

(c) Find the median temperature.

(d) Copy the table, add a third column headed '*Temperature* × *Frequency*' and hence calculate the mean temperature.

Temperature (°C)	Frequency
−3	6
−2	3
−1	6
0	7
1	3
2	3
3	2

5 The numbers of tacks in 20 boxes were counted and the results graphed as in the following diagram:

Calculate the mean number of tacks per box.

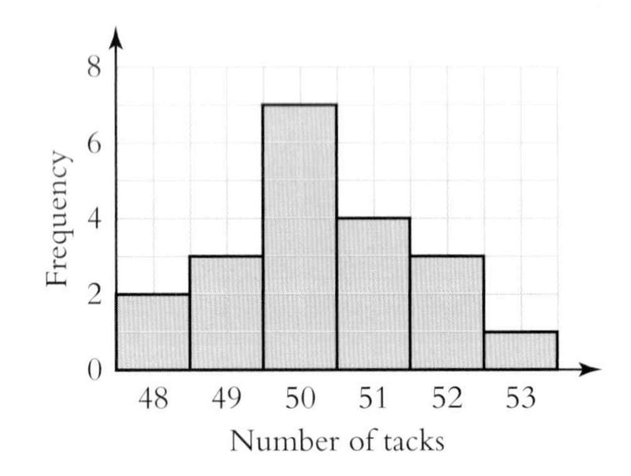

6 The frequency table below shows the number of textbooks carried by a group of college students:

Number of textbooks	0	1	2	3	4	5	6
Frequency	6	8	14	9	6	4	3

Calculate the mean number of textbooks carried per student.

Comparison of Data

Now that you are familiar with the meanings of mean, median, mode, and Range, you should be able to make sensible judgements about the nature of one set of data, and sensible comparisons between two or more sets of data.

We begin by interpreting a single set of data and we look at which of the 'averages' might be more typical of the data than the others.

Example Which is the most typical 'average' in each of the following data sets – the mean, the median, or the mode?

(a) The attendances at five Premier League football matches on Saturday were:

10 800, 9500, 11 000, 8900, 58 500.

(b) The marks of a group of students who sat a test were:

45, 53, 56, 57, 60, 64, 72, 75, 78, 78, 82, 84.

(c) The numbers of goals scored by the teams in the Italian Football League one week were:

0, 1, 0, 0, 1, 2, 0, 1, 0, 3, 1, 1, 0, 2, 0, 0, 1, 0, 4, 0.

Solution

(a) Check that the mean = 19 740 and the median = 10 800. There is no mode.

The mean has been affected by one very high atypical attendance, therefore the **median** is most typical of the data.

(b) Check that the mean = 67, the median = 68, and the mode = 78.

The **mean** and **median** both give a truer picture of the average than the mode.

(c) Check that the mean = 0·85, the median = 0·5, and the mode = 0.

The **mode** is most typical, because half of the teams playing scored no goals. This clearly gives a lot of information about the League.

Exercise 3

1 Six students in an Intermediate 1 class sat the same test, and their marks were:

35, 37, 40, 42, 43 and 79.

(a) Calculate the mean mark.

(b) How many pupils scored below the mean mark?

(c) Find the median mark.

(d) Which average gives a true picture of the marks – the mean or the median?

2 The 'Pizza Cake' Home Delivery Company took the following times (in minutes) to deliver the first eight orders to its customers last Saturday evening.

13 18 20 12 8 14 10 17.

Its rival 'Pizzas4U' was also timed for the first eight orders on the same Saturday evening. Its times (in minutes) were:

20 22 19 20 17 21 22 19.

(a) Calculate the mean time for each company.

(b) Which company tends to deliver the pizzas more quickly?

(c) Calculate the Range for each company.

(d) Which company provides the more consistent service?

(e) Comment on the advantages and disadvantages of the two Companies.

3 Tom and Dick are playing skittles. The following figures show the numbers of skittles that each player knocks down in six shots:

Tom 1 6 9 8 10 4
Dick 5 8 9 6 5 3.

(a) Find the median number of skittles knocked down by each player.

(b) Calculate the Range for each player.

(c) Which player (i) scored better?

(ii) was most consistent?

Give reasons for your answer.

Exercise 3 continued

4 The values of the five houses in Southfork Drive are:

£75 000 £70 000 £95 000 £80 000 £375 000.

(a) Calculate the mean value of the houses.

(b) Find the median value of the houses.

(c) Which of these 'averages' gives a truer picture of the values of these houses? Explain your answer.

5 A group of children was asked to keep a record of how many portions of fruit each ate during one week. The results are shown in the following stem and leaf diagram.

Number of Portions of Fruit

Stem	Leaf
0	0 0 0 2 3 5
1	0 2 3 6 6 7
2	1 3 5 5 8 8
3	3 5 8

(n = 21) (3 | 5 means 35)

(a) Write down the mode.

(b) Find the median.

(c) Which of these 'averages' gives a better picture of the results?

6 Desta is training for an athletics championship. Her coach times her over 25 laps of the running track.

(a) What is the modal lap time?

(b) Calculate the mean lap time.

(c) Which of the two 'averages' is more typical of her overall performance?

Lap time (seconds)	Frequency	Lap time × Frequency
78	6	
79	5	
80	5	
81	5	
82	4	
	(Total = 25)	(Total =)

Probability

Probability is a measure of chance on a scale between 0 and 1. An event which is impossible has a probability of 0 and an event which is absolutely certain has a probability of 1. Thus the probability that I will live for ever is 0 and the probability of scoring '1', '2', '3', '4', '5', or '6' with one roll of a standard die is 1. Most situations we meet have probabilities between 0 and 1, and in this section you should be able to state numerically the probability of a simple outcome.

Probability is defined as follows:

$$\text{Probability} = \frac{\textit{Number of favourable outcomes}}{\textit{Total number of outcomes}}.$$

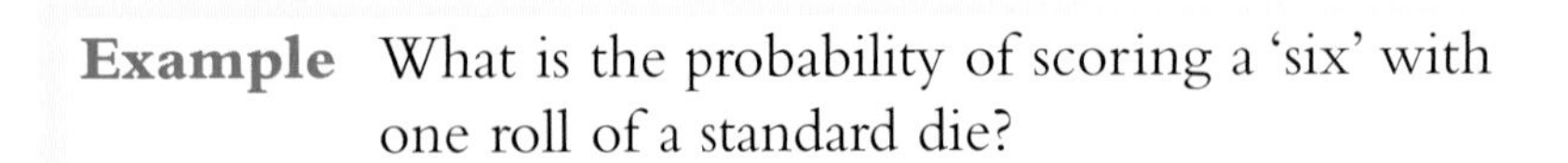

Example What is the probability of scoring a 'six' with one roll of a standard die?

Solution

The *total* number of outcomes is 6 (1, 2, 3, 4, 5 or 6).

The number of *favourable* outcomes is 1 (a 'six').

Hence probability $= \frac{1}{6}$.

Example A letter is chosen at random from the word INTERMEDIATE. What is the probability it is a vowel?

Solution

Intermediate has 12 letters. 6 of them are vowels (A, E, I, O or U).

The probability is therefore $\frac{6}{12}$ or $\frac{1}{2}$.

The probability of winning the lottery is very low

Exercise 4

1 A standard die numbered from 1 to 6 is rolled. What is the probability that it shows

(a) a four

(b) an even number?

Exercise 4 continued

2 A letter is chosen at random from the word STATISTICS. What is the probability that it is (a) S (b) A (c) I?

3 A bag contains 5 red sweets, 3 blue sweets and 2 yellow sweets. If a sweet is chosen at random, what is the probability that it is blue?

4 A spinner numbered from 1 to 5 is spun and allowed to come to rest.

What is the probability that it comes to rest on (a) 1 (b) an odd number?

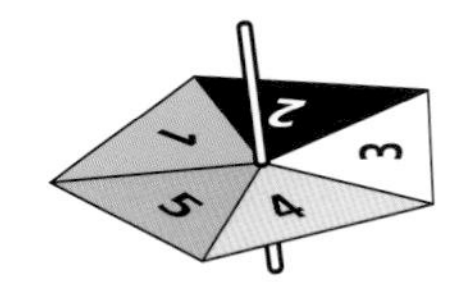

5 There are 300 tickets sold in a raffle. Dale buys 10. What is the probability that he has bought the winning ticket?

6 Five cards – Ace, King, Queen, Jack, and Ten – are laid face down on a table.

(a) Omar chooses a card. What is the probability that he chooses the Ace?

(b) He chooses the Ace and does not replace it. What is the probability that the next card he chooses is the King?

7 ANTIDISESTABLISHMENTARIANISM is the longest real word in the English language. If a letter is chosen at random what is the probability that it is a vowel?

8 A bag contains 6 blue beads, 4 red beads and 3 green beads.

(a) If one bead is picked out of the bag, what is the probability that it is green?

(b) If a red bead is chosen and not replaced, what is the probability that the next bead picked is also red?

9 Terry has eight bags of sweeties. The bags contain 14, 15, 15, 16, 16, 16, 17, and 18 sweeties. If a bag is chosen at random, what is the probability that it contains 15 sweeties?

10 The numbers of pencils carried by all the girls of class 4A were counted. The results were as follows:

Number of pencils	0	1	2	3	4
Frequency	3	8	6	2	1

If one girl is chosen at random, what is the probability that she is carrying 1 pencil?

Exercise 5 – Revision of Chapter 9

1 Ian was writing a book. He kept a record of how many pages he wrote each week. Over a ten week period, he wrote the following numbers of pages:

17 21 24 16 25 18 20 18 23 18.

Calculate (a) the mean (b) the median (c) the mode (d) the Range.

2 Three judges awarded Tatiana marks of 9·3, 9·5 and 9·0 in a gymnastics competition.

After a fourth judge awarded her mark, Tatiana's mean score was 9·4. What mark did the fourth judge award?

3 The marks of a fifth year class in an October test were as follows:

50 57 78 43 59 60 66 89 54 43

64 73 65 56 73 58 81 77 56 43.

(a) Illustrate this data in a stem and leaf diagram.

(b) Write down the mode.

(c) Find the median.

(d) Which of the mode and median is more typical of the marks? Explain.

4 Magnus times himself each day for 50 days, while he does his favourite crossword puzzle.

Time taken (minutes)	Frequency	Time taken × Frequency
12	10	120
13	15	195
14	11	
15	10	
16	4	

(Total = 50)

Exercise 5 – Revision of Chapter 9 continued

(a) What is the modal time?

(b) Find the median time.

(c) Copy and complete the table and hence find the mean time to complete the crossword puzzle.

(d) In what percentage of his attempts does Magnus complete the puzzle in less than quarter of an hour?

5 A letter is chosen at random from the word ASSESSMENT. What is the probability that it is S?

Summary

(Calculated statistics provide us with accurate information in our interpretation of data. You should now be familiar with and confident about the following items.)

1 The *mean*, *mode* and *median* are three useful 'averages'.

- *mean* = $\dfrac{\textit{sum of values}}{\textit{number of values}}$
- *median* = *middle* value in an ordered set
- *mode* = most *frequent* value.

The *Range* of a data set = highest value – lowest value.

Thus for the data set 5, 7, 4, 8, 8, 10:

mean = $\dfrac{5 + 7 + 4 + 8 + 8 + 10}{6} = \dfrac{42}{6} = 7$,

median = 7·5 (middle value of 4, 5, 7, 8, 8, 10),

mode = 8,

and *Range* = 10 – 4 = 6.

Summary continued

2 The three 'averages' and the Range can be found from a *frequency table*.

Thus:

Marks	Frequency	Marks × Frequency
14	7	$14 \times 7 = 98$
15	12	$15 \times 12 = 180$
16	23	$16 \times 23 = 368$
17	8	$17 \times 8 = 136$
	(Total = 50)	(Total = 782)

mean $= \frac{782}{50} = 15{\cdot}64$; mode (most frequent mark) $=16$; median (middle mark) $= 16$;
Range $= 17–14 = 3$.

3 Probability $= \dfrac{\textit{Number of favourable outcomes}}{\textit{Total number of outcomes}}$.

The probability of drawing a King from a pack of playing cards is $\frac{4}{52}$ or $\frac{1}{13}$.

End Of Unit Tests

Test One (Non-Calculator)

1 Work out:

(a) $-50 + 15$ (b) $25 - 40$ (c) $-22 - 17$.

2 Work out:

(a) -8×9 (b) $7 \times (-50)$ (c) $-40 \div 8$.

3 Three corners of a parallelogram have coordinates A (–5, –4), B (3, –2), and C (5, 4).

Plot A, B, and C, and hence find the coordinates of the fourth corner D.

4 (a) A bus left Glasgow at 9.35 pm and reached Aberdeen at 1.05 am the following day. How long did the journey take?

(b) The average speed of the bus was 66 kilometres per hour. Find the distance travelled from Glasgow to Aberdeen.

5 The mean temperature in a Scottish town for the month of July was calculated in degrees Celsius over a number of years. The results are shown in the table.

Year	2001	2002	2003	2004	2005	2006
Temperature (°C)	16	17	17	19	20	23

(Mean temperature in July)

(a) Draw a line graph on squared paper to illustrate this data.

(b) What is the trend of the graph?

6 Henry is given a maths homework exercise each week. He records his percentage mark each week for 20 weeks as follows:

56 66 75 58 66 73 80 66 77 69

60 71 82 66 70 82 91 69 75 88.

(a) Construct a stem and leaf diagram to show his marks.

(b) What is the modal mark?

7 At Abermuchty High School, 12 pupils sit both chemistry and physics exams, and their percentage marks are as follows:

Students	Abi	Ben	Cleo	Don	Eli	Fred	Gill	Herb	Iona	Jack	Kay	Len
Chemistry	72	88	53	69	73	90	47	50	32	46	78	59
Physics	65	78	49	62	65	80	43	45	30	40	70	53

(a) Illustrate this data on a scattergraph.

(b) Draw a best-fitting line on your graph.

(c) Use your line to estimate the physics mark of a thirteenth student who scored 55% for chemistry, but missed the physics exam.

8 Find (a) the mean (b) the median (c) the mode for the following set of data:

2 3 3 7 8 8 9 10 11 11 11 16 18.

9 A letter is chosen at random from the alphabet.

What is the probability that it is a vowel?

Test Two (Calculator)

1 Calculate the size of x in each of the following triangles.

(a)

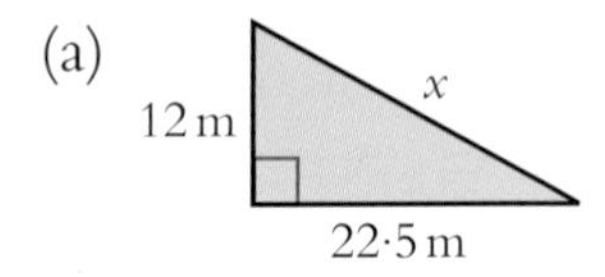

(b)

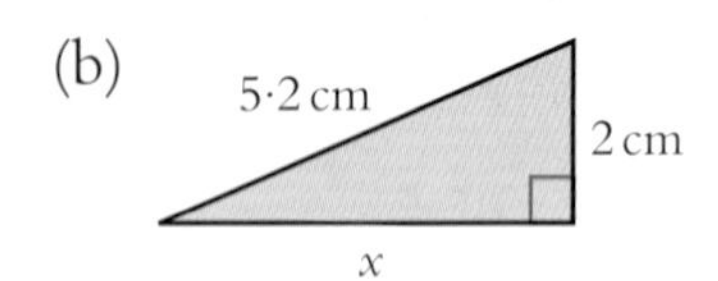

2 Copy and complete the following magic square so that all the rows, columns, and diagonals add up to the same total.

9	−12	
−6		
−3		

3 (a) Alison drives for 3 hours 15 minutes at an average speed of 36 km/h. How far does she travel?

(b) Melissa drives a distance of 399 kilometres at an average speed of 84 km/h. How long did the journey take? Give your answer in hours and minutes.

(c) Jagtar drives 252 miles in 3 hours 30 minutes. Calculate his average speed.

4 A train began its journey at 21 25 hours, and arrived at its destination at 05 10 hours the following day. If it travelled at an average speed of 104 km/h, what distance did it travel?

5 The distances between Moorhill and two other towns are shown in the following diagram.

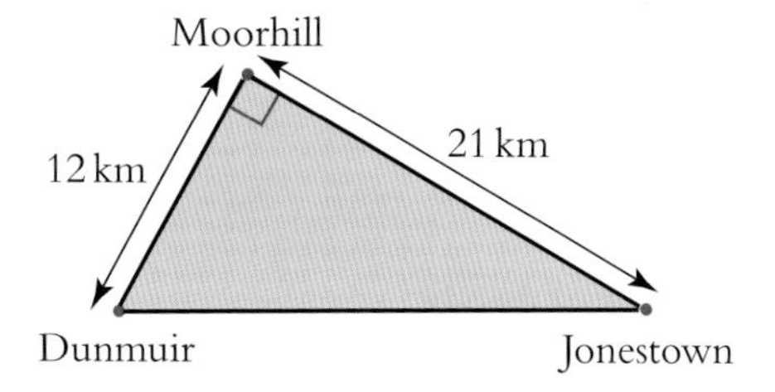

Calculate the distance between Dunmuir and Jonestown.

(Give your answer correct to one decimal place.)

6 Members of a group of 80 students were asked which flavour of crisps they preferred.

The results are shown in the following pie chart.

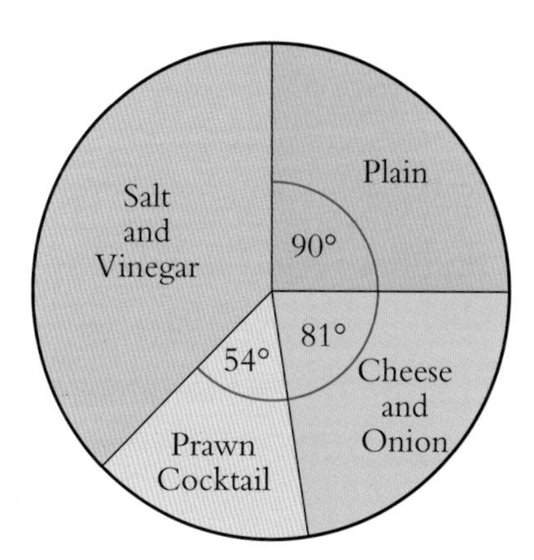

How many students preferred 'salt and vinegar' flavour?

7 The temperature at 8 am was recorded each day for a week at Aviemore during February. The results were as follows:

−3°C, 5°C, −1°C, 2°C, −8°C, 0°C, −2°C.

(a) Calculate the mean temperature.

(b) Find the median temperature.

8 Joan drives from her home to work. The following graph shows her journey.

She leaves home at 8 am and arrives back at 6.30 pm.

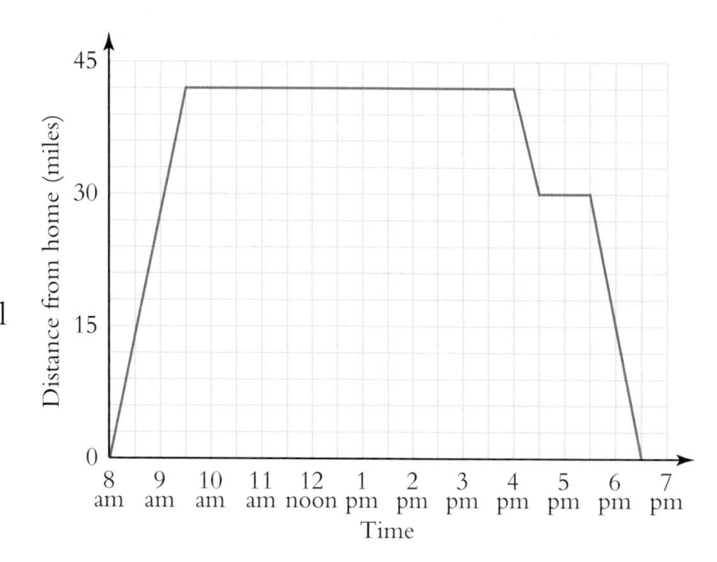

(a) How long did she spend at work?

(b) Calculate the average speed, in miles per hour, of her journey to work.

9 The length of each telephone call received at a call centre was recorded over a one hour period. The results were displayed in a frequency table as follows.

Length of call (to nearest minute)	Frequency	Length of call × Frequency
0	14	0
1	35	35
2	21	42
3	18	
4	9	
5	3	

(Total = 100)

(a) Copy and complete the frequency table.

(b) Calculate the mean length of a call.

(c) Write down the modal length of call.

Test Three (A/B Content)

1 Calculate:

(a) $12 - (-5)$ (b) $-3 \times (-7)$

(c) $-20 \div (-5)$ (d) $-3 \times 5 \times (-4)$.

2 The temperature at Glasgow Airport was measured as 3°C at 5 pm one evening. By 1 am the following morning, it had fallen to –6°C. By how many degrees had the temperature fallen?

3 Julie plans to arrive at Prestwick Airport at 13 20 hours to catch a flight.

She lives 70 miles from the airport. If she drives at an average speed of 40 mph, what is the latest time she can set out from home to reach the airport on time?

4 EFGH is a rhombus.

Find the length of side EF.

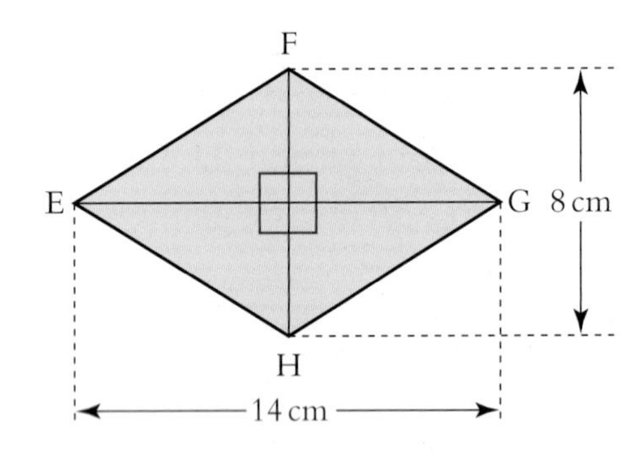

(Give your answer correct to one decimal place.)

5 The seven staff who attended the annual office golf outing recorded the following scores:

84 80 79 81 82 77 119.

(a) Calculate the mean score.

(b) Calculate the median score.

(c) Which 'average' is more typical of the scores – the mean or the median?

Explain your answer.

6 The following diagram shows a 'number cell'.

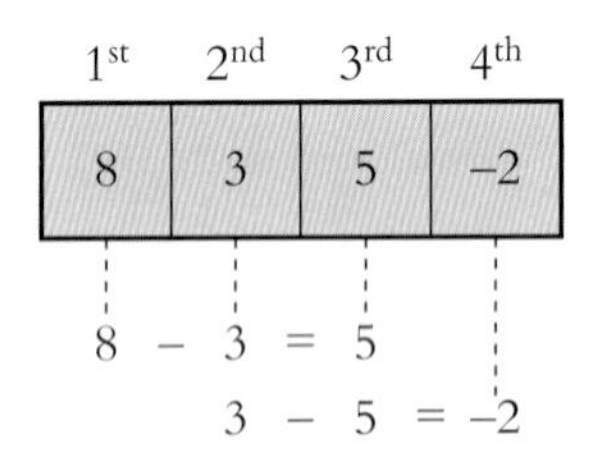

From the diagram you will see that:

- 1st number – 2nd number = 3rd number 8 – 3 = 5
- 2nd number – 3rd number = 4th number 3 – 5 = – 2

Copy and complete the following 'number cells'.

(a)

6	2		

(b)

		7	–9

(c)

3			5

UNIT 3

10 Simple Algebraic Operations

We return to Algebra in this chapter with a view to solving equations and inequalities. These occur regularly in Mathematics and in many other subjects too. Before solving equations however, we look in a little more detail at some processes with which you should be reasonably familiar.

Evaluating Formulae

In Chapter 3 we evaluated simple expressions and formulae, some of which were in words. Here we deal purely with *symbols*, and the expressions and formulae are a little more difficult. You may care to look back to Chapter 3 and revise the correct order for operations – 'BOMDAS'.

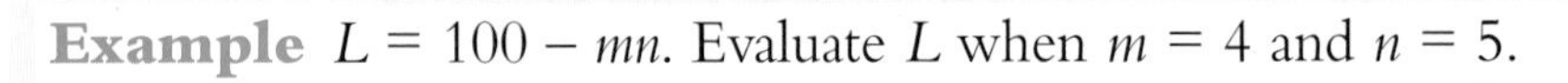

Example $L = 100 - mn$. Evaluate L when $m = 4$ and $n = 5$.

Solution $L = 100 - mn = 100 - 4 \times 5 = 100 - 20 = 80$.

Example The radius of a circle may be calculated from the formula

$r = \sqrt{\dfrac{A}{\pi}}$ where A is the area of the circle.

Find the radius of a circle whose area, A, is 154 square centimetres.

Solution $r = \sqrt{\dfrac{A}{\pi}} = \sqrt{\dfrac{154}{\pi}} = \sqrt{49{\cdot}0197} = 7{\cdot}00$.

Hence the radius is 7 cm.

Take care

When evaluating a formula which involves a square root sign, you should do any calculations within the square root sign before pressing the square root key on your calculator.

Example The formula $s = \frac{1}{2}ft^2$ is used in physics to describe accelerated motion. Find s when $f = 8$ and $t = 5$.

Solution $s = \frac{1}{2}ft^2 = \frac{1}{2} \times 8 \times 5^2 = \frac{1}{2} \times 8 \times 25$
$= 0{\cdot}5 \times 8 \times 25 = 100.$

Exercise 1

1 If $P = 3x + 5$, evaluate P when $x = 5$.

2 If $P = 5q - 2r$, evaluate P when $q = 4$ and $r = 3$.

3 The circumference of a circle can be found by using the formula $C = 2\pi r$.
Find C when $r = 6$. (Give your answer correct to one decimal place.)

4 If $S = 8\sqrt{d}$, calculate S when $d = 100$.

5 Find the radius, r, of a circle with area, A, of 314 cm^2, given that $r = \sqrt{\frac{A}{\pi}}$.
(Give your answer to the nearest whole number.)

6 The formula $h = \frac{V}{lb}$ may be used to find the height of a cuboid.
Calculate:

(a) h when $V = 72$, $l = 6$ and $b = 4$

(b) h when $V = 100$, $l = 4$ and $b = 2$.

7 If $Q = \frac{p}{t^2}$, calculate Q when $p = 80$ and $t = 5$.

Exercise 1 continued

8 Evaluate the expression $3pq - r$ when

(a) $p = -1$, $q = 4$, and $r = 5$

(b) $p = -2$, $q = -4$, and $r = -8$.

9 Given the formula $V = \sqrt{gr}$, calculate V when $g = 9{\cdot}81$ and $r = 8$.
(Give your answer to 2 decimal places.)

10 The volume of a cone may be found using the formula $V = \frac{1}{3}\pi r^2 h$, where r is the base radius, and h is the vertical height of the cone.

Calculate V when $r = 5$ and $h = 12$. (Give your answer to the nearest whole number.)

11 The following diagram shows a cuboid whose volume is V cubic centimetres.

Its height may be found from the formula $h = \frac{V}{l^2}$.

(a) calculate h when $V = 252$ and $l = 6$

(b) calculate V when $h = 7{\cdot}5$ and $l = 8$.

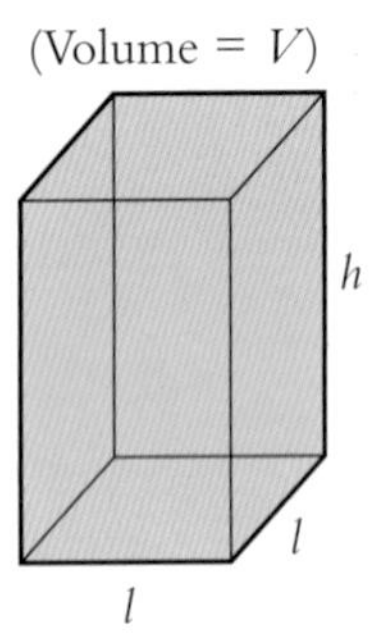

12 The volume of a cylinder may be found from the formula $V = \pi r^2 h$.
Calculate V when $r = 7$ and $h = 8$. (Give your answer to the nearest ten units.)

13 If $V = \sqrt{a^2 - b}$, calculate V when $a = 8$ and $b = 15$.

14 If $S = \frac{1}{2}at^2$, calculate S when $a = 9{\cdot}81$ and $t = 10$.

15 Given the formula $A = 4xy^2$, calculate A when $x = 2{\cdot}5$ and $y = 4$.

16 Using the formula $b = \sqrt{(c^2 - a^2)}$,

(a) calculate b when $c = 13$ and $a = 5$

(b) calculate b when $c = 221$ and $a = 220$.

Exercise 1 continued

17 A can of soup is in the shape of a cylinder as shown.

The surface area of a cylinder may be found using the formula $A = 2\pi rh + 2\pi r^2$.

Calculate the surface area of the can when $r = 3$ and $h = 10$.

(Give your answer to the nearest whole number.)

18 Evaluate the expression x^2y when $x = -4$ and $y = -3$.

19 The Electric Power of a device (P) can be calculated from the current (I) and the resistance (R) by the formula $P = I^2R$.

Calculate P when $I = 3{\cdot}6$ and $R = 75$.

20 Evaluate the expression $\frac{3ab}{c}$ when

(i) $a = -5$, $b = 4$ and $c = 2$

(ii) $a = 2$, $b = -10$ and $c = -15$.

21 Use the formula $M = 3ab^2$ to calculate M when $a = 5$ and $b = 2$.

Simplifying Algebraic Expressions

Some expressions you will come across may be extremely simple, such as $4x$. Others, such as $5e + f - 2e - 3f$ may appear a lot more difficult.

In this section, you should be able to simplify an expression by:

- collecting like terms
- writing it in a shorter form.

Key words and definitions

Terms such as $5x$, $2x$, $7x$, are called **like terms** and can be collected together in expressions such as $5x + 4x - 2x$, leading to $7x$. Terms such as $5x$ and $2y$ are **not** like terms and so an expression such as $5x + 2y$ cannot be simplified. A term like x or y means the same as $1x$ or $1y$.

$3a + 4b$

Example Simplify

(a) $7p + 3p - 2p$

(b) $7d + 5 - d$

(c) $4a + 6b + 3a - 8b$.

Solution (a) $8p$

(b) $6d + 5$

(c) $7a - 2b$.

Example Write in shorter form

(a) $t \times 7$

(b) $4m \times 5n$

(c) $5a \times 2a$.

Solution (a) $7t$

(b) $20mn$

(c) $10a^2$.

Exercise 2

No calculator allowed

1 Simplify:

(a) $5a + 2a$ (b) $8f - 3f$ (c) $7y + y$ (d) $7w + 3w - 4w$

(e) $4x - x$ (f) $4u - 3u$ (g) $2e - 5e$ (h) $4d - 3d + 5d$.

2 Simplify by collecting like terms:

(a) $6r + 3r + 4t$ (b) $6y + 3z - 2y$ (c) $9a + 5b - a$ (d) $4h + 7 + h$.

Exercise 2 continued

No calculator allowed 

3 Simplify by collecting like terms:

(a) $5 + 9c - 8$ (b) $11k + 15 - 7k$ (c) $7b + 4a - 9b$ (d) $5u - 9v + 3v$.

4 Simplify by collecting like terms:

(a) $5p + 6q + 2p + 3q$ (b) $3a + 5b - a + 4b$

(c) $7x - 4y + x + 2y$ (d) $3c + 7d - c - 7d$

(e) $8u - 6v - 2u + 7v$ (f) $5e + f - 2e - 3f$.

5 Write the following expressions in a simpler shorter form:

(a) $f \times 2$ (b) $5 \times k$ (c) $y \times y$

(d) $h \times 10$ (e) $b \times 1$ (f) $m \times m$.

6 Write the following expressions in a simpler shorter form:

(a) $7 \times 2m$ (b) $4 \times 2b$ (c) $6y \times 5$

(d) $9 \times 3e$ (e) $2 \times (-5e)$ (f) $3 \times (-2u)$.

7 Simplify the following expressions:

(a) $3x \times x$ (b) $2p \times p$ (c) $t \times 5t$ (d) $4m \times 3m$

(e) $7y \times 7y$ (f) $3k \times 3k$ (g) $2a \times b$ (h) $y \times 2z$

(i) $v \times 5u$ (j) $3a \times 5b$ (k) $4m \times 3n$ (l) $2e \times 5f$.

Multiplying out Brackets

Sometimes when simplifying expressions, we come across terms which have been grouped within brackets. It is essential that you can expand brackets when required, and then simplify the results.

Consider an expression such as $4(x + 2)$. We can think of this as follows:

$4(x + 2) = 4 \times (x + 2) = (x + 2) + (x + 2) + (x + 2) + (x + 2)$
$= 4x + 8$.

We get the same result if we multiply each term inside the bracket by the term in front of the bracket. Thus:

$4(x + 2) = 4 \times x + 4 \times 2 = 4x + 8$. This is a much easier approach.

Example Multiply out these brackets:
(a) $6(x + 5)$
(b) $3(2y - 8z)$.

Solution (a) $6x + 30$
(b) $6y - 24z$.

When you are asked to 'multiply out brackets and simplify', you have to multiply out the brackets as shown, and then collect any like terms in the resulting expression.

Example Multiply out the brackets and simplify:
(a) $6(e + 2f) - 5f$
(b) $8 + 3(4p - 3)$
(c) $7(3 - u) + 10u$.

Solution (a) $6(e + 2f) - 5f$
$= 6e + 12f - 5f$
$= 6e + 7f$

(b) $8 + 3(4p - 3)$
$= 8 + 12p - 9$
$= 12p - 1$

(c) $7(3 - u) + 10u$
$= 21 - 7u + 10u$
$= 21 + 3u$.

Exercise 3

1 Multiply out the brackets:

(a) $3(a + 2)$ (b) $5(b - 4)$ (c) $7(p + 1)$ (d) $8(c + 6)$

(e) $2(d - 9)$ (f) $4(e - 5)$ (g) $6(4 + f)$ (h) $9(3 - g)$

(i) $10(1 - x)$ (j) $5(h - 2)$ (k) $7(n - 4)$ (l) $12(3 - p)$.

Exercise 3 continued

2 Multiply out the brackets:

(a) $5(4a + 3)$ (b) $6(2p - 3)$ (c) $7(3x - 2)$ (d) $8(4 - 3h)$

(e) $9(5 - 2y)$ (f) $3(4x + 1)$ (g) $4(2g - 9)$ (h) $12(3 - 5p)$.

3 Multiply out the brackets:

(a) $3(2a + b)$ (b) $2(4p - 5q)$ (c) $6(7y - z)$ (d) $5(d + 2e)$

(e) $4(m - 5n)$ (f) $8(3p - 4r)$ (g) $5(3x + 7y)$ (h) $6(3t + 9u)$.

4 Multiply out the following brackets and simplify:

(a) $6(a + 5) + 3$ (b) $4(b + 2) - 5$ (c) $5(3c + 2) - 12$ (d) $2(d - 5) + 13$

(e) $8h + 3(2 - h)$ (f) $5t + 3(4 - t)$ (g) $7y + 2(5 - y)$ (h) $5 + 4(3b + 2)$

(i) $4 + 3(5y - 2)$ (j) $7 + 2(3a - 8)$ (k) $4(x + 3y) - 7y$ (l) $5(p + 4r) - 11r$

(m) $4(3 - h) + 7h$ (n) $5a + 3(4 - a)$ (o) $16k + 2(5 - 3k)$ (p) $3(2x + 5y) - 9x$.

5 Multiply out the brackets:

(a) $a(b + c)$ (b) $p(p - q)$ (c) $2y(y + z)$

(d) $3b(2b - 5c)$ (e) $7p(p - 2q)$.

6 Multiply out the brackets and simplify:

(a) $3(x + 5) + 2(x - 4)$ (b) $4(5 - x) + 2(3x - 2)$.

Factorisation

In this section, you will learn how to factorise an algebraic expression using a common factor.

We know now how to multiply out a bracket.
For example $4(x + 2) = 4x + 8$.

The reverse process to multiplying out a bracket is called *factorisation*. If we were asked to factorise $4x + 8$ we would write $4(x + 2)$. The number before the bracket, 4 in this case, is called the **common factor**.

To factorise an expression, you must find the **highest common factor** of the terms in the expression, and then take this 'out' of the terms.

Example Factorise:

(a) $7m + 21$
(b) $45 - 9a$
(c) $12 + 8m$.

Solution (a) $7(m + 3)$
(b) $9(5 - a)$
(c) $4(3 + 2m)$.

Hints and tips

- After you have factorised an expression, you can always check if you are correct by multiplying out the brackets in your answer. This should give you what you started with.
- Always make sure you have taken the **highest** common factor outside the bracket. For example $2(6 + 4m)$ is an incomplete factorisation of $12 + 8m$, because 2 is not the **highest** common factor of 12 and $8m$.

Exercise 4

1 Factorise the following expressions:

(a) $5d + 10$ (b) $8w - 16$ (c) $5c + 25$ (d) $6t - 18$
(e) $10p - 90$ (f) $4x + 16$ (g) $9v - 45$ (h) $2u + 14$
(i) $10y + 10$ (j) $3w + 24$.

Exercise 4 continued

2 Factorise the following expressions:

(a) $12 + 4b$ (b) $10 - 5q$ (c) $8 - 2d$ (d) $27 - 9u$
(e) $64 + 8x$ (f) $60 - 10l$ (g) $9 - 3m$ (h) $6 - 6z$
(i) $20 + 5n$ (j) $24 - 12a$.

3 Factorise the following expressions:

(a) $8a + 10$ (b) $4p - 6$ (c) $10n + 15$ (d) $14x + 21$
(e) $10y + 25$ (f) $9u - 12$ (g) $10t + 35$ (h) $6p + 15$
(i) $8y - 20$. (j) $18d + 30$.

4 Factorise the following expressions:

(a) $10 + 6x$ (b) $12 - 8m$ (c) $50 + 20u$ (d) $10 + 8e$
(e) $35 + 14b$ (f) $24 - 16x$ (g) $8 + 6z$ (h) $20 - 18q$
(i) $45 - 25a$ (j) $36 + 27x$.

5 Factorise the following expressions:

(a) $7e + 28$ (b) $14 - 2v$ (c) $16w + 20$ (d) $30 - 20f$
(e) $5a - 5b$ (f) $4a + 12b$ (g) $8m - 64$ (h) $40 - 30x$
(i) $64 - 8w$ (j) $12x + 30y$.

Solving Equations

When we need to know the value of an unknown quantity, we set up an *equation* involving the quantity. We then apply the rules for solving an equation.

In this section, you should be able to solve a simple equation, such as $x + 4 = 11$.

What this means is 'what number or value does x represent?' The term x is called a *variable* and to solve the equation involving x, we have to 'undo' any operations carried out on x.

Main points

An equation can be thought of as a set of balanced scales.

To keep a set of scales balanced, then whatever we add or subtract on one side, we must add or subtract on the other side.

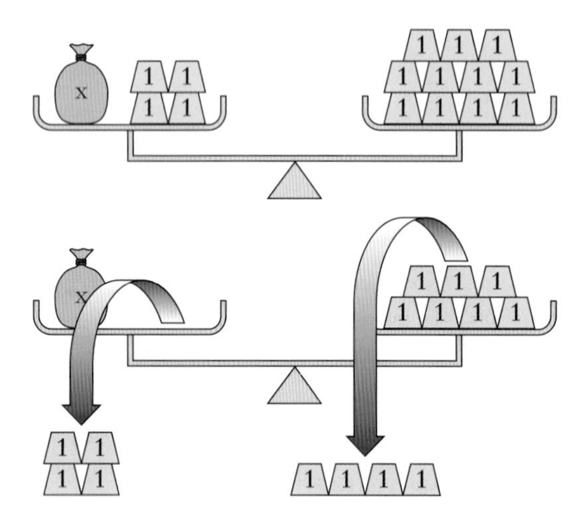

Thus to solve $x + 4 = 11$ we proceed as follows:

$$x + 4 = 11$$
$$(-4) \quad (-4) \quad \text{(subtract 4 from **each** side)}$$

Hence $x = 7$.

Example Solve the following equations:

(a) $x + 5 = 2$ (b) $y - 7 = 2$

(c) $7a = 21$ (d) $\frac{1}{4}x = 5$.

Solution (a) $x + 5 = 2$

$(-5) \quad (-5)$

Hence $x = -3$

(b) $y - 7 = 2$

$(+7) \quad (+7)$

Hence $y = 9$

(c) $7a = 21$

$(\div 7) \quad (\div 7)$

Hence $a = \frac{21}{7} = 3$

(d) $\frac{1}{4}x = 5$

$(\times 4) \quad (\times 4)$

Hence $x = 20$.

Exercise 5

Do not use a calculator 

1 Solve the following equations:

(a) $x + 3 = 9$ (b) $h + 7 = 12$ (c) $b + 3 = 14$

(d) $6 + k = 12$ (e) $5 + y = 3$.

2 Solve the following equations:

(a) $x - 5 = 4$ (b) $y - 7 = 1$ (c) $p - 4 = -2$

(d) $y - 6 = 0$ (e) $m - 2 = -6$.

3 Solve the following equations:

(a) $4q = 12$ (b) $5w = 35$ (c) $8e = 40$

(d) $3r = 7{\cdot}5$ (e) $2t = 9$ (f) $9y = 72$.

4 Solve the following equations:

(a) $\frac{1}{2}x = 4$ (b) $\frac{1}{3}x = 8$ (c) $\frac{1}{4}y = 3$

(d) $\frac{1}{9}p = 0$ (e) $\frac{1}{6}a = 10$ (f) $\frac{1}{5}d = 2$.

5 Solve the following equations:

(a) $f - 5 = 9$ (b) $g + 7 = 3$ (c) $10h = 80$ (d) $\frac{1}{2}k = 15$

(e) $l - 5 = -4$ (f) $9z = 54$ (g) $6 + x = -2$ (h) $2c = 7$

(i) $23 + v = 32$ (j) $\frac{1}{3}b = 7$.

More Difficult Equations

The equations so far have been simple and straightforward. In more complicated situations however, more difficult equations may arise. These more difficult equations may require several operations before we can solve them.

Example Solve the following equations algebraically:

(a) $3n + 5 = 23$
(b) $3m - 5 = m + 11$
(c) $4(y - 5) = 28$.

Solution

(a)
$$\begin{aligned} 3n + 5 &= 23 \\ (-5) \quad &\quad (-5) \\ \Rightarrow 3n &= 18 \\ (\div 3) \quad &\quad (\div 3) \\ \Rightarrow n &= \frac{18}{3} \\ \Rightarrow n &= 6 \end{aligned}$$

(b)
$$\begin{aligned} 3m - 5 &= m + 11 \\ (-m) \quad &\quad (-m) \\ \Rightarrow 2m - 5 &= 11 \\ (+5) \quad &\quad (+5) \\ \Rightarrow 2m &= 16 \\ (\div 2) \quad &\quad (\div 2) \\ \Rightarrow m &= 8 \end{aligned}$$

(c)
$$\begin{aligned} 4(y - 5) &= 28 \\ \Rightarrow 4y - 20 &= 28 \\ (+20) \quad &\quad (+20) \\ 4y &= 48 \\ (\div 4) \quad &\quad (\div 4) \\ \Rightarrow y &= \frac{48}{4} \\ \Rightarrow y &= 12. \end{aligned}$$

When you are asked to solve an equation **algebraically**, you should show the sequence of operations in your working. If you simply write down the answer, even though it is correct, you may risk getting no marks.

Exercise 6

1 Solve the following equations algebraically:

(a) $2x + 6 = 14$ (b) $3y - 5 = 1$ (c) $6a + 3 = 9$ (d) $3q + 2 = 17$
(e) $5 + 7x = 19$ (f) $5m - 3 = 32$ (g) $8f + 3 = 51$ (h) $7 + 2x = 10$
(i) $10a + 2 = 2$ (j) $4v - 3 = 25$.

2 Solve the following equations algebraically:

(a) $4x + 3 = 2x + 11$ (b) $7y + 2 = 2y + 12$ (c) $8t + 3 = 4t + 23$
(d) $3d - 5 = d + 13$ (e) $5y - 4 = y + 24$ (f) $7y - 6 = 3y + 38$
(g) $9c + 3 = 27 + c$ (h) $10i - 7 = 29 + 4i$.

Exercise 6 continued

3 Solve the following equations algebraically:

(a) $12 + 5x = 2x + 15$ (b) $z + 7 = 4z - 5$

(c) $x + 35 = 4x + 8$ (d) $2w + 8 = 7w - 7$

(e) $p + 12 = 7p - 6$ (f) $4f + 5 = 6f - 17$.

4 Solve the following equations algebraically:

(a) $4(x + 2) = 20$ (b) $3(w + 1) = 12$ (c) $2(h - 4) = 10$ (d) $5(y - 6) = 15$

(e) $3(b + 5) = 21$ (f) $7(k - 4) = 7$ (g) $5(3 + x) = 55$ (h) $6(p - 7) = 18$.

Solving Inequalities

An *equation* can be thought of as a set of balanced scales. Provided we do the same to each side, the 'balance' is maintained both for the equation and the scales.

An inequality can be thought of as a set of scales in an 'out-of-balance' condition. Provided we do the same to each side, the out-of-balance condition is maintained in the same way.

Main points

You must be familiar with the following inequality symbols:

- $x > 5$ means x is greater than 5
- $y < 8$ means y is less than 8
- $a \geqslant 3$ means a is greater than or equal to 3
- $b \leqslant 4$ means b is less than or equal to 4.

Exercise 7

1 Write in either $>$ or $<$ between each pair of numbers to make a true statement.

(a) 3…6 (b) 8…7 (c) 0…−3 (d) −5…−4

(e) −1…−6 (f) $\frac{1}{4}\ldots\frac{1}{3}$.

2 By choosing x only from the numbers −2, −1, 0, 1, 2, 3, 4 or 5, find all the solutions of:

(a) $x > 2$ (b) $x < 2$ (c) $x \geqslant 4$ (d) $x \leqslant 0$

(e) $x > 4$ (f) $x \leqslant -2$ (g) $x \geqslant -1$ (h) $x < -1$.

Remember

To *solve* inequalities, we use exactly the same methods as we used to solve equations. Whatever we do to one side of the inequality, we must do exactly the same to the other side.

Example Solve the inequalities

(a) $7x - 5 > 23$

(b) $2t + 3 \leqslant 10$.

Solution

(a)

$$7x - 5 > 23$$
$$(+5)\quad(+5)$$
$$\Rightarrow 7x > 28$$
$$(\div 7)\quad(\div 7)$$
$$\Rightarrow x > \frac{28}{7}$$
$$\Rightarrow x > 4$$

(b)

$$2t + 3 \leqslant 10$$
$$(-3)\quad(-3)$$
$$\Rightarrow 2t \leqslant 7$$
$$(\div 2)\quad(\div 2)$$
$$\Rightarrow t \leqslant \frac{7}{2}$$

Exercise 8

1 Solve the following inequalities:

(a) $x + 2 \leqslant 6$ (b) $x + 5 < 9$ (c) $y - 3 \geqslant 5$

(d) $p - 6 > 5$ (e) $h + 5 \geqslant 7$.

2 Solve the following inequalities:

(a) $5f < 25$ (b) $7t \leqslant 42$ (c) $9m \geqslant 45$

(d) $3u > 7{\cdot}5$ (e) $2b \leqslant 13$.

3 Solve the following inequalities:

(a) $5y + 4 \leqslant 19$ (b) $7f - 2 > 47$ (c) $6p + 3 \geqslant 9$ (d) $8y - 2 < 30$

(e) $6y + 5 < 17$ (f) $4w - 7 \leqslant 1$ (g) $10b + 3 < 23$ (h) $5t - 9 \geqslant 26$.

4 Solve the following inequalities:

(a) $2y + 1 < 10$ (b) $2w + 3 > 11$ (c) $4b + 3 \leqslant 27$ (d) $8j - 2 > 70$

(e) $8p - 3 \geqslant 37$ (f) $3u + 4 < 28$ (g) $7m + 1 > 1$ (h) $5r - 7 < 38$.

5 Solve the following inequalities:

(a) $6q - 3 > 15$ (b) $2 + 4t \geqslant 30$ (c) $2w + 1 > 14$ (d) $3e - 2 > 2{\cdot}5$.

Exercise 9 — Revision of Chapter 10

1 Given the formula $v = \sqrt{ar}$, find v when $a = 12$ and $r = 3$.

2 Use the following formula to find the value of Q when $p = 243$ and $r = 3$.

$$Q = \sqrt{\frac{p}{r}}.$$

Exercise 9 – Revision of Chapter 10 continued

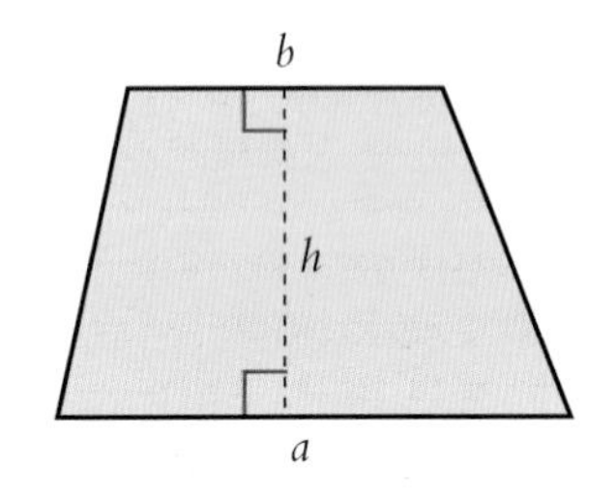

3 The area of a trapezium may be found using the formula $A = \frac{1}{2}h(a + b)$.

Calculate the area of a trapezium in which $h = 5$, $a = 12$, and $b = 8$.

4 Evaluate the expression $2mn^2$ when $m = 3{\cdot}5$ and $n = 6$.

5 Multiply out the brackets and simplify:

(a) $3(p + 5) + 2$ (b) $8x + 2(3 - x)$ (c) $7 + 2(3a - 1)$ (d) $4(x + 3y) - 7y$.

6 Factorise:

(a) $5x + 20$ (b) $7t - 14$ (c) $15 + 5m$

(d) $8z + 12$ (e) $15 - 10y$.

7 Solve the following equations algebraically:

(a) $6x + 3 = 21$ (b) $2y - 3 = 6$ (c) $6x + 2 = 2x + 18$

(d) $4z - 5 = z + 16$ (e) $7 + 4b = b + 16$ (f) $10x - 3 = 15 + 4x$

(g) $y + 5 = 6y - 25$ (h) $\frac{1}{3}x = 6$.

8 Solve the following inequalities:

(a) $5x + 3 > 18$ (b) $6y - 4 < 8$ (c) $2v - 1 < 8$ (d) $7x - 3 < 60$.

9 The time taken, T seconds, for one complete swing of a pendulum of length L (metres) is given by the formula

$$T = 2\pi\sqrt{\frac{L}{g}}, \quad \text{where } g = 9{\cdot}81.$$

Calculate T for a pendulum of length 0·95m. (Give your answer correct to two decimal places.)

Summary

(Check that you are familiar with and confident about the following items before leaving this chapter. If necessary you should re-trace your steps by practising again some questions from the Exercises.)

1 *Evaluating a formula*

For example, if $A = \sqrt{\dfrac{W}{xy}}$, find A when $W = 144$, $x = 12$ and $y = 3$.

Then $A = \sqrt{\dfrac{144}{12 \times 3}} = \sqrt{\dfrac{144}{36}} = \sqrt{4} = 2$.

2 *Multiplying out brackets*

For example, $5(6 + 3a) - 7a = 30 + 15a - 7a = 30 + 8a$.

3 *Factorising using a common factor*

For example, $18 - 12w = 6(3 - 2w)$.

4 *Solving simple equations algebraically*

For example,

$$
\begin{array}{lrcl}
 & 2p + 6 & = & 7p - 14 \\
 & (-2p) & & (-2p) \\
\Rightarrow & 6 & = & 5p - 14 \\
 & (+14) & & (+14) \\
\Rightarrow & 20 & = & 5p \quad \text{or} \quad 5p = 20 \\
 & & & (\div 5) \quad (\div 5) \\
 & & & p = 4.
\end{array}
$$

5 *Solving inequalities*

Use the same methods for solving inequalities as for equations. For example,

$$
\begin{array}{lll}
 & 6x - 2 & < 16 \\
 & (+2) & (+2) \\
 & 6x & < 18 \\
 & (\div 6) & (\div 6) \\
\Rightarrow & x & < \dfrac{18}{6} \\
\Rightarrow & x & < 3.
\end{array}
$$

Graphical Relationships

In Chapter 8 we looked at the graphical presentation of some statistics before moving on to some calculated statistics in Chapter 9. As we saw, a graphical presentation of data conveys a 'picture' which a table of results does not.

Now that we have looked at algebraic expressions in some detail, we can apply some of this knowledge to describing graphical relationships algebraically.

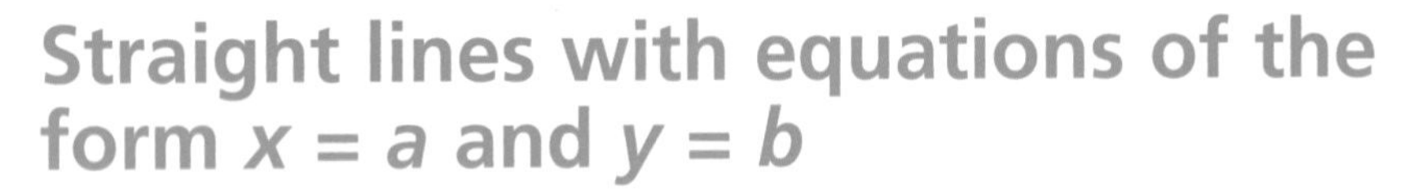

Straight lines with equations of the form $x = a$ and $y = b$

In this section, you will learn how to recognise and describe straight line graphs which are parallel to the x-axis or the y-axis. These graphs have equations of the form $x = a$ and $y = b$. A graph of the form $y = b$ might be used to show the motion of an aircraft flying at steady speed.

Consider the straight line passing through the points (3, 0), (3, 4) and (3, –2).

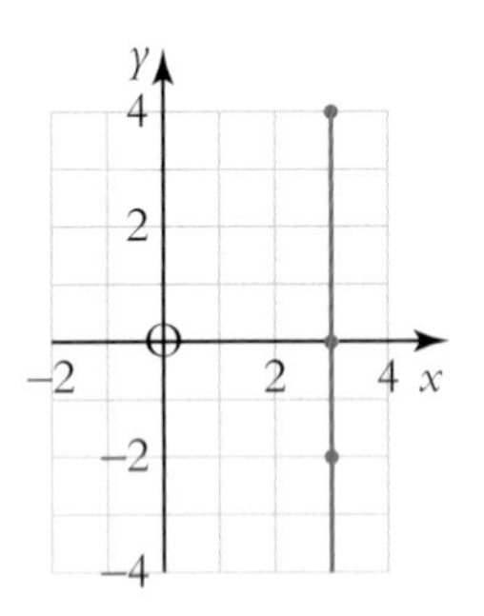

This straight line can be extended in both directions. Every point on this line has its x-coordinate equal to 3, and we say that the equation of this line is $x = 3$.

Now consider the straight line passing through the points (0, 4), (5, 4) and (–3, 4).

Again this line can be extended in both directions. Every point on this line has its y-coordinate equal to 4, and we say that the equation of this line is $y = 4$.

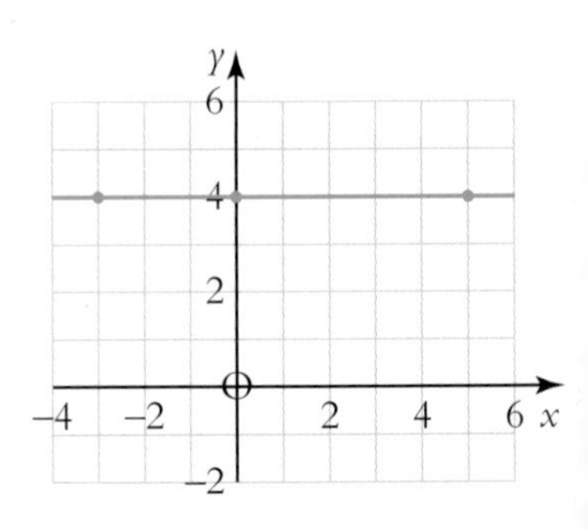

Remember

An equation of the form $x = a$ is the equation of a vertical line parallel to the y-axis, and passing through the point $(a, 0)$.

An equation of the form $y = b$ is the equation of a horizontal line parallel to the x-axis, and passing through the point $(0, b)$.

Example (a) Draw the lines with equations $x = 2$ and $y = 3$.

(b) Write down the coordinates of the point where these lines intersect.

Solution (a)

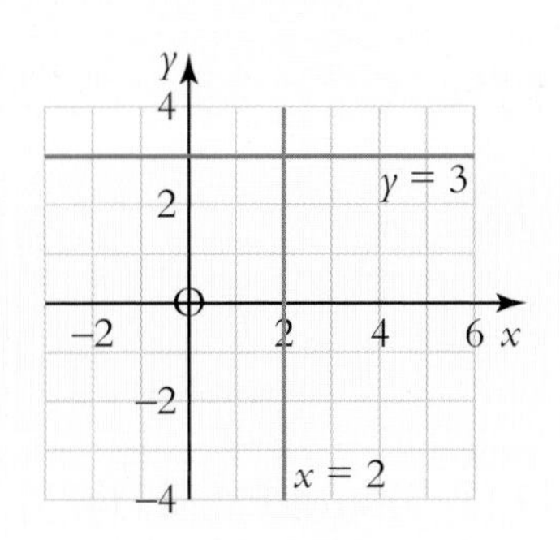

(b) (2, 3)

Exercise 1

1 (a) Plot the points (4, 0), (4, 2) and (4, 4).

(b) Draw a straight line passing through these points.

(c) What can you say about the x-coordinate of every point on this line?

(d) Write down the equation of this line.

2 (a) Draw a straight line passing through the points (0, 3), (4, 3) and (–2, 3).

(b) What can you say about the y-coordinate of every point on this line?

(c) Write down the equation of this line.

Exercise 1 continued

3 (a) Draw the lines with equations $x = 5$ and $y = 4$.

(b) Write down the coordinates of the point where these lines intersect.

4 Repeat question 3 for the lines with equations $x = 1$ and $y = 2$.

5 Write down the equations of the lines a, b, c, and d in the diagram.

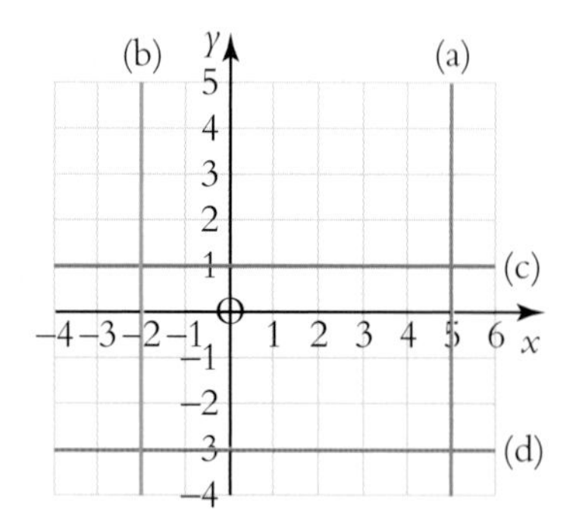

6 (a) Draw the line with equation $x = 3$.

(b) Plot the points A (4, 3), B (5, –1), C (2, 0), D (1, 3) and E (–2, 1).

(c) State whether each of A, B, C, D, and E is to the right or left of $x = 3$.

7 (a) Draw the line with equation $y = 5$.

(b) Plot the points F (4, 6), G (5, 2), H (–3, 6), I (0, 4), and J (2, 7).

(c) State whether each of F, G, H, I and J is above or below $y = 5$.

8 Without drawing, state whether each of the following points is above, below or on the line with equation $y = 1$.

K (5, 2), L (3, 0), M (–2, 2), N (6, 1), O (0, 0).

9 A designer draws a square bounded by the lines $x = 1$, $y = 2$, $x = 4$, and $y = 5$ as shown in the diagram.

Calculate the area of the square.

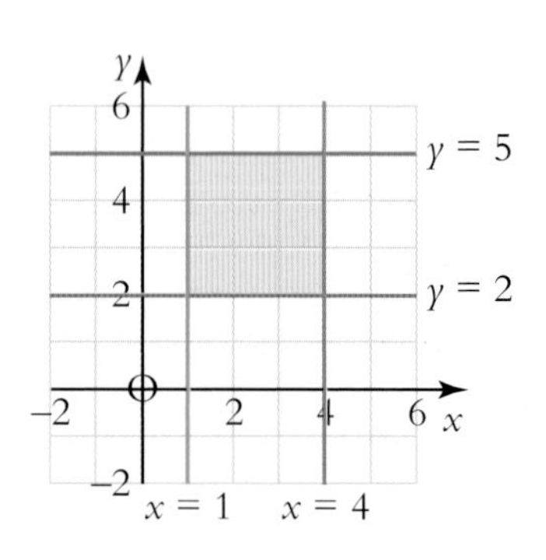

Exercise 1 continued

10 (a) Draw the lines with equations $x = 4$, $x = -1$, $y = 5$, and $y = -2$.

(b) Find the area of the rectangle enclosed by these four lines.

11 Answer **true** or **false**:

(a) the point (9, 3) lies on the line $x = 9$

(b) the point (5, 6) lies on the line $y = 5$

(c) the lines $x = 3$ and $y = 3$ are parallel

(d) the point (–2, 5) lies to the left of the line $x = 1$

(e) the line $y = 7$ is parallel to the y-axis

(f) the point (95, 93) lies above the line $y = 94$

(g) the equation of the x-axis is $y = 0$.

12 Write down the coordinates of the point of intersection of the straight lines with equations $x = 5$ and $y = -2$.

Straight lines with equations of the form $y = ax$

Now that we are familiar with lines which are parallel to the axes, we can consider lines which are at an angle to the axes. We begin with angled lines which pass through the Origin.

In this section, you will learn how to recognise and draw such lines.

These lines have equations of the form $y = ax$.

A graph of the form $y = ax$ might be used to show how the temperature of water (originally ice cold) rises, when it is heated at a steady rate.

Consider the straight line passing through the points (0, 0), (1, 1), (2, 2), (3, 3), and (4, 4).

This straight line can be extended in both directions.

Every point on the line has its x-coordinate equal to its y-coordinate, and we say that the equation of this line is $y = x$.

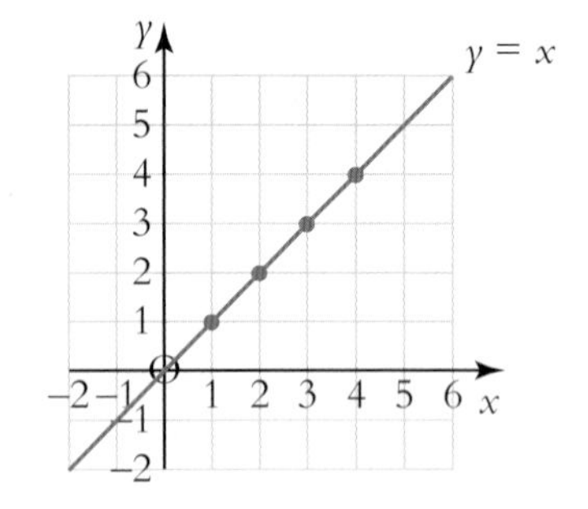

Technique

Now consider a line with equation $y = \frac{1}{2}x$. To draw this line, we choose at least three values for x, and find the corresponding values for y. Suppose we choose x values of -4, 0, and 6.

When $x = -4$, $y = \frac{1}{2}$ of $-4 = -2$, therefore $(-4, -2)$ lies on the line.

When $x = 0$, $y = \frac{1}{2}$ of $0 = 0$, therefore $(0, 0)$ lies on the line.

When $x = 6$, $y = \frac{1}{2}$ of $6 = 3$, therefore $(6, 3)$ lies on the line.

This information may be summarised as in the table.

x	-4	0	6
y	-2	0	3

By plotting the three points, we can now draw the line with equation $y = \frac{1}{2}x$.

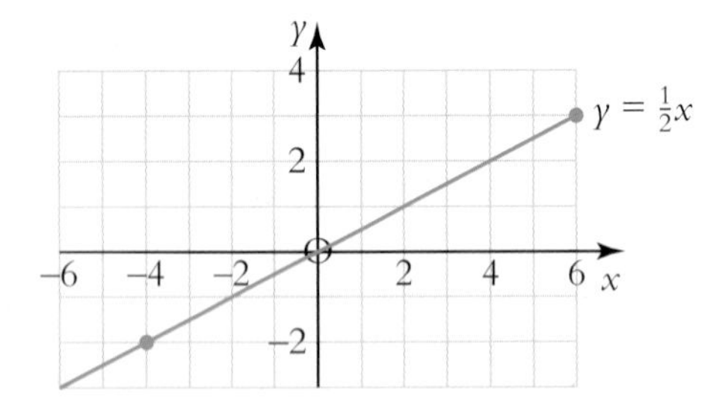

Remember

Straight lines with equations of the form $y = ax$ all pass through the Origin, $(0, 0)$.

Exercise 2

1 (a) Draw the line with equation $y = x$. Use squared paper.
(b) On your diagram, plot the points A (3, 2), B (3, 5), C (0, 2), and D (4, 3).
(c) State whether each of A, B, C, and D lie above or below the line $y = x$.

2 State whether each of the following points lies above, below, or on the line $y = x$:
E (8, 8), F (10, 11), G (15, 14), H (–5, –4).

3 (a) Copy and complete the table for the line with equation $y = 2x$.

x	–3	0	5
y			

(b) Draw the line with equation $y = 2x$ on a piece of squared paper.

4 (a) Copy and complete the table for the line with equation $y = \frac{1}{4}x$.

x	–4	0	8
y			

(b) Draw the line with equation $y = \frac{1}{4}x$ on a piece of squared paper.

5 (a) Draw the lines with equations $y = x$ and $x = 3$.
(b) Write down the coordinates of the point where these lines intersect.

6 (a) Copy and complete the table for the line with equation $y = –3x$.

x	–2	0	3
y			

(b) Draw the line with equation $y = –3x$ on a piece of squared paper.

7 What do the lines $y = x$, $y = \frac{1}{2}x$, $y = 2x$, $y = \frac{1}{4}x$, and $y = –3x$ have in common?

8 By drawing both lines, find the point of intersection of the lines with equations $y = 2x$ and $y = 6$.

9 Answer **true** or **false**:
(a) (6, 12) lies on the line $y = \frac{1}{2}x$
(b) the line with equation $y = 100x$ passes through the Origin
(c) (–4, 8) lies on the line $y = –2x$
(d) the line $y = 5x$ passes through the point (20, 4)
(e) (100, 24) lies below the line $y = \frac{1}{4}x$.

Exercise 2 continued

10 (a) Draw the lines whose equations are $y = 2x$ and $x = 5$.

(b) Shade in the triangle bounded by $y = 2x$, $x = 5$ and the x-axis.

(c) Calculate the area of this triangle.

Straight lines with equations of the form $y = ax + b$

We have already looked at lines which are parallel to the x-axis and the y-axis, and lines which pass through the Origin. All other lines have an equation of the form $y = ax + b$.

This equation would describe the speed–time graph say for an accelerating car, whose speed was greater than zero when first observed.

In this section, you should be able to

- recognise the equation $y = ax + b$ as the general equation of a straight line
- draw a straight line with equation of the form $y = ax + b$, after completing a table of values.

Example (a) Copy and complete the table for the line whose equation is $y = \frac{1}{2}x + 3$.

x	-2	0	4
y			

(b) Hence draw the line whose equation is $y = \frac{1}{2}x + 3$.

Solution (a) When $x = -2$, $y = \frac{1}{2}$ of $-2 + 3 = -1 + 3 = 2$.

When $x = 0$, $y = \frac{1}{2}$ of $0 + 3 = 0 + 3 = 3$.

When $x = 4$, $y = \frac{1}{2}$ of $4 + 3 = 2 + 3 = 5$.

x	−2	0	4
y	2	3	5

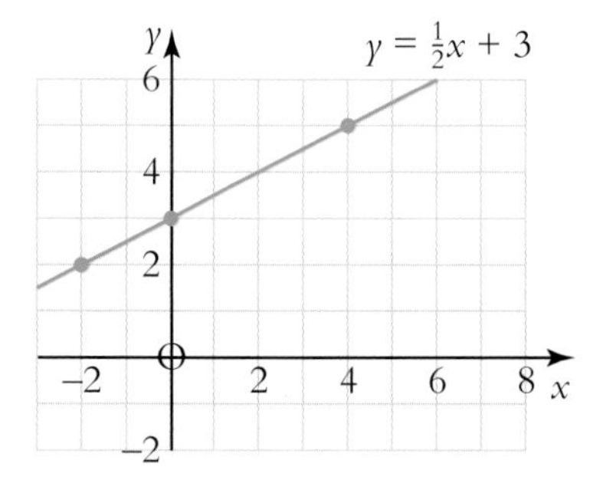

(b) This line passes through the points (−2, 2), (0, 3), and (4, 5).

Exercise 3

(Do not use a calculator)

In each of the following questions

(a) copy and complete the table for the given line

(b) draw the line on a piece of squared paper.

1 $y = x + 2$

x	−6	0	5
y			

2 $y = 2x + 3$

x	−4	0	3
y			

3 $y = -2x + 3$

x	−3	0	4
y			

4 $y = 3x - 4$

x	−2	0	4
y			

5 $y = -3x + 4$

x	−2	0	4
y			

6 $y = \frac{1}{4}x + 1$

x	−8	0	4
y			

7 $y = 5 - 2x$

x	−2	0	3
y			

Exercise 4 – Revision of Chapter 11

1 (a) Plot the points P (5, –2), Q (5, 0) and R (5, 3).

(b) Write down the equation of the straight line passing through P, Q and R.

2 (a) Use one diagram to draw the lines with equations $x = 4$ and $y = 3$.

(b) Write down the coordinates of the point where these lines intersect.

3 (a) Copy and complete the table for the line whose equation is $y = 3x$.

x	–3	0	3
y			

(b) Hence draw the line with equation $y = 3x$.

4 (a) Use one diagram to draw the lines with equations $y = x$ and $y = 5$.

(b) Write down the coordinates of the point where these lines intersect.

5 Answer **true** or **false**:

(a) (50, 25) lies on the line $y = 2x$

(b) the line $y = x + 5$ passes through the point (–8, 3)

(c) the lines $y = x$ and $y = 2x$ intersect at the Origin

(d) (20, 61) lies below the line $y = 3x$.

6 (a) Copy and complete the table for the line whose equation is $y = 2x + 1$.

x	–4	0	3
y			

(b) Hence draw the line with equation $y = 2x + 1$.

7 (a) Copy and complete the table for the line whose equation is $y = -2x - 4$.

x	–5	0	2
y			

(b) Hence draw the line with equation $y = -2x - 4$.

8 (a) Copy and complete the table for the line whose equation is $y = \frac{1}{2}x + 1$.

x	–6	0	8
y			

(b) Draw the line $y = \frac{1}{2}x + 1$.

Summary

(You should now be able to describe straight line graphs algebraically, as described by the following points.)

1 An equation of the form $x = a$ represents a vertical line parallel to the y-axis passing through the point $(a, 0)$.

An equation of the form $y = b$ represents a horizontal line parallel to the x-axis passing through the point $(0, b)$.

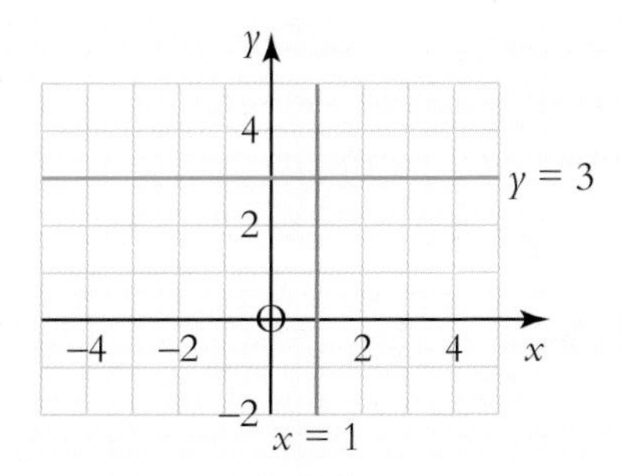

2 An equation of the form $y = ax$ represents a straight line passing through the Origin (0, 0).

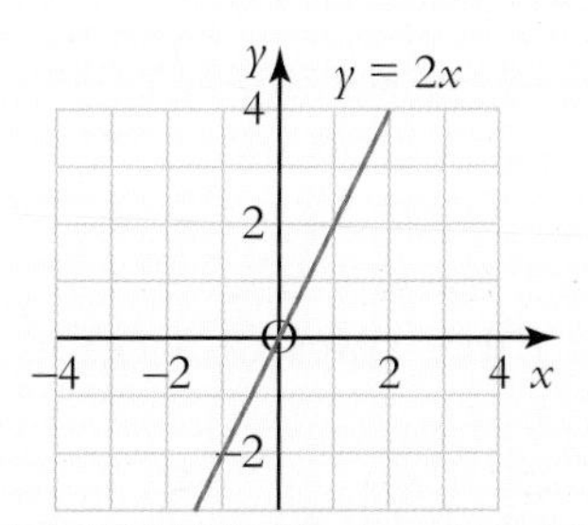

3 The equation $y = ax + b$ is the general equation of a straight line. You should draw such a line after completing a table of values.

Thus to plot the line $y = \frac{1}{3}x + 1$, choose x values such as –3, 0, 6. These produce y values of 0, 1, 3. Thus plot the points (–3, 0), (0, 1), and (6, 3), and draw the line.

Trigonometry in a Right-angled Triangle

Trigonometry is the branch of mathematics which is used to solve problems involving the lengths of sides and sizes of angles in triangles. The triangle is the simplest straight-sided geometric shape, and it can be isosceles, equilateral, right-angled, or scalene. In this chapter, we concentrate on problems with right-angled triangles.

Important Definitions

In this section, you should be able to:

- name the three sides in a right-angled triangle
- use the **sin**, **cos** and **tan** keys on your calculator.

The Three Sides

We use special names for each of the three sides of a right-angled triangle.

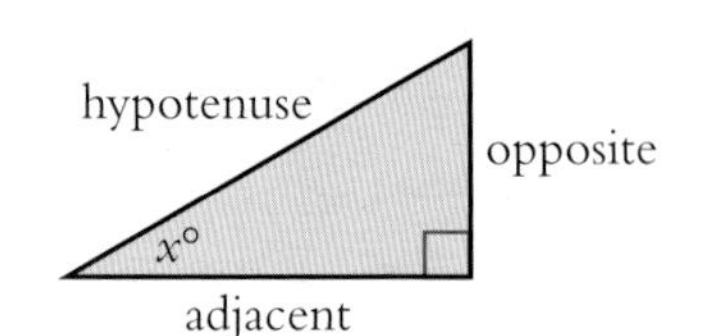

The side opposite the right angle is called the **hypotenuse**.

The side opposite the angle marked $x°$ is called the **opposite**.

The remaining side is called the **adjacent**.

(Adjacent means 'next to', and this side is next to the angle $x°$.)

Example Copy the triangle shown and name the three sides.

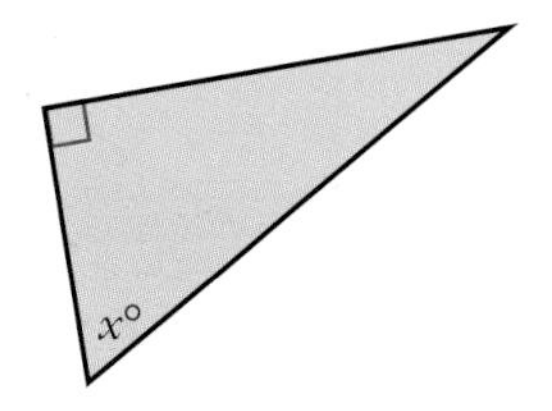

Solution The *hypotenuse* is opposite the right angle. The *opposite* is opposite the angle marked $x°$, and the *adjacent* is the remaining side. Hence the labelled triangle is as shown.

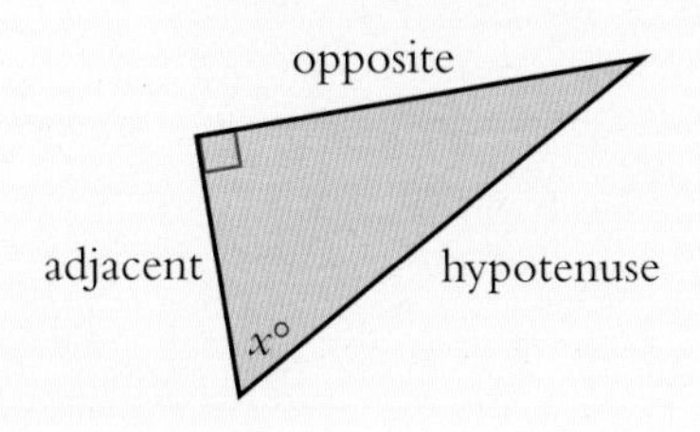

Exercise 1

1 Copy the triangles shown, and write on each diagram the names of the three sides – hypotenuse, opposite and adjacent.

(a) 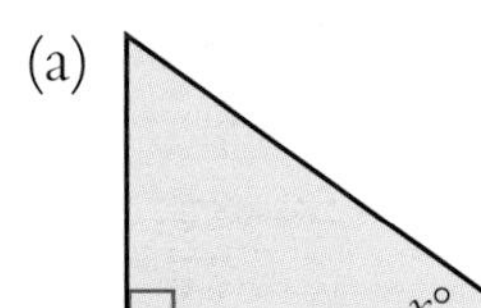

(b) 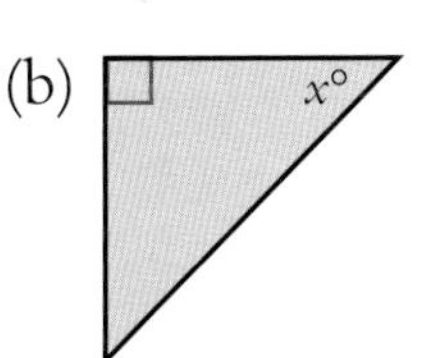

(c) 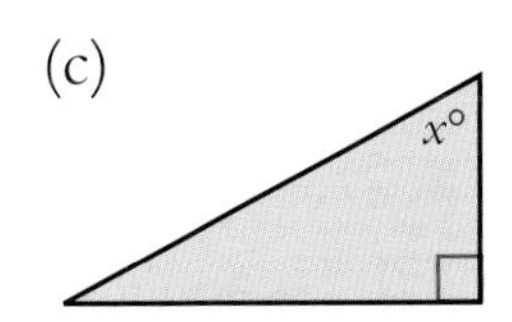

(d) 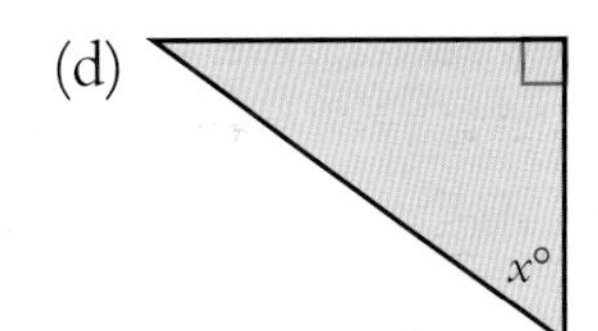

The three ratios

Copy and complete the following table for the right-angled triangles shown overleaf.

Triangle	side opposite 31°	side adjacent to 31°	$\frac{opposite}{adjacent}$
(a)	3 cm	5 cm	$\frac{3}{5} = 0{\cdot}6$
(b)	6 cm	10 cm	$\frac{6}{10} =$
(c)	4·5 cm	7·5 cm	
(d)			

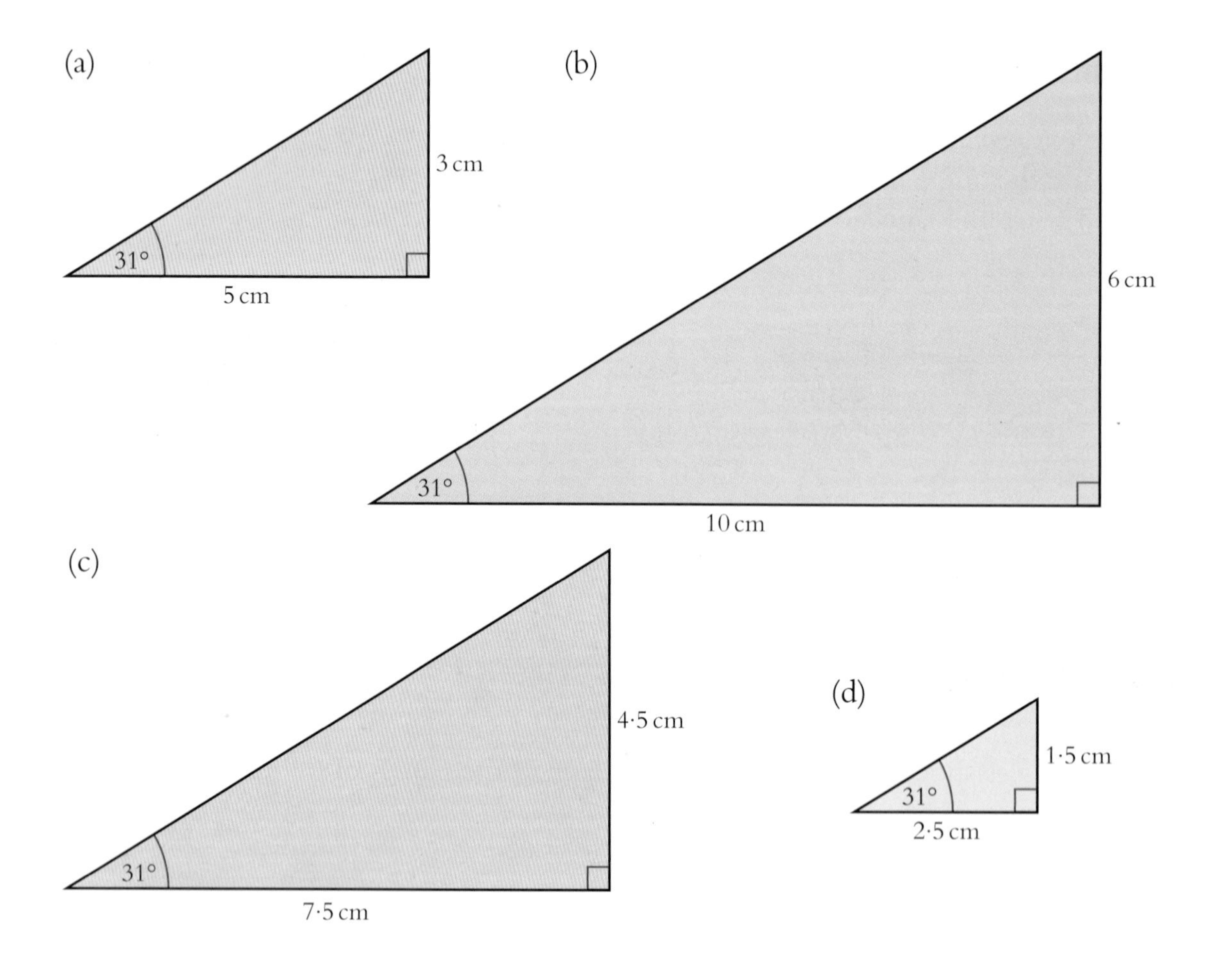

In every right-angled triangle with an angle of 31°, the ratio $\frac{\text{opposite}}{\text{adjacent}}$ is equal to 0·6.

This ratio $\frac{\text{opposite}}{\text{adjacent}}$ is called the **tangent** of the angle.
Thus the tangent of 31° is 0·6. We write this as tan 31° = 0·6.

Exercise 2

1 Find the tangents of the angles shown in the following triangles.

(a)

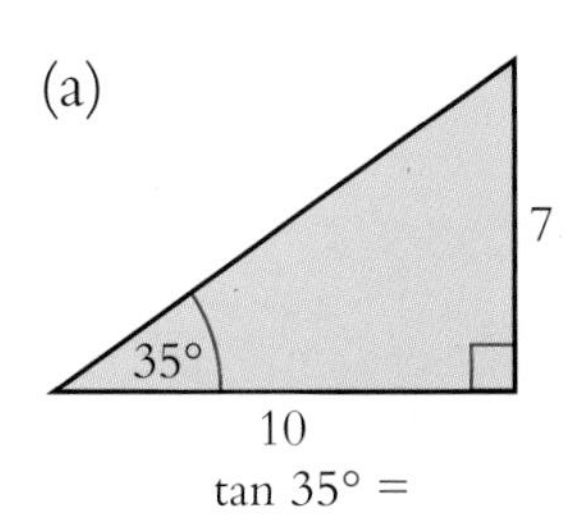

tan 35° =

(b)

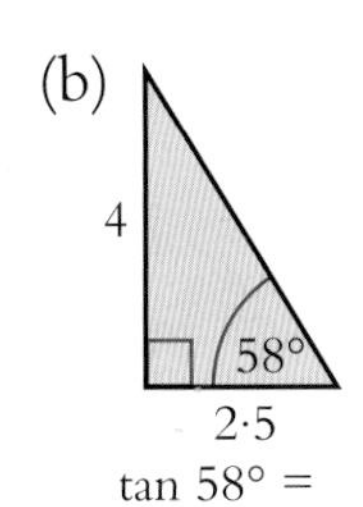

tan 58° =

(c)

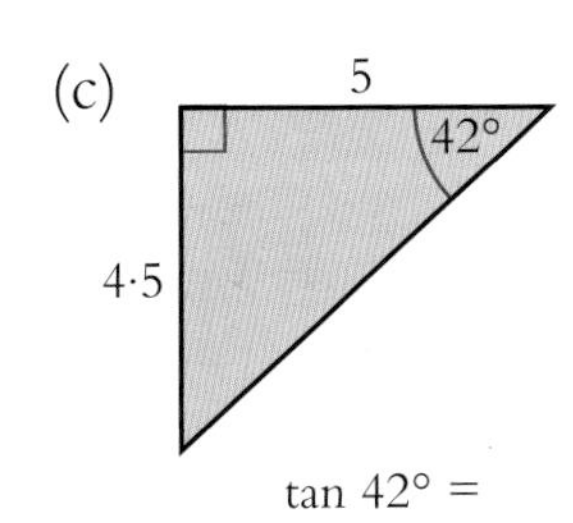

tan 42° =

(d)

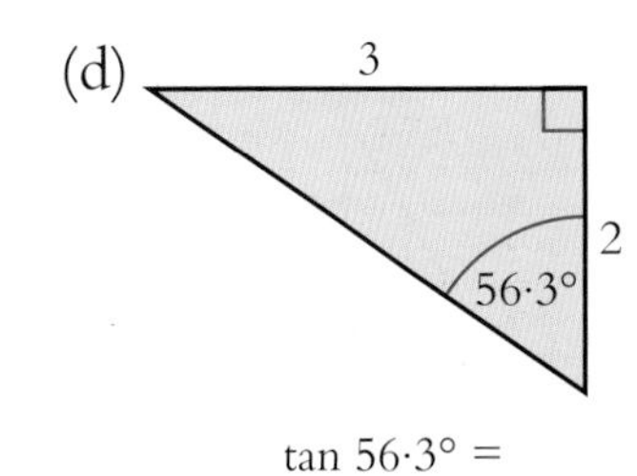

tan 56·3° =

Key words and definitions

We can define two other ratios in a right-angled triangle.

The ratio $\dfrac{\text{opposite}}{\text{hypotenuse}}$ is called the **sine** of the angle. We write **sin** for short.

The ratio $\dfrac{\text{adjacent}}{\text{hypotenuse}}$ is called the **cosine** of the angle. We write **cos** for short.

Example Use the triangle shown to find the values of sin 53·1° and cos 53·1°.

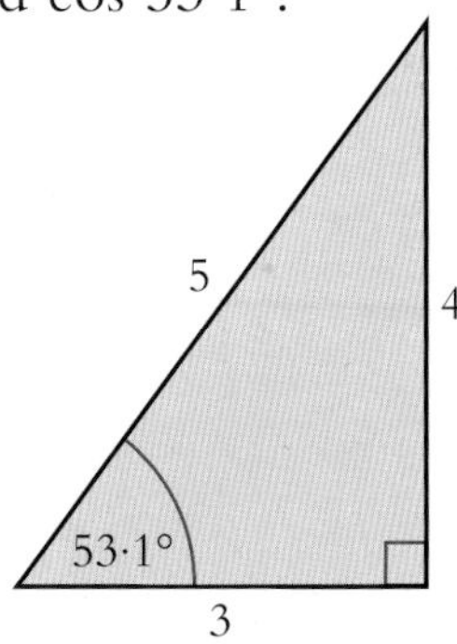

Solution

$$\sin 53{\cdot}1° = \frac{\text{opposite}}{\text{hypotenuse}} = \frac{4}{5} = 0{\cdot}8$$

$$\cos 53{\cdot}1° = \frac{\text{adjacent}}{\text{hypotenuse}} = \frac{3}{5} = 0{\cdot}6.$$

Exercise 3

1 Find the sine, cosine, or tangent of the angle marked in each of the following triangles.

(a)
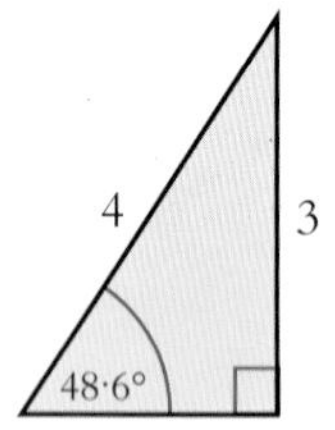

(b)
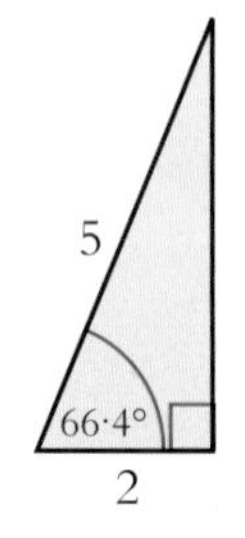

(c)
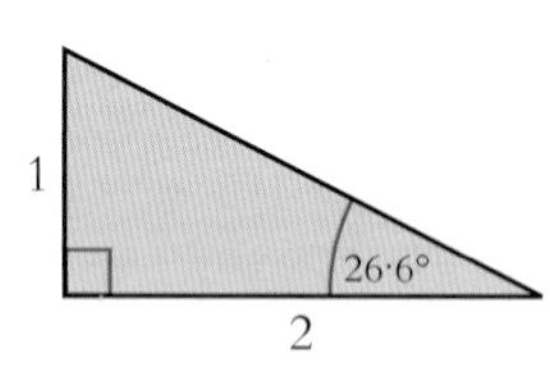

(d)
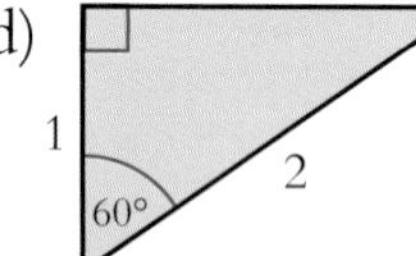

(e)
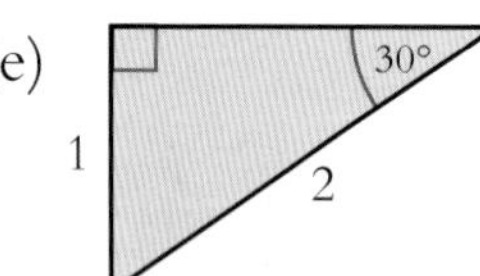

(f)
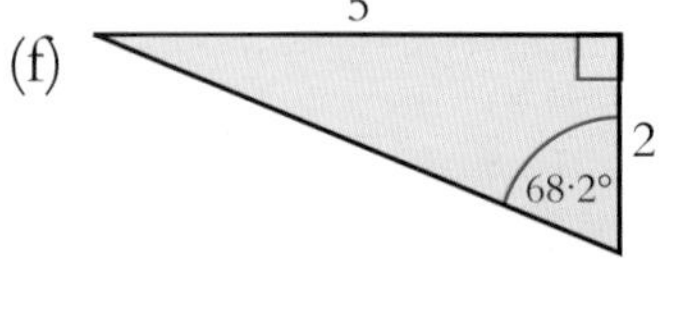

(g)
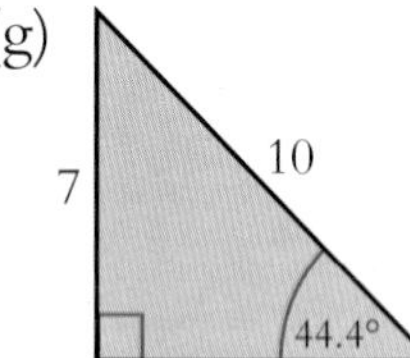

(h)
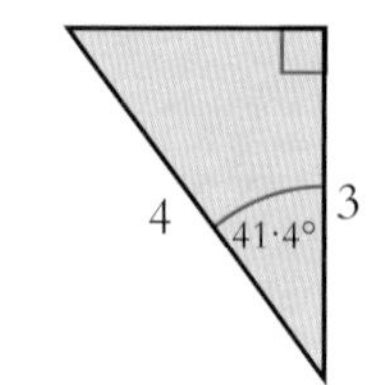

(i)
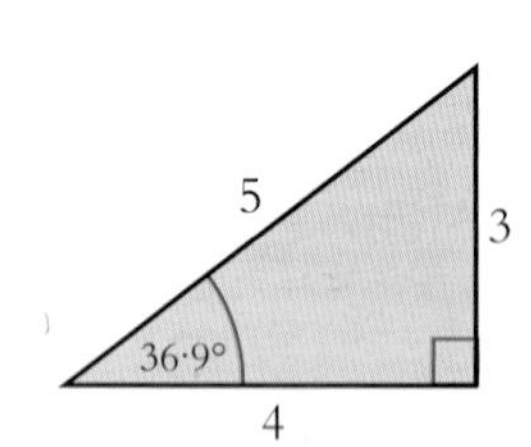

Find sin 36·9°
cos 36·9°
and tan 36·9°

Using the Calculator

The values of trigonometric ratios can be found using a scientific calculator. This saves having to draw a triangle, measure the lengths of its sides and then calculate a particular ratio.

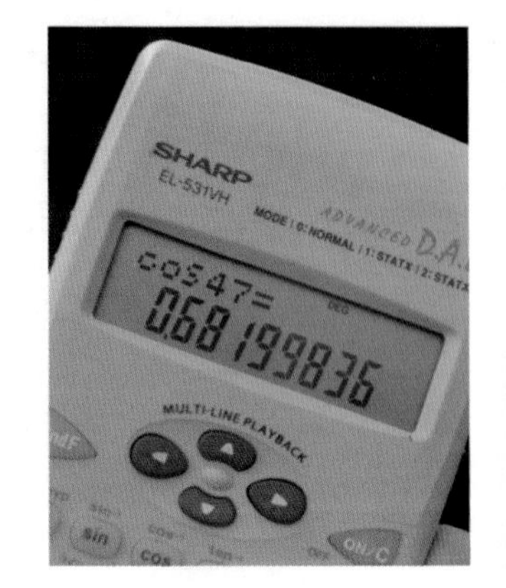

Example Find (a) sin 30°

(b) cos 47° (to three decimal places).

Solution (a) Press 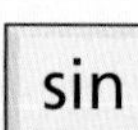   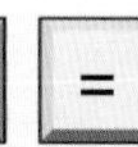

The calculator will show 0·5, so sin 30° = 0·5.

(b) Press    

The calculator will show 0·68199..., so cos 47° = 0·682 (to 3 d.p.).

Take care

You must always make sure that your calculator is in **degree** mode, usually indicated by **DEG** or **D** on the screen.

Technique

To find the size of an angle for which we already know the sine, cosine, or tangent, we use the **2nd F** key on the calculator.

(Your calculator may have a **SHIFT** or **INV** key instead of a **2nd F** key.)

Example Given that $\tan x° = 0{\cdot}625$, find x correct to one decimal place.

Solution Press

[2nd f] [tan] [0] [·] [6] [2] [5] [=]

The calculator will show $\tan^{-1} 0{\cdot}625 = 32{\cdot}0053$, so $x = 32{\cdot}0$ (to 1 d. p.).

Exercise 4

1 Find the value of each of the following trigonometric ratios correct to three decimal places.

(a) sin 35° (b) cos 47° (c) tan 21° (d) sin 77°

(e) tan 15° (f) cos 7·5° (g) tan 45° (h) cos 2°

(i) sin 14·8° (j) tan 72° (k) cos 45° (l) tan 89°.

Exercise 4 continued

2 Find the value of x in each of the following cases.
(Give your answer correct to one decimal place.)

(a) $\sin x° = 0{\cdot}475$ (b) $\cos x° = 0{\cdot}671$ (c) $\tan x° = 0{\cdot}435$ (d) $\tan x° = 3{\cdot}125$

(e) $\cos x° = 0{\cdot}6$ (f) $\cos x° = 0{\cdot}714$ (g) $\sin x° = 0{\cdot}877$ (h) $\tan x° = 0{\cdot}4$

(i) $\cos x° = 0{\cdot}543$ (j) $\sin x° = 0{\cdot}866$ (k) $\tan x° = 1{\cdot}45$ (l) $\sin x° = 0{\cdot}619$.

Finding the Length of a Side in a Right-angled Triangle

Suppose a ladder which is 5m long is placed against a vertical wall, and that the ladder makes an angle of 60° with the ground. Do you think you could find how high up the wall the ladder reaches?

In this section, you should be able to calculate the length of a side in a right-angled triangle when you know the length of another side and the size of an angle.

To solve such a problem, you will need to use the following formulae list.

$$\tan x° = \frac{\textbf{opposite}}{\textbf{adjacent}}$$

$$\sin x° = \frac{\textbf{opposite}}{\textbf{hypotenuse}}$$

$$\cos x° = \frac{\textbf{adjacent}}{\textbf{hypotenuse}}$$

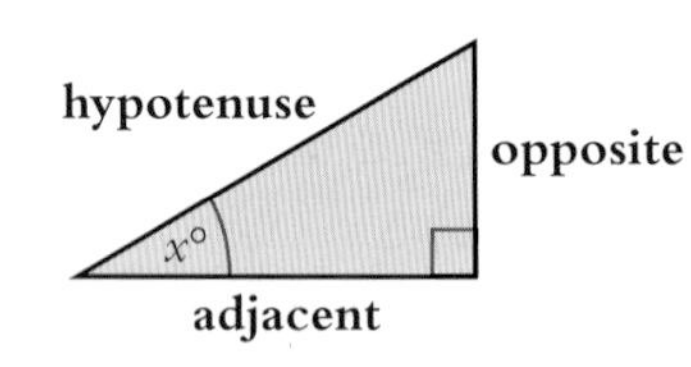

Example Find the length of the side marked L in the following diagram.

(Give your answer correct to one decimal place.)

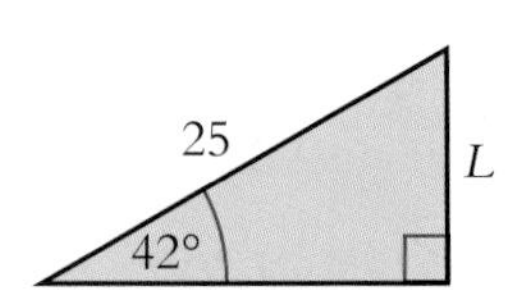

Solution We start by naming the three sides – hypotenuse, opposite, adjacent.

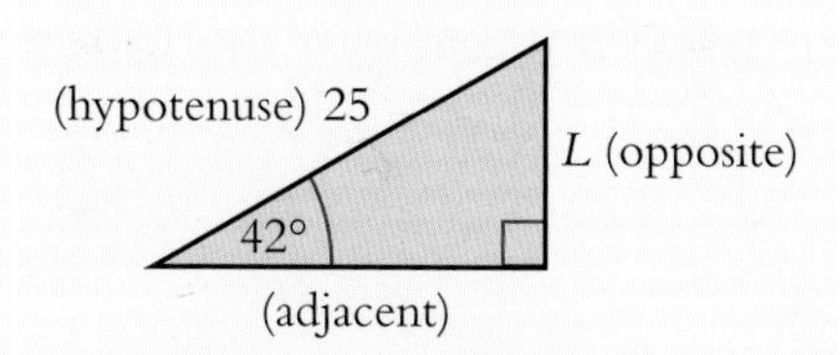

The side given is the **hypotenuse** and side L is the **opposite**, and so from the formulae list, we must use **sine**.

Hence $\sin 42° = \dfrac{\text{opposite}}{\text{hypotenuse}} = \dfrac{L}{25}$

$\Rightarrow L = 25 \times \sin 42° = 16{\cdot}7$ (to 1 d.p.)

Example Find the length of the side marked a in the triangle shown. (Give your answer correct to one decimal place.)

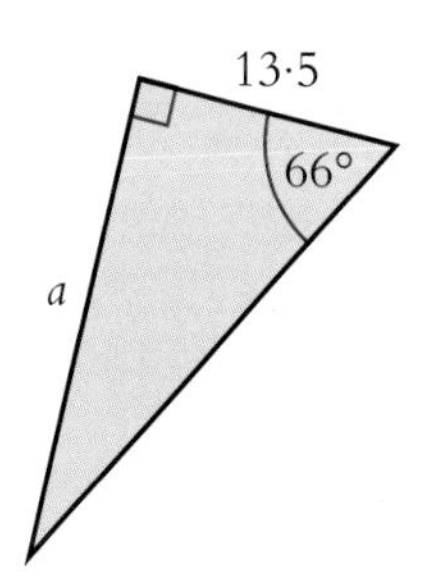

Solution Again we name the three sides – hypotenuse, opposite, adjacent.

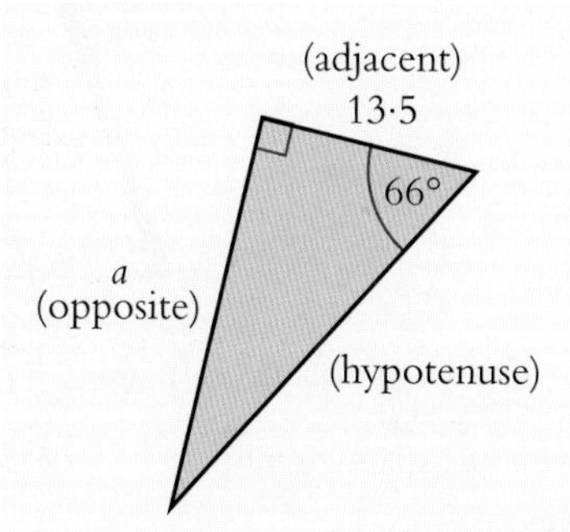

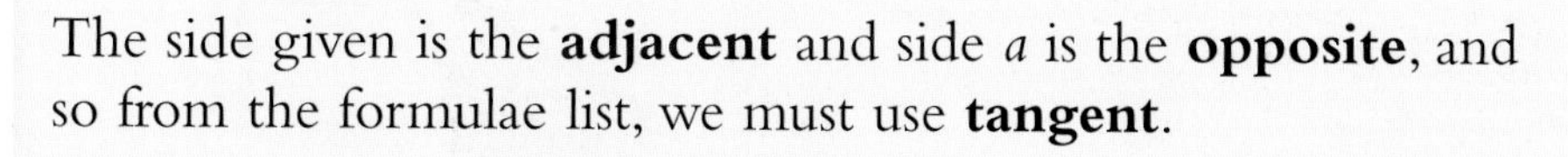

The side given is the **adjacent** and side a is the **opposite**, and so from the formulae list, we must use **tangent**.

Hence $\tan 66° = \dfrac{\text{opposite}}{\text{adjacent}} = \dfrac{a}{13{\cdot}5}$

$\Rightarrow a = 13{\cdot}5 \times \tan 66° = 30{\cdot}3$ (to 1 d.p.)

Exercise 5

Give all your answers correct to one decimal place

1 Calculate the length of the unknown side marked h in each of the following triangles.

(a)

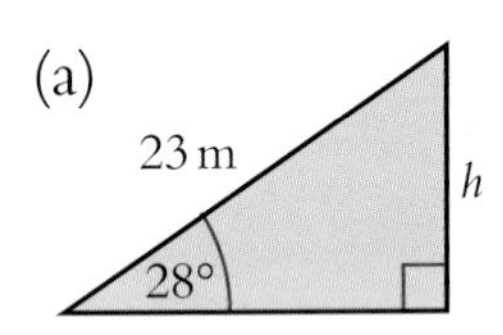

(b)

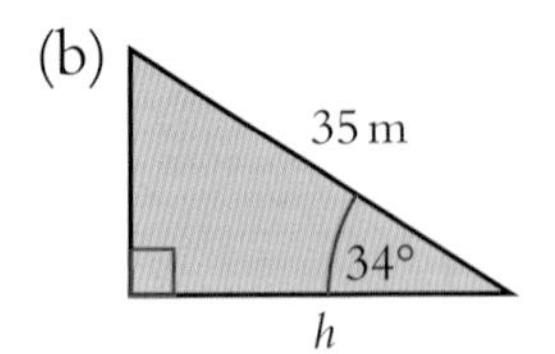

(c)

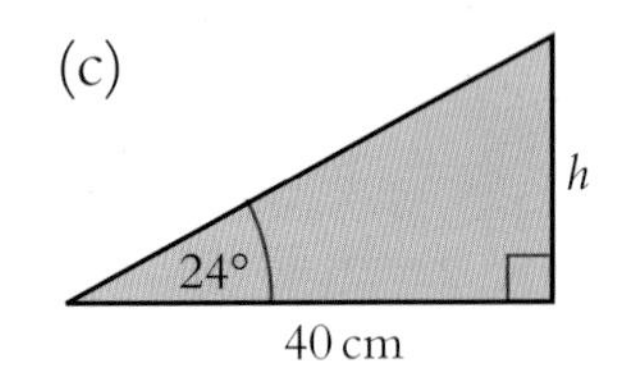

(d)

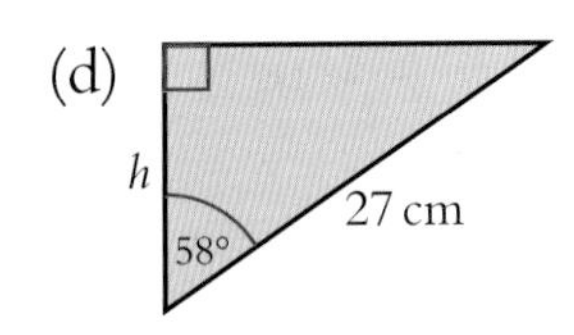

(e)

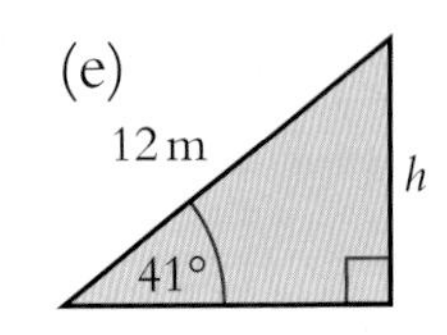

(f)

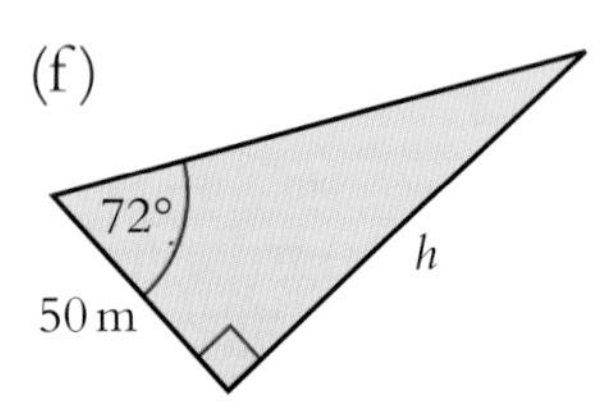

2 A ladder rests against a wall. The ladder is 4 metres long, and it makes an angle of 78° with the ground.

Calculate the distance, d metres, from the foot of the ladder to the wall.

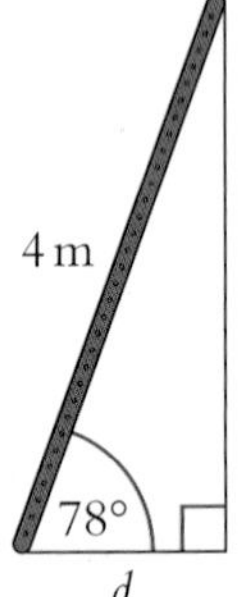

3 The slide in a children's play park makes an angle of 40° with the ground. The length of the slide is 6·5 metres.

Calculate the height, h metres, of the slide above the ground.

6·5 m

h

40°

4 A shop sign of width w metres is supported by a metal bar. The bar is 1·5 metres long. The bar makes an angle of 27° with the shop sign, as shown in the diagram.

Find the width, w, of the shop sign.

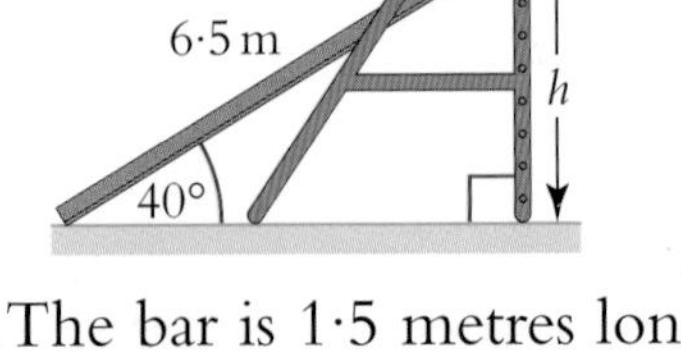

Exercise 5 continued

Give all your answers correct to one decimal place

5 A surveyor wishes to find the height of a tower. He stands 30 metres from the base of the tower. He measures the angle of elevation to the top of the tower to be 15°, as shown.

Calculate the height, h (metres), of the tower.

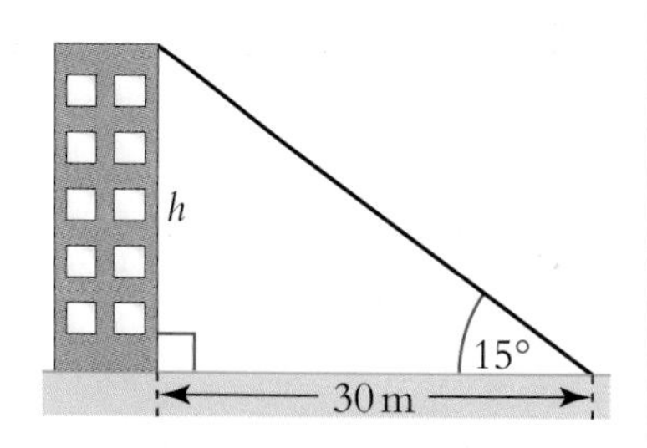

6 A wire 15 metres long supports a flagpole. This wire makes an angle of 62° with the ground, as shown in the diagram.

Calculate the height, f (metres), of the flagpole.

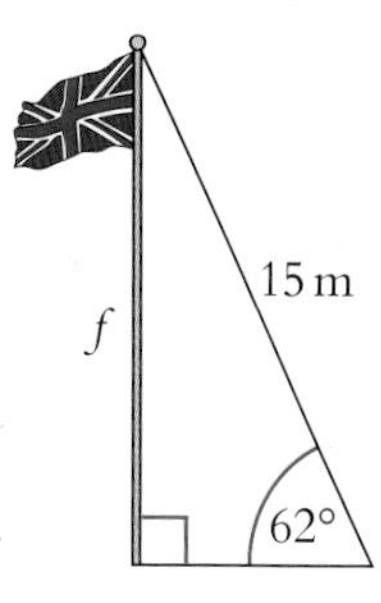

7 A ramp allows wheelchair access to a building. The ramp is 3 metres long and makes an angle of 4° with the horizontal, as shown in the diagram.

Calculate the height, h (metres), of the ramp.

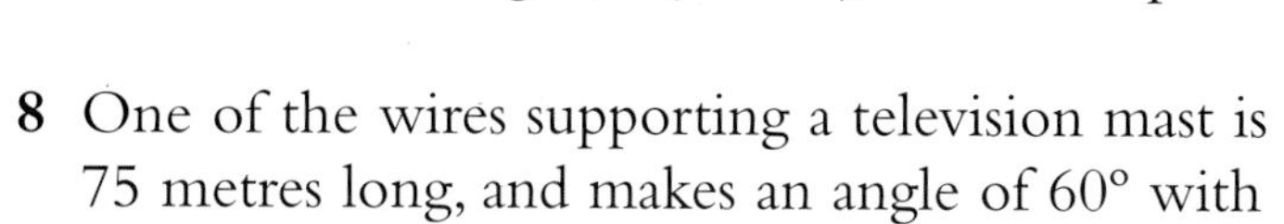

8 One of the wires supporting a television mast is 75 metres long, and makes an angle of 60° with the ground.

If the wire is attached to the mast at a point 10 metres from the top, calculate the height, m (metres), of the mast.

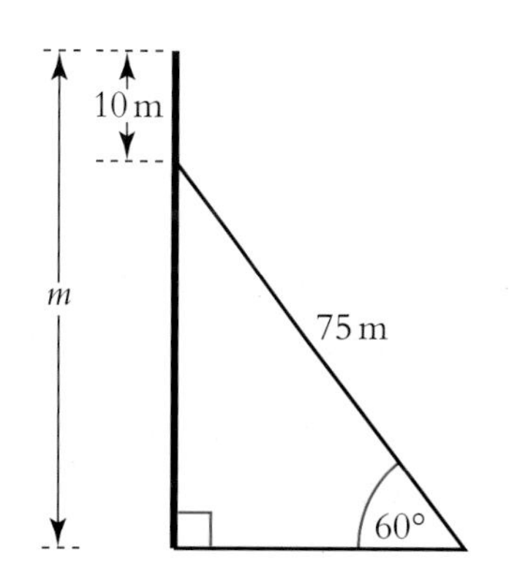

9 During an orienteering competition, Liz ran 3 kilometres from points P to Q on a bearing of 036°, as shown in the diagram.

Calculate y to find how far north she is from her starting point.

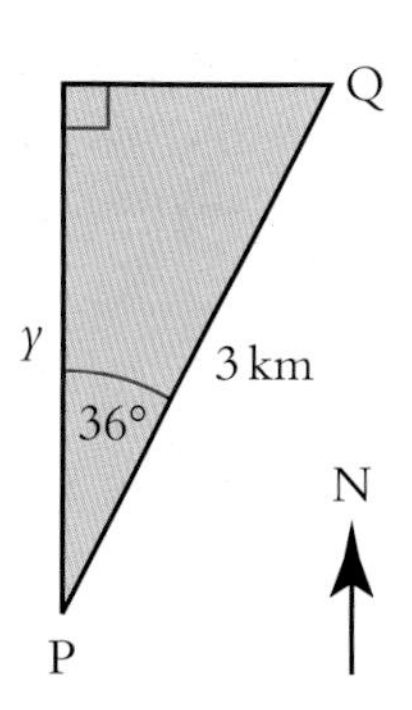

Finding the Size of an Angle in a Right-angled Triangle

Sometimes we have the sizes of the sides of a right-angled triangle but we have no information about the sizes of the angles, other than the right-angle.

In this section, you should be able to find the size of an angle in a right-angled triangle when you know the sizes of two of the sides.

To solve such a problem, you will need to use the same formulae list which was quoted in the last section.

$$\tan x^\circ = \frac{\textbf{opposite}}{\textbf{adjacent}}$$

$$\sin x^\circ = \frac{\textbf{opposite}}{\textbf{hypotenuse}}$$

$$\cos x^\circ = \frac{\textbf{adjacent}}{\textbf{hypotenuse}}$$

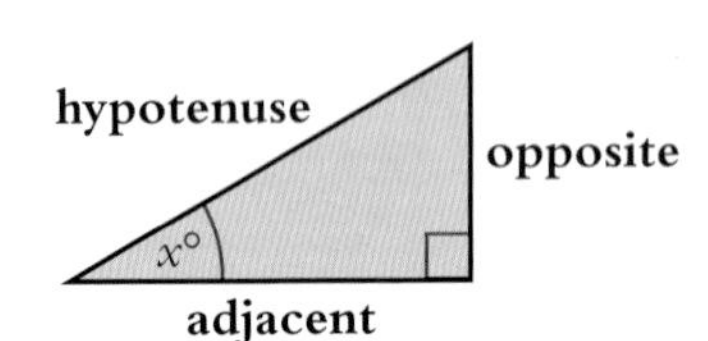

Example Find the size of the angle marked $a°$ in the diagram.

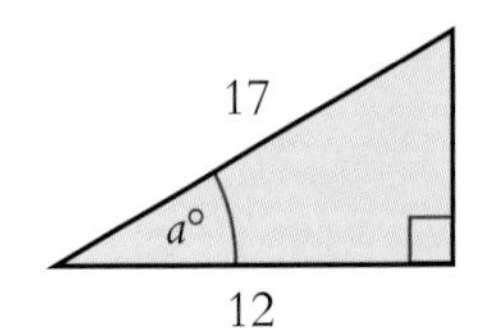

Solution We start by naming the three sides – hypotenuse, opposite, adjacent.

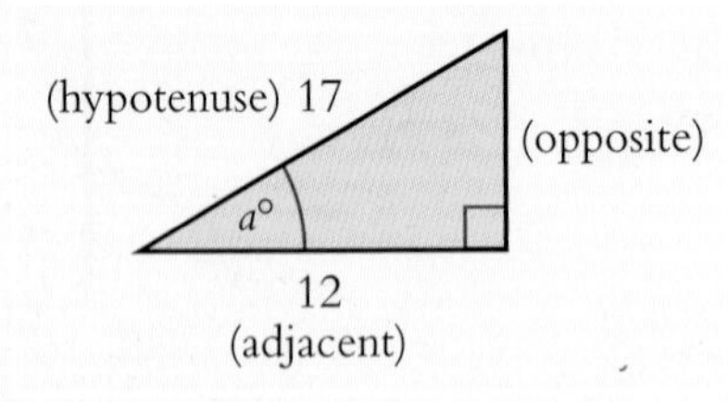

Here we know the sizes of the **hypotenuse** and the **adjacent**, and so from the formulae list, we must use **cosine**.

Hence $\cos a° = \dfrac{\text{adjacent}}{\text{hypotenuse}} = \dfrac{12}{17}$.

Now Press **2nd F cos (12 ÷ 17) =**

The calculator will show $\cos^{-1} (12 \div 17) = 45.099\ldots$, so $a = 45{\cdot}1$ (to 1 d.p.).

Example A horizontal flagpole is supported by a bracket as shown.

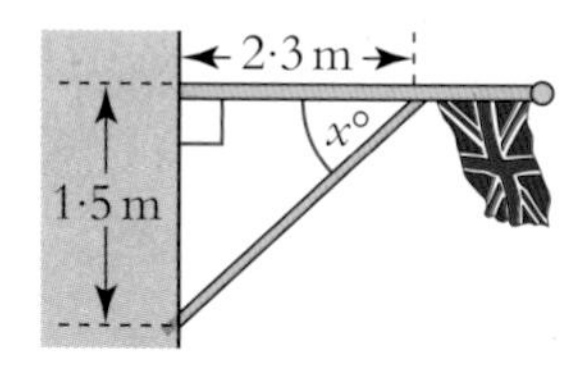

The bracket reaches 2·3 metres along the flagpole and is fixed to the wall 1·5 metres below the flagpole. Calculate the angle $x°$ between the bracket and the flagpole.

Solution By naming the sides you will see that we are given the sizes of the **opposite** and **adjacent** sides, and so from the formulae list, we must use **tangent**.

Hence $\tan x° = \frac{\text{opposite}}{\text{adjacent}} = \frac{1{\cdot}5}{2{\cdot}3}$.

Now Press **2nd F tan (1·5 ÷ 2·3) =**

The calculator will show $\tan^{-1}(1{\cdot}5 \div 2{\cdot}3) = 33{\cdot}1113...$, so the angle the flagpole makes with the bracket is 33·1° (to 1 d.p.).

Exercise 6

Give all answers correct to 1 decimal place unless otherwise stated

1 Calculate the size of the unknown angle marked $x°$ in each of the following triangles.

(a)

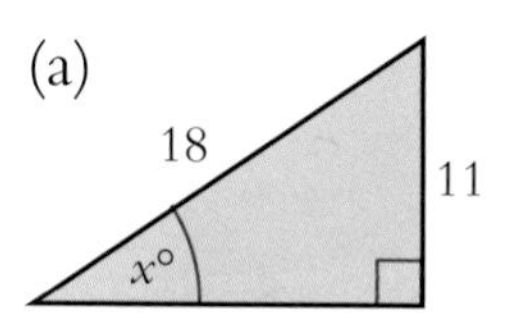

(b)

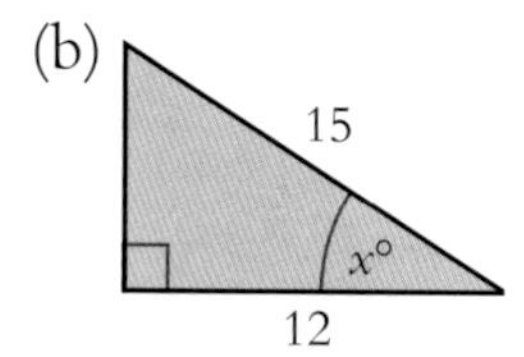

(c)

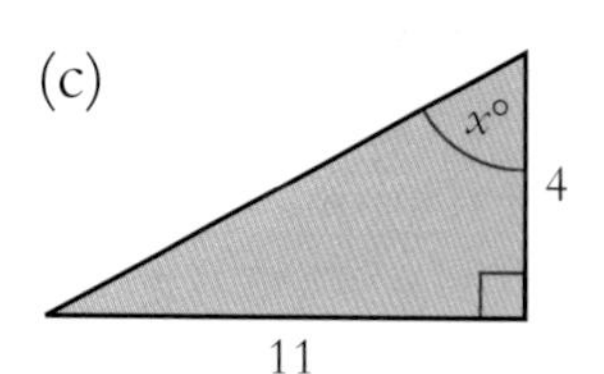

(d)

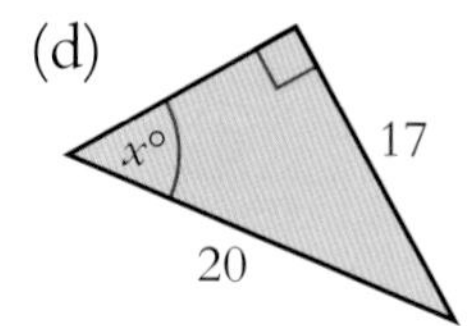

(e)

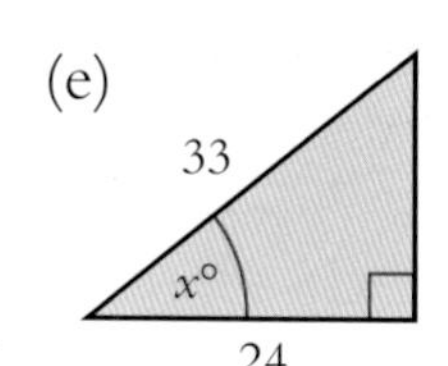

(f)

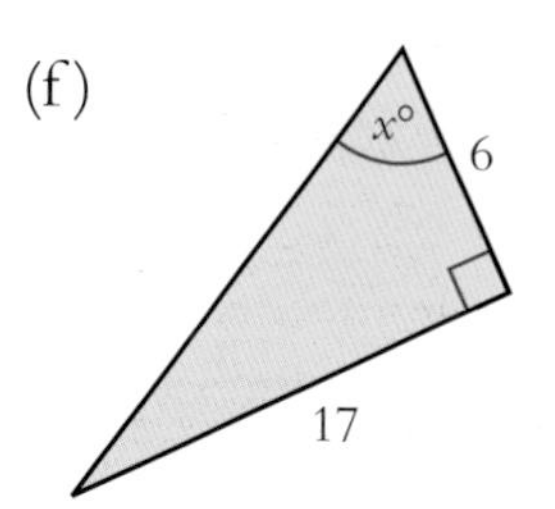

Exercise 6 continued

Give all answers correct to 1 decimal place unless otherwise stated

2 In triangle ABC, AC = 12 cm, AB = 15 cm.

Calculate the size of angle ABC.

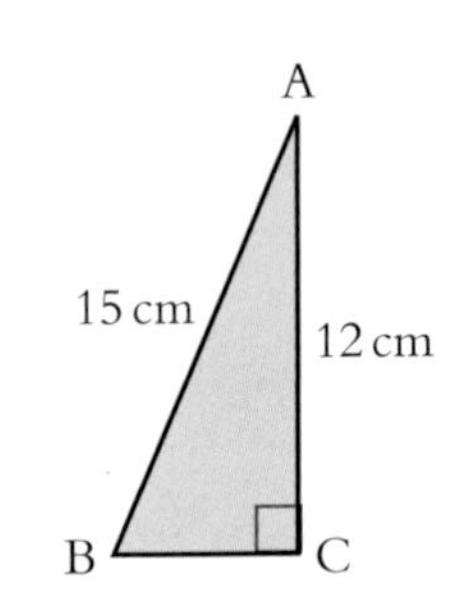

3 An aeroplane flies 2000 metres in a straight line after take-off. It reaches a height of 450 metres.

Calculate the angle $x°$ to the horizontal at which it is flying.

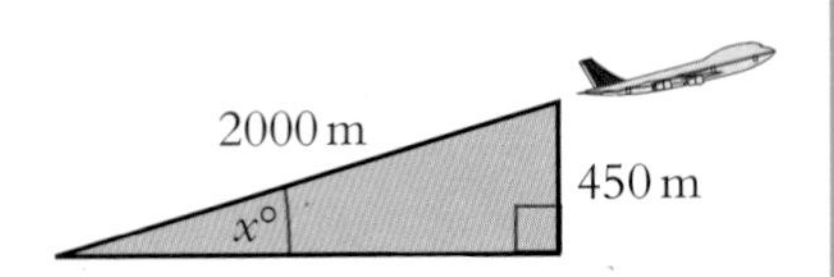

4 Christopher sails west from a port P for 15 kilometres, then south for 10 kilometres.

Find the bearing of his return journey to port P, represented by angle b in the diagram. (Give your answer to the nearest degree.)

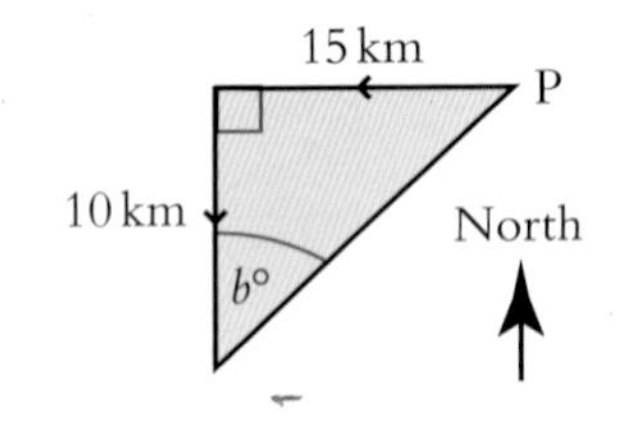

5 ABCD is a rectangle measuring 13 cm by 8 cm.

Calculate the size of angle BDC.

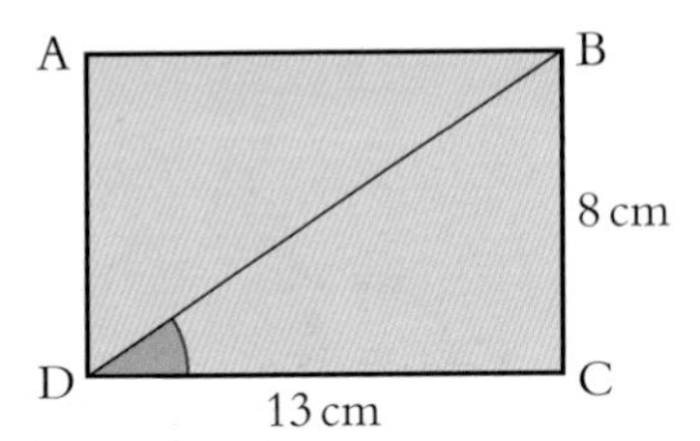

6 The Pilatus Railway in Switzerland is the steepest rack railway in the world. At one stage the train gains 48 metres vertically for every 100 metres travelled.

Calculate the size of angle r to the horizontal at which the train is travelling.

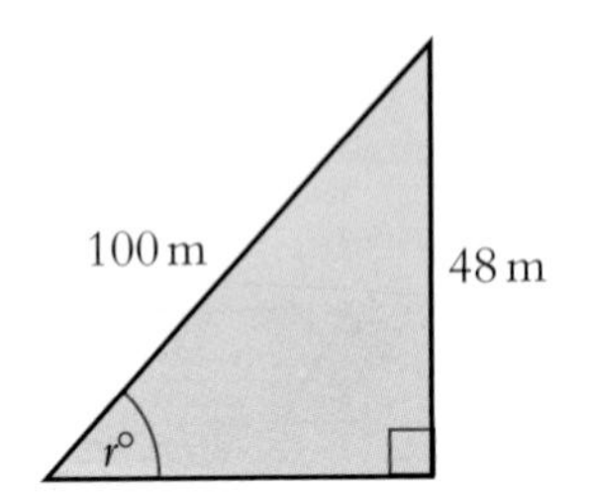

Exercise 6 continued

Give all answers correct to 1 decimal place unless otherwise stated

7 Justin is building a stairway in his new house. The vertical height of the stairway is 125 cm and the horizontal extent is 216 cm.

Calculate the angle, $a°$, the stairway makes with the horizontal.

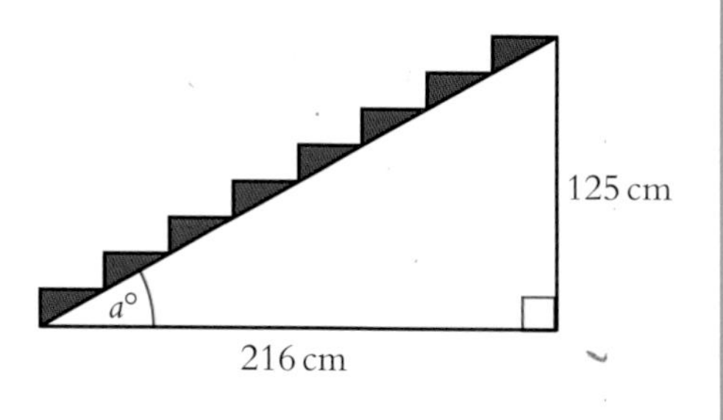

8 A small launch ramp in a skateboard park is as shown.

Calculate the angle the ramp makes with the ground.

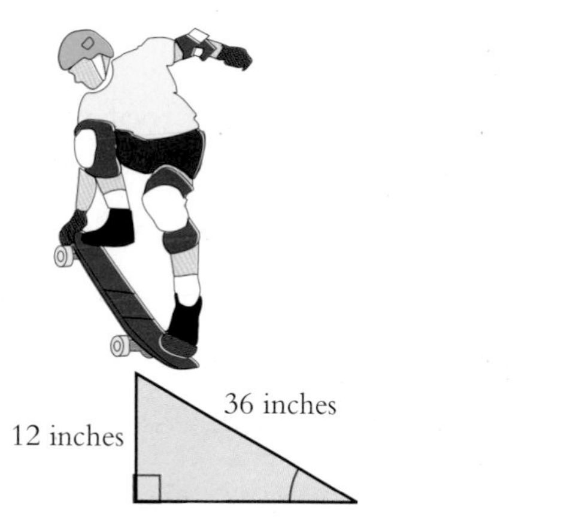

Mixed Examples

We have now used trigonometry both to find the length of a side and to calculate the size of an angle in a right-angled triangle. In the following exercise, there is a mixture of questions testing both skills.

Exercise 7

Give all answers correct to 1 decimal place

1 Calculate the size of the unknown angle marked $z°$ in the following triangles.

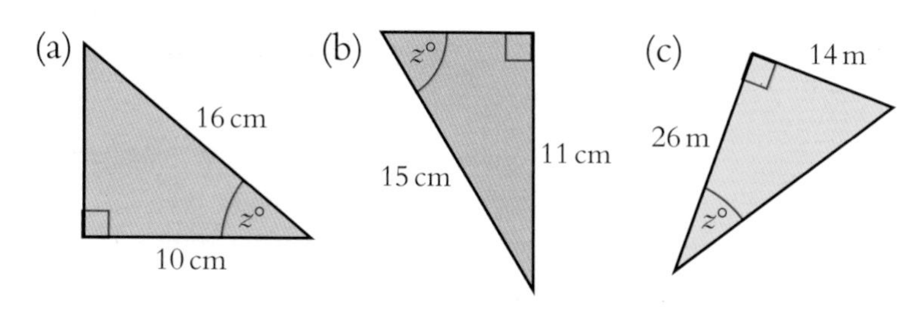

Exercise 7 continued

Give all answers correct to 1 decimal place

2 Calculate the length of the unknown side marked d in the following triangles.

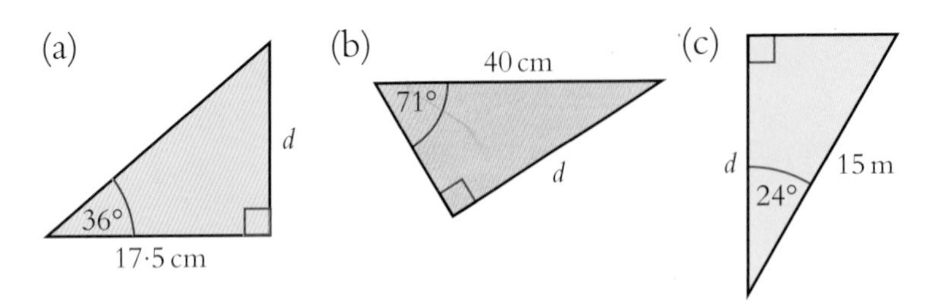

3 Yasar has a sloping roof on his house. The diagram shows the shape and sizes of part of the side view of the house.

Calculate the angle, $x°$, which the sloping roof makes with the horizontal.

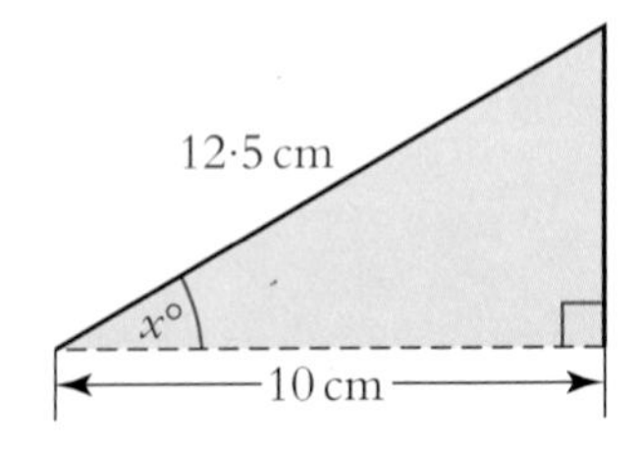

4 Scarlett has a garden in the shape of a right-angled triangle.

Calculate the length of side QR.

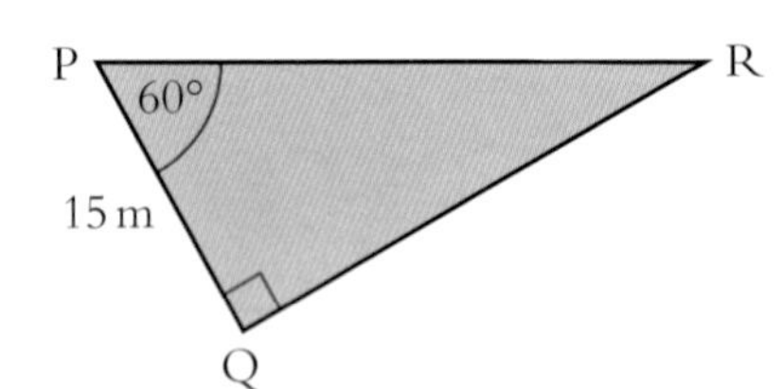

5 Find the value of x in each of the rectangles shown.

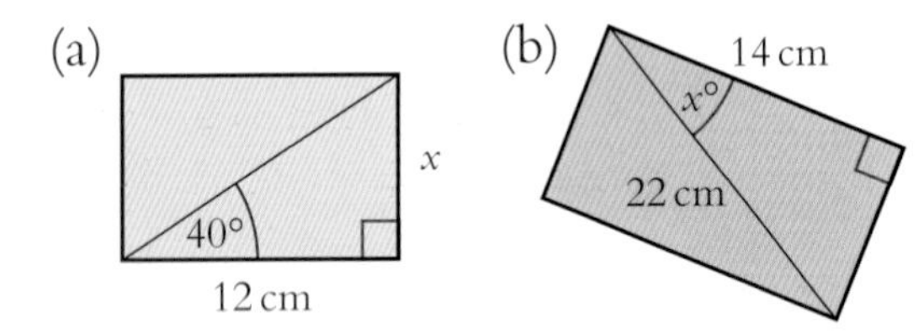

6 Frank is attempting to measure the height of a skyscraper. At a distance of 50 metres from the foot of the building he measures an angle of elevation of 76° to the top of the building.

Calculate the height, h metres, of the skyscraper.

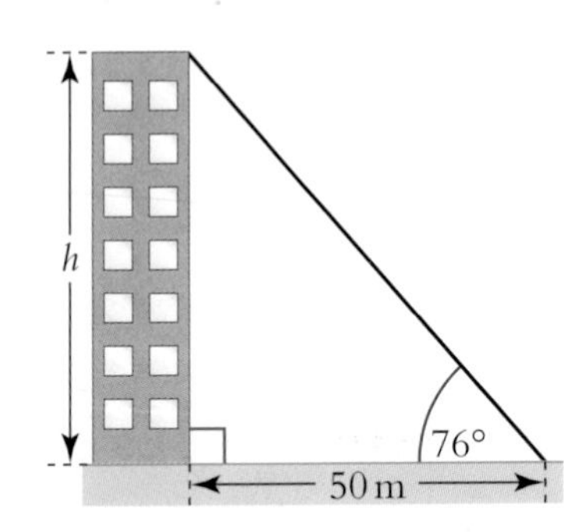

Exercise 7 continued

Give all answers correct to 1 decimal place

7 In triangle DEF, angle DEF = 90°, DF = 25 cm, EF = 18 cm.

Calculate

(a) the length of side DE

(b) the size of angle DFE

(c) the size of angle EDF.

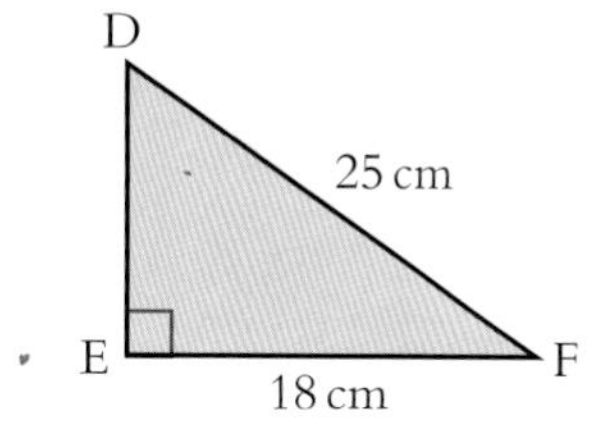

8 Plot the points A (2, 2), B (6, 2), and C (6, 5), on a piece of squared paper.

Draw triangle ABC and calculate the size of angle CAB.

9 Tom's cat is stuck in a tree.

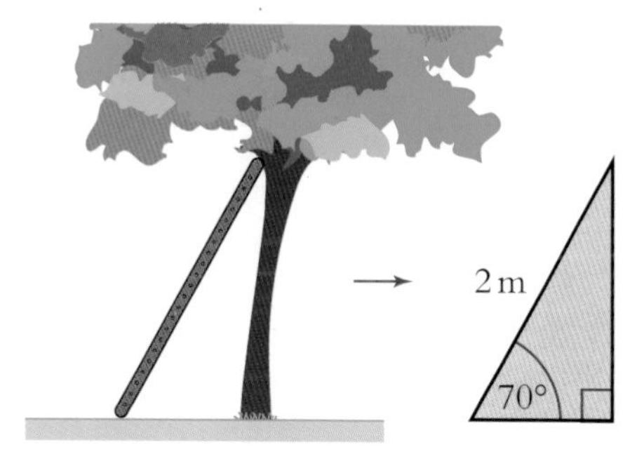

He rests his ladder against the tree. The ladder is 2 metres long, and makes an angle of 70° with the ground. Tom can safely reach 1·6 metres above the top of the ladder.

If his cat is 3·4 metres above the ground, can Tom rescue it safely?

Give a reason for your answer.

More Advanced Problems

We finish by applying what we have learned so far in more difficult practical situations.

In this section, you should be able to identify *appropriate* right-angled triangles in a problem situation and then use trigonometry with these triangles to solve the problem.

Example Darren has gone camping. A view of the front of his tent is shown, together with its sizes.

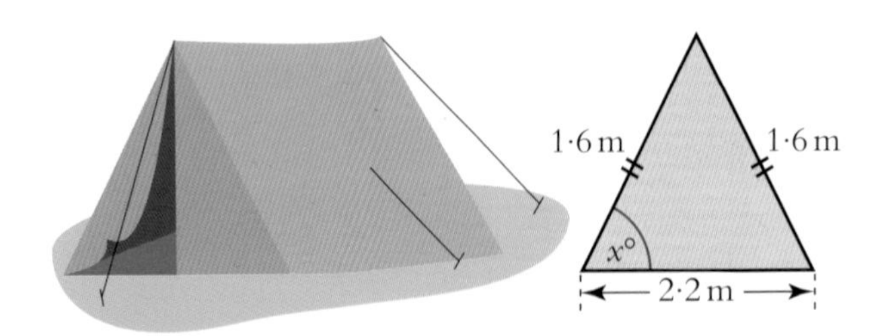

Calculate the size of the angle marked $x°$.

Solution We split the main triangle into two equal right-angled triangles, as shown in the diagrams.

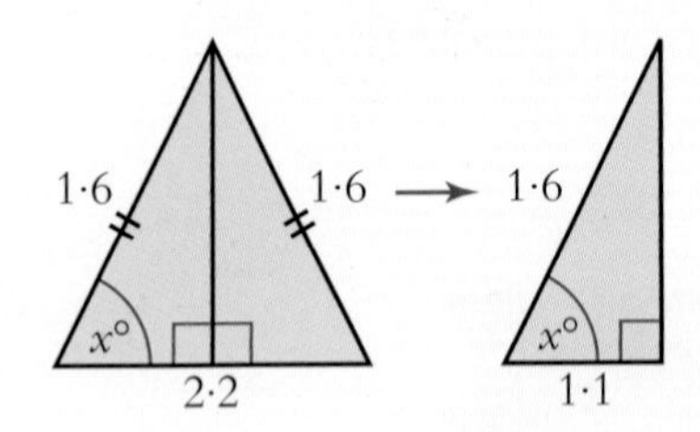

Hence $\cos x° = \dfrac{\text{adjacent}}{\text{hypotenuse}} = \dfrac{1{\cdot}1}{1{\cdot}6}$.

Now press **2nd F cos (1·1 ÷ 1·6) =**

The calculator will show $\cos^{-1}(1{\cdot}1 \div 1{\cdot}6) = 46{\cdot}5674...$,

Hence $x = 46{\cdot}6$ (to 1 d.p.).

Example A detailed side view of a lean-to conservatory is shown in the diagram.

If the sloping part of the roof makes an angle of 17° with the horizontal, calculate the overall height, h (metres), of the conservatory.

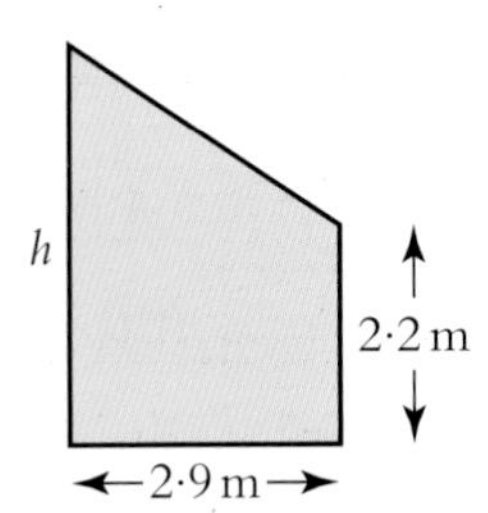

Solution We first split the outline into a rectangle and a right-angled triangle as shown.

Let the size of the vertical side of the triangle be x (metres).

In the triangle, $\tan 17° = \dfrac{\text{opposite}}{\text{adjacent}} = \dfrac{x}{2{\cdot}9}$.

$$\Rightarrow x = 2{\cdot}9 \times \tan 17° = 0{\cdot}89$$
$$\Rightarrow h = 2{\cdot}2 + 0{\cdot}89 = 3{\cdot}09$$

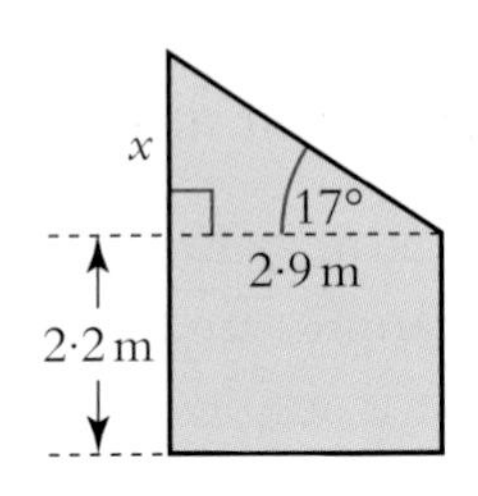

The overall height of the conservatory is therefore 3·09 metres.

Exercise 8

1 The diagram shows a trapezium, ABCD. Find the size of angle ADC.

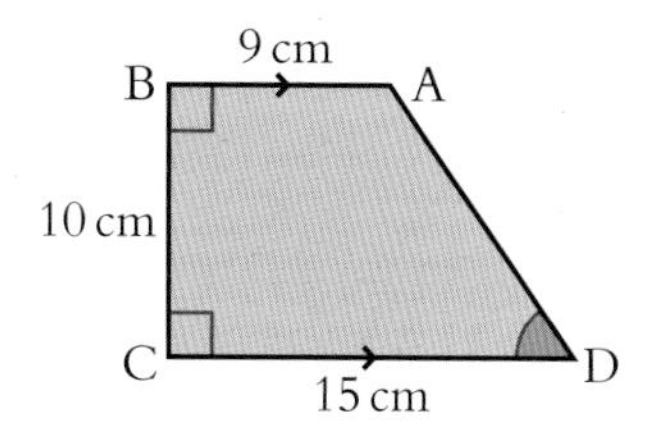

2 The triangle ABC is isosceles. Calculate the size of the angle $y°$.

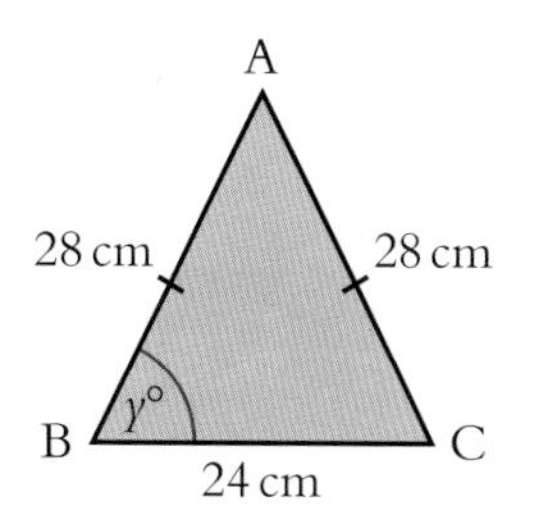

3 The diagram shows a trapezium, PQRS. Calculate the length of side QR.

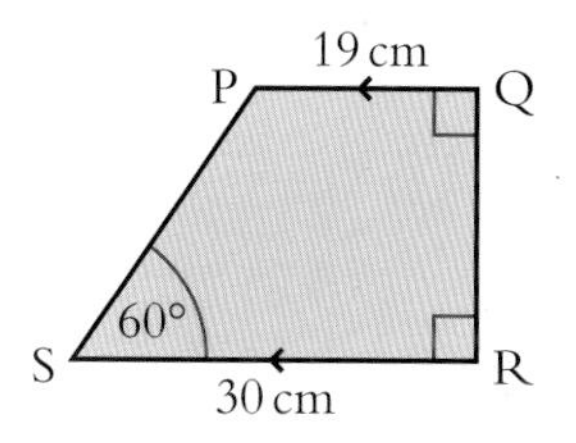

4 Two points A and B, are at the opposite sides of a city.

A by-pass around the outskirts of the city links up A and B as shown.

Calculate the distance AB.

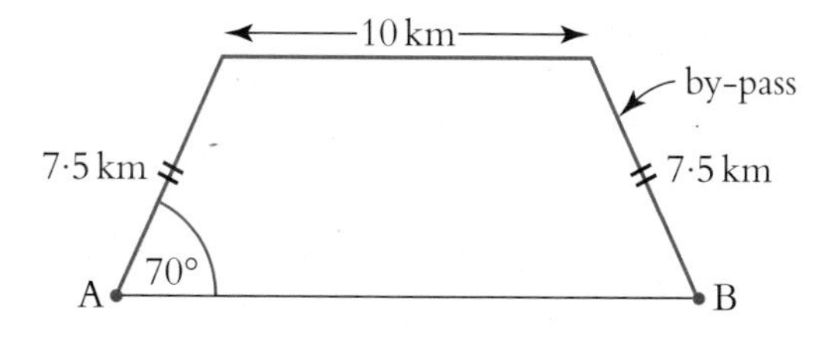

5 Three towns are represented by the points P, Q, and R in the diagram.

Town Q is due east of town P. Town R is 25 km south of town Q. Town P is 35 km from town R. Calculate the bearing, $x°$, of R from P.

(Give your answer to the nearest degree.)

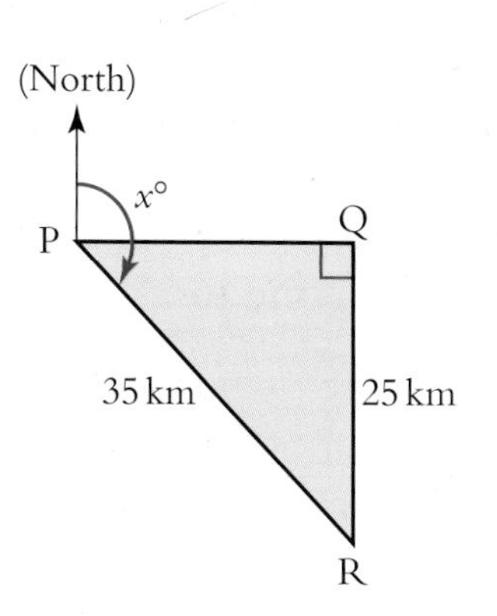

Exercise 8 continued

6 A telegraph pole, 8 metres tall, is supported by two wires, one on each side, as shown.

Calculate the distance, d (metres), between the anchor points of the two wires.

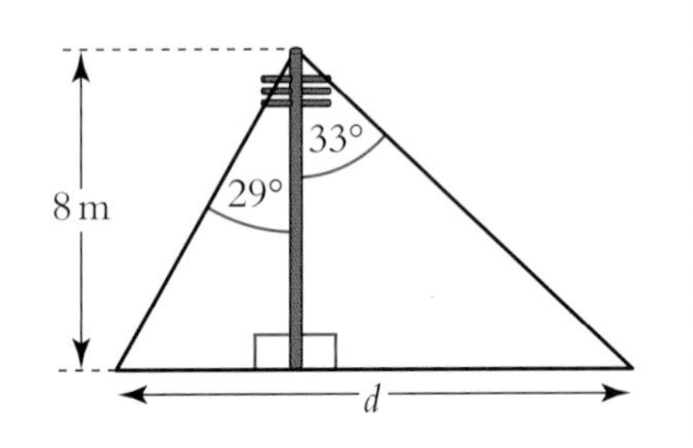

Exercise 9 – Revision of Chapter 12

1 Calculate the value of x in each of the following triangles.

(a)

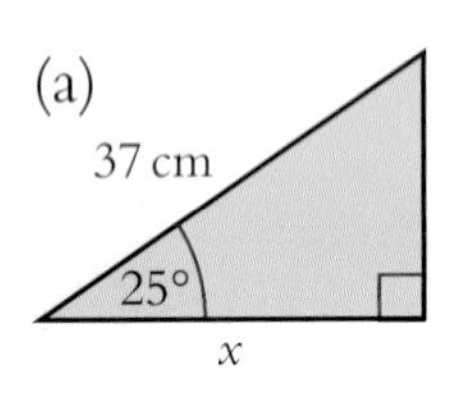

(b) 12 m, 23 m, $x°$

(c)

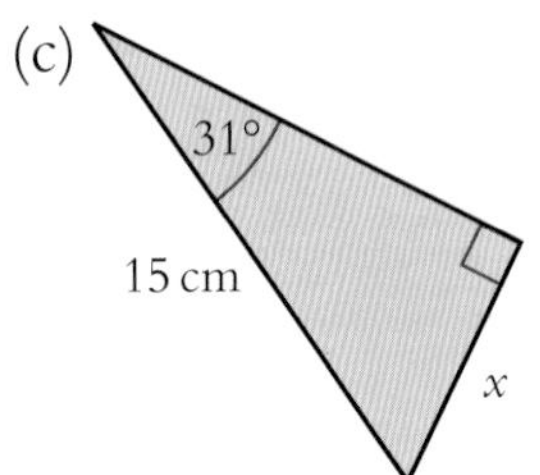

(d) 13 cm, 12 cm, $x°$

2 Triangle DEF is right-angled at E, as shown.

In this triangle, DE = 17 cm, and DF = 30 cm. Calculate the size of angle DFE.

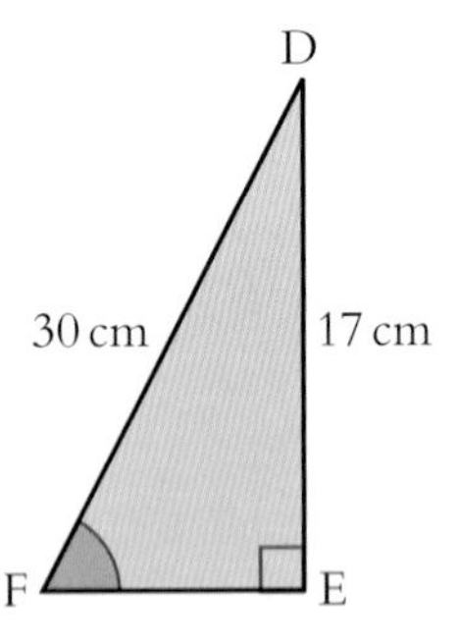

3 Edmond is attempting to calculate the height of his school building.

From a distance of 50 metres, he measures the angle of elevation to the top to be 21°.

What does he calculate the height, h (metres), of the school to be?

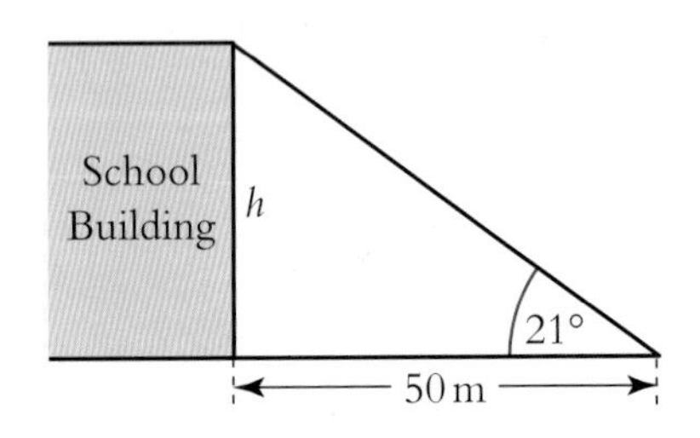

Exercise 9 – Revision of Chapter 12 continued

4 A window designed for a new car is in the shape of a trapezium as shown.

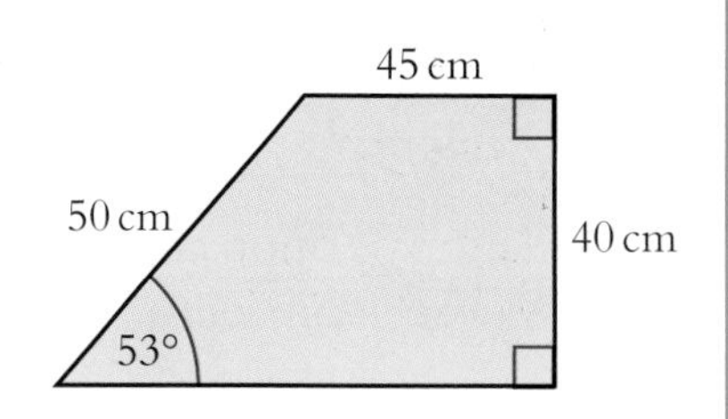

Rubber edging is to be fitted around the perimeter of the window. Find the length of rubber edging required. (Give your answer correct to the nearest centimetre.)

5 The diagram shows a rhombus KLMN.

Calculate the size of the shaded angle, KLM.

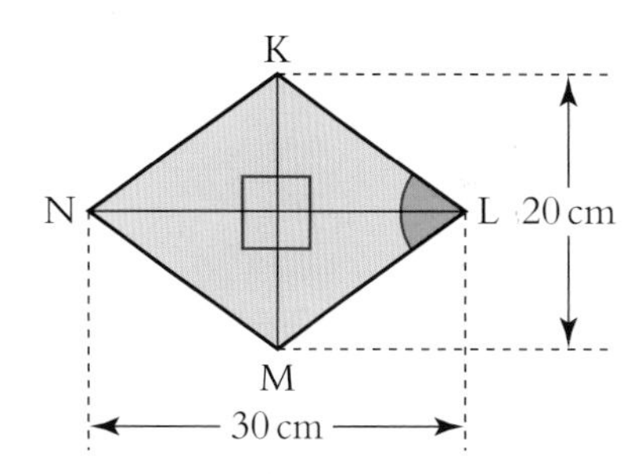

6 A detailed sketch of a bridge is as shown.

Calculate the overall length of the bridge (represented by AB in the diagram).

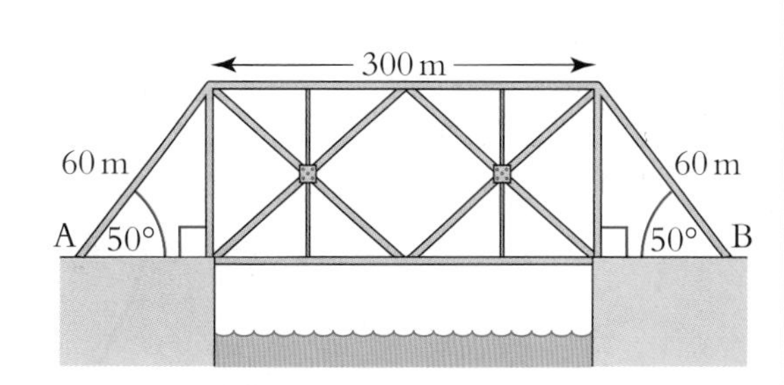

Summary

(Trigonometry is the study of triangles. When dealing with a right-angled triangle, the following points should come to mind quickly.)

1 ***Formulae*** (Trigonometric ratios in a right-angled triangle)

$\tan x^\circ = \dfrac{\text{opposite}}{\text{adjacent}}$,

$\sin x^\circ = \dfrac{\text{opposite}}{\text{hypotenuse}}$,

$\cos x^\circ = \dfrac{\text{adjacent}}{\text{hypotenuse}}$

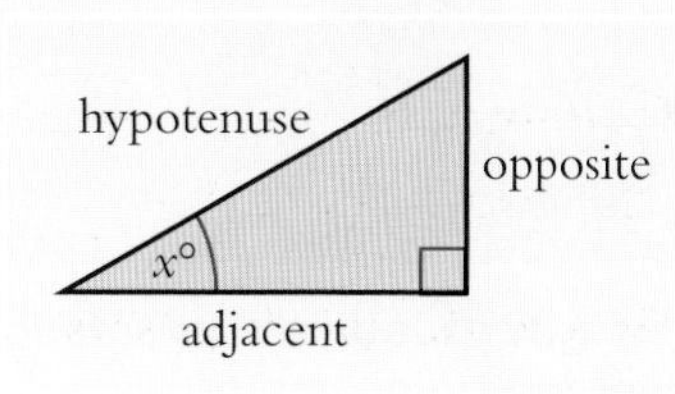

Summary continued

2 ***Finding the length of a side.***

To calculate y:

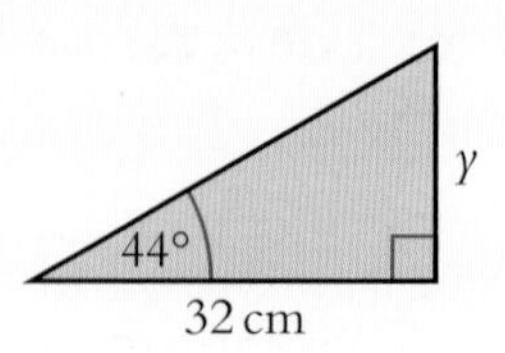

$$\tan 44° = \frac{\text{opposite}}{\text{adjacent}} = \frac{y}{32}$$

$\Rightarrow y = 32 \times \tan 44° = 30{\cdot}9$ (cm)

3 ***Calculating the size of an angle.***

To calculate x:

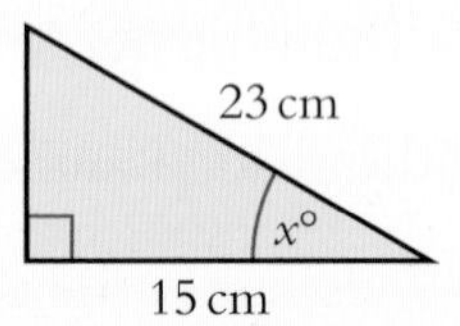

$$\cos x° = \frac{\text{adjacent}}{\text{hypotenuse}} = \frac{15}{23}$$

Now Press **2nd F cos (15 ÷ 23) =**

The calculator will show $\cos^{-1}(15 \div 23) = 49{\cdot}2942...$,

so $x = 49{\cdot}3$ (to 1 d.p.).

The position of this astronaut on the International Space Station is determined by calculations involving trigonometry

13 Standard Form

The Sun is the nearest star to Earth, but its light takes just over 8 minutes to reach us, even though light travels at 186 000 miles per second!

The next nearest star is called 'Proxima Centauri', and its light takes $4\frac{1}{2}$ years to reach us!! This star is therefore over twenty five million million miles away. That is, more than 25 000 000 000 000 miles!

The enormity of the solar system, including three possible 'new' planets

An extremely large number like this is not only difficult to write, but also difficult to deal with in this form. So too is an extremely small number, such as the wavelength of red light, which is approximately 0.000 000 7 metres. We deal with such numbers by writing them in *Standard Form*.

Writing normal numbers in Standard Form

In this first section, you should be able to write any large or small number in Standard Form. Standard form is also called **scientific notation** and is a way of writing very large or very small numbers in a much shorter and more convenient form.

Dealing with large numbers

The number of bits on a computer hard disk is often written in standard form

Large numbers such as (a) 50 000 (b) 7400 may be dealt with as follows:

(a) $50\,000 = 5 \times 10 \times 10 \times 10 \times 10$ which can be written as 5×10^4

(b) $7400 = 7{\cdot}4 \times 10 \times 10 \times 10$ which can be written as $7{\cdot}4 \times 10^3$.

The expressions 5×10^4 and $7{\cdot}4 \times 10^3$ are examples of numbers written in Standard Form.

When a number has been written in standard form, it may be expressed in the form

$a \times 10^n$ where a is a number between 1 and 10 and n is an integer.

Technique

To write a number such as 7400 in Standard Form, we proceed as follows:

(a) 7400 may be written as 7400·0

(b) We now move the decimal point *to the left* until we have a number between 1 and 10 (7·4).

(c) We now count how many places we have moved the point (3).

(d) We now raise 10 to this power (10^3).

(e) 7400 may therefore be written as $7{\cdot}4 \times 10^3$.

Example Write in standard form
(a) 7 million
(b) 69 350 000 000 000
(c) 675 000.

Solution (a) 7 million = 7 000 000 = 7×10^6

(b) 69 350 000 000 000 = $6{\cdot}935 \times 10^{13}$

(c) 675 000 = $6{\cdot}75 \times 10^5$.

Exercise 1

1 Write the following large numbers in standard form. (For example, $7800 = 7{\cdot}8 \times 10^3$.)

(a) 700 (b) 9000 (c) 8 million (d) 700 000

(e) 750 (f) 12 000 000 (g) 725 000 (h) 7 856 000 000

(i) 4876 (j) 32 000 000 000 000 000.

Dealing with Small numbers

Small numbers such as 0·000 32 may be dealt with as follows:

$0{\cdot}000\,32 = \dfrac{3{\cdot}2}{10 \times 10 \times 10 \times 10}$, which can be written as $3{\cdot}2 \times 10^{-4}$.

The expression $3{\cdot}2 \times 10^{-4}$ is another example of Standard Form.

Technique

To write any number such as 0·000 32 in Standard Form, we proceed as follows:

(a) We move the decimal point *to the right* until we have a number between 1 and 10 (3·2).

(b) We now count how many places the point has been moved (4).

(c) We now write this as a negative power of 10 (10^{-4}).

(d) 0·000 32 may therefore be written as $3{\cdot}2 \times 10^{-4}$.

Example Write in standard form

(a) 0·0004

(b) 0·000 000 057

(c) 0·000 000 000 000 543.

Solution (a) 4×10^{-4}

(b) $5{\cdot}7 \times 10^{-8}$

(c) $5{\cdot}43 \times 10^{-13}$.

Exercise 2

1 Write the following small numbers in standard form. (For example, $0{\cdot}025 = 2{\cdot}5 \times 10^{-2}$.)

(a) 0·002 (b) 0·05 (c) 0·000 006 (d) 0·7

(e) 0·000 005 6 (f) 0·271 (g) 0·000 005 67 (h) 0·023

(i) 0·000 045 (j) 0·000 000 000 000 001 69.

2 Write the following numbers in Scientific Notation:

(a) 756 000 (b) 0·000 098 (c) 65 million (d) 0·000 000 15

(e) 7394 (f) 0·000 045 (g) 89 000 (h) 71 200

(i) 0·000 000 006 (j) 541 000.

3 Re-write the following sentences with the numbers in standard form.

(a) The distance from the Earth to the Sun is 93 000 000 miles.

(b) Light travels at 186 000 miles per second.

(c) The mass of a dust particle is 0·000 000 000 753 kilograms.

(d) The area of Africa is 30 293 000 square kilometres.

(e) The Pacific Ocean has an area of 70 million square miles.

(f) The probability of winning the lottery is 0·000 000 071.

(g) There are 0·000 015 78 miles to an inch.

(h) The population of the Earth is 6 525 000 000.

Expressing numbers in Standard Form as Normal Numbers

In this section, we look at reversing the processes we met in the last section. You should then be able to write a number expressed in standard form as a normal number.

Remember that we wrote 7400 as $7{\cdot}4 \times 10^3$.

This means that $7{\cdot}4 \times 10^3$ can be written as 7400 or 7400·0. That is, the decimal point has been moved 3 places *to the right*.

Remember too that we wrote $0{\cdot}000\,32$ as $3{\cdot}2 \times 10^{-4}$. This means that $3{\cdot}2 \times 10^{-4}$ can be written as $0{\cdot}000\,32$. That is, the decimal point has been moved 4 places *to the left*.

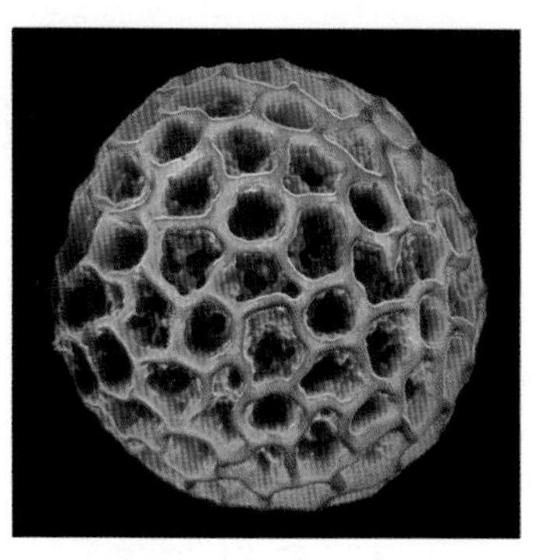

The weight of this pollen grain can be written in standard form

Example Express the following as numbers in normal form:

(a) 6×10^5 (b) $2{\cdot}03 \times 10^{12}$

(c) $1{\cdot}9 \times 10^{-3}$ (d) $4{\cdot}651 \times 10^{-6}$

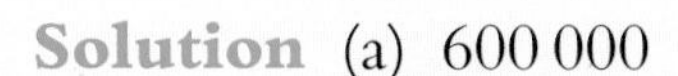

Solution (a) 600 000

(b) 2 030 000 000 000

(c) 0·001 9

(d) 0·000 004 651

Exercise 3

(Do not use a calculator)

1 Write the following as normal numbers:

(a) 5×10^4 (b) 7×10^2 (c) $5{\cdot}8 \times 10^5$ (d) $2{\cdot}51 \times 10^3$

(e) $9{\cdot}03 \times 10^6$ (f) $1{\cdot}59 \times 10^7$ (g) $6{\cdot}543 \times 10^4$ (h) $3{\cdot}875 \times 10^9$.

2 Write the following as normal numbers:

(a) 8×10^{-3} (b) $2{\cdot}0 \times 10^{-5}$ (c) 9×10^{-1} (d) $5{\cdot}7 \times 10^{-3}$

(e) $7{\cdot}15 \times 10^{-4}$ (f) $1{\cdot}178 \times 10^{-6}$ (g) $4{\cdot}4 \times 10^{-8}$ (h) $9{\cdot}62 \times 10^{-2}$.

3 Write the following as normal numbers:

(a) $1{\cdot}245 \times 10^2$ (b) $3{\cdot}54 \times 10^{-2}$ (c) $2{\cdot}7 \times 10^{-7}$ (d) $4{\cdot}215 \times 10^5$

(e) $1{\cdot}9 \times 10^3$ (f) $9{\cdot}65 \times 10^6$ (g) $3{\cdot}14 \times 10^{-5}$ (h) $7{\cdot}133 \times 10^{-4}$.

4 Re-write the following sentences with the numbers as normal numbers.

(a) A light year is $9{\cdot}46 \times 10^{12}$ kilometres.

(b) A grain of pollen weighs $2{\cdot}2 \times 10^{-3}$ grams.

(c) There are 1×10^{-6} kilometres to a millimetre.

(d) Approximately $6{\cdot}1 \times 10^7$ people live in the United Kingdom.

(e) The maximum break in snooker is $1{\cdot}47 \times 10^2$.

(f) Visible green light has a wavelength of $5{\cdot}1 \times 10^{-4}$ millimetres.

Exercise 3 continued

(Do not use a calculator) 

5 By writing the following as normal numbers, put them in ascending order (lowest to highest):

4×10^{-2}, 5×10^{-3}, $4{\cdot}5 \times 10^{-4}$.

6 (a) Express $\frac{3}{1000}$ as a decimal number.

(b) Write $3{\cdot}2 \times 10^{-4}$ as a normal number.

(c) Hence write the following in ascending order: $\frac{3}{1000}$, $0{\cdot}028$, $3{\cdot}2 \times 10^{-4}$.

7 (One billion = 1000 million.)

During an election, a politician promises to spend £3×10^{10} on defence.
By writing this amount as a normal number, find how many billion pounds this is.

8 (a) Given that one trillion = 1×10^{12}, write this as a normal number.

(b) Express 154 trillion in Standard Form.

Using a Calculator for Standard Form

In this section, you should be able to:

- interpret a display involving standard form on a calculator
- perform simple calculations using standard form on a calculator.

Many of the numbers we have met in this chapter have too many digits to fit into a calculator display. However, if you have a scientific calculator, there is a way round the problem. We can use the **Exp** key on the calculator to express numbers in Standard Form. For example if for $6{\cdot}12 \times 10^{17}$, we key in **6·12 Exp 17** to the calculator, we should see one of the following.

6.12 17   6.12 x10¹⁷   6.12 E 17

If however the power of 10 is negative, we use the (–) key on the calculator to enter a negative power. For example, if for $3{\cdot}2 \times 10^{-9}$, we key in **3·2 Exp (–) 9**, we should see one of the following.

On some calculators the +/– key must be used to enter a negative power.

With the aid of a calculator we can easily perform simple calculations using standard form.

Example The orbit path of a satellite around a small planet is approximately $4{\cdot}7 \times 10^4$ kilometres in length. How far does the satellite travel in eight orbits?

Solution Distance travelled = $8 \times (4{\cdot}7 \times 10^4)$

Key in **8 × 4·7 Exp 4**

Hence distance travelled = 376 000 kilometres.

(If the answer is required in standard form, $376\,000 = 3{\cdot}76 \times 10^5$.)

Exercise 4

1 The following displays appear on a calculator. Express each as a normal number.

(a) 

(b) 

(c) 

(d) 

(e) 

(f) 

(g) 

(h) 

(i) 5.3 E -05

Exercise 4 continued

2 The orbit path of a satellite is $5{\cdot}4 \times 10^4$ kilometres in length. How far does the satellite travel in 7 orbits? (Give your answer in standard form.)

3 Calculate each of the following, giving your answer in scientific notation:

(a) $7 \times (3{\cdot}2 \times 10^6)$ (b) $5 \times (4{\cdot}2 \times 10^5)$ (c) $8 \times (3{\cdot}05 \times 10^7)$

(d) $2 \times (8{\cdot}4 \times 10^3)$ (e) $25 \times (3{\cdot}8 \times 10^4)$ (f) $18 \times (4{\cdot}5 \times 10^9)$.

4 Calculate each of the following, giving your answer in scientific notation:

(a) $(7{\cdot}2 \times 10^6) \div 12$ (b) $(8 \times 10^{12}) \div 500$ (c) $(3{\cdot}6 \times 10^8) \div 8000$.

5 The average crowd in a football stadium is $5{\cdot}75 \times 10^4$ spectators. If there are 24 matches played at the stadium during the season, find the total number of spectators for the season. Express your answer as a normal number.

6 Given that one light year $= 9{\cdot}46 \times 10^{12}$ kilometres, find how many kilometres there are in 15 light years. Give your answer in standard form.

7 Calculate each of the following, giving your answer in standard form:

(a) $5 \times (3{\cdot}12 \times 10^{-4})$ (b) $8 \times (1{\cdot}6 \times 10^{-12})$ (c) $45 \times (3{\cdot}6 \times 10^{-7})$.

8 An atom of hydrogen weighs $1{\cdot}6735 \times 10^{-24}$ grams. Calculate the weight of five million hydrogen atoms. Give your answer in standard form.

9 The total prize fund for the lottery one week was £$8{\cdot}76 \times 10^6$. If the prize was divided equally between 32 winners, how much did each winner receive? Give your answer as a normal number.

10 A star football player earns £$9{\cdot}5 \times 10^4$ each week. How much will he earn in a year?

11 The orbit of a planet around a star is circular.

The radius of the orbit is $1{\cdot}5 \times 10^8$ kilometres. By taking $\pi = 3{\cdot}14$, calculate the circumference of the orbit. Give your answer in standard form.

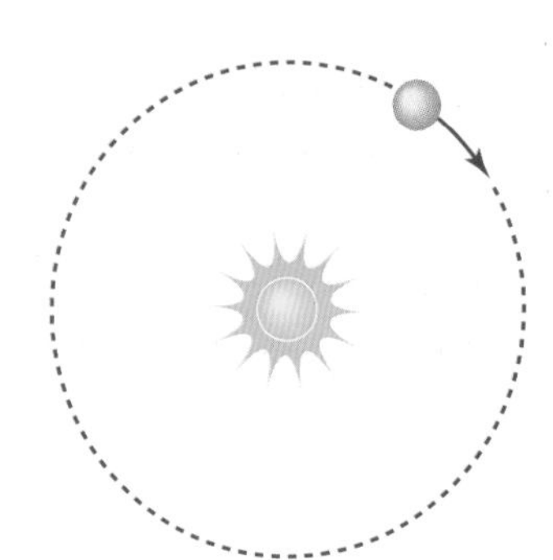

Exercise 5 – Revision of Chapter 13

1 Write the following numbers in standard form:

(a) 760 000 (b) 575 000 (c) 72 million (d) 0·000 005

(e) 0·000 27 (f) 21 830 (g) 0·000 000 28 (h) 53 000 000 000.

2 Write the following numbers in normal form:

(a) $5{\cdot}1 \times 10^4$ (b) $8{\cdot}32 \times 10^6$ (c) $4{\cdot}9 \times 10^{-2}$ (d) $7{\cdot}5 \times 10^{-4}$

(e) 6×10^{-3} (f) $6{\cdot}59 \times 10^2$ (g) $5{\cdot}12 \times 10^5$ (h) $2{\cdot}7 \times 10^{-3}$.

3 Russian billionaire Alexi Romanov has a personal fortune of £$2{\cdot}54 \times 10^{11}$. Write this as a normal number.

4 The Avogadro Constant is a very large number used in chemistry. Its value is 602 213 670 000 000 000 000 000. Write this number in scientific notation.

5 The length of a virus is 0.0005 millimetres. Write this length in scientific notation.

6 The following displays appear on calculators. Write each as a normal number.

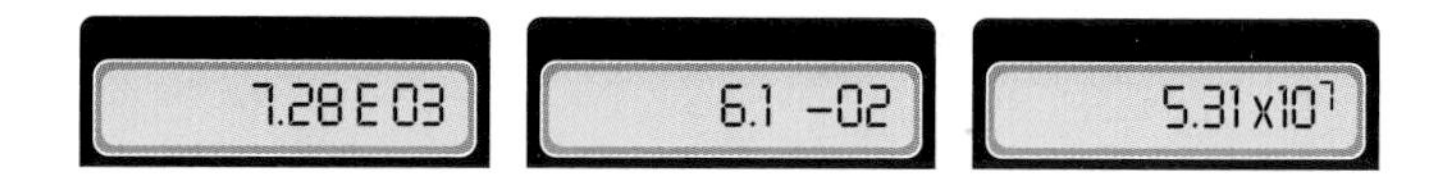

7 Calculate each of the following, giving your answer in scientific notation:

(a) $5 \times (3{\cdot}24 \times 10^6)$ (b) $(8{\cdot}4 \times 10^{12}) \div 12$.

8 Write the following in ascending order.

$\frac{9}{100}$, 0·089, $9{\cdot}1 \times 10^{-3}$

9 Light travels 3×10^5 kilometres in 1 second. How far will light travel in 1 minute? Give your answer in standard form.

10 There are $2{\cdot}24 \times 10^3$ pounds in a ton. How many pounds are in 5 tons? Give your answer in standard form.

Summary

(You should now be familiar with each of the following five items. These should remind you of the ways in which we deal with numbers which are either very large or very small.)

1 ***Standard Form*** (or Scientific Notation)

A number written in standard form is expressed in the form $a \times 10^n$, where a is between 1 and 10 and n is an integer.

2 ***Writing a number in standard form***

$725\,000\,000 = 7{\cdot}25 \times 10^8$

$0{\cdot}000\,023 = 2{\cdot}3 \times 10^{-5}$.

3 ***Writing a number in normal form***

$6{\cdot}14 \times 10^7 = 61\,400\,000$

$1{\cdot}6 \times 10^{-4} = 0{\cdot}000\,16$.

4 ***Using a calculator.***

To show $5{\cdot}7 \times 10^{14}$ on your calculator, key in **5·7 Exp 14.**

A display of 2·1 E –04 means $2{\cdot}1 \times 10^{-4}$ or 0·000 21.

5 ***Performing calculations using a calculator***

For $8 \times (3{\cdot}25 \times 10^6)$, key in 8 × 3·25 Exp 6. The result is 26 000 000, or $2{\cdot}6 \times 10^7$.

The speed of light can be expressed in standard form

End Of Unit Tests

Test One (Non-Calculator)

1 The formula $v = u + at$ is used frequently in calculations in physics. Evaluate v when $u = 12$, $a = 5$ and $t = 7$.

2 Multiply out the brackets and simplify:

(a) $2p + 3(p - 4q)$ (b) $3(2a - 5) + 7$.

3 Factorise:

(a) $7y - 21$ (b) $10x - 35$.

4 Solve the following equations:

(a) $3(2a + 4) = 21$ (b) $5t - 3 = t + 13$.

5 Solve the inequality $4z - 6 > 18$.

6 (a) On a piece of squared paper, draw the straight line with equation $x = 5$.

(b) On the same paper, draw the line with equation $y = -2$.

(c) Write down the coordinates of the point where the lines intersect.

7 (a) Copy and complete the table for the line whose equation is $y = \frac{1}{3}x + 5$.

x	-6	0	3
y			

(b) Hence draw the line with equation $y = \frac{1}{3}x + 5$ on a piece of squared paper.

8 Express the following numbers in standard form:

(a) 530 000 (b) 0·000 002 71.

9 Write the following standard form numbers as normal numbers:

(a) $9{\cdot}75 \times 10^6$ (b) $4{\cdot}03 \times 10^{-5}$.

Test Two (Calculator)

1 The formula $C = \frac{5F - 160}{9}$ may be used to convert temperatures from the Fahrenheit scale (F) to the Celsius scale (C). Convert 50 degrees Fahrenheit to degrees Celsius.

2 When £P is invested in a bank for T years at a rate of interest of R% per annum, the simple interest due, £I, may be found using the formula $I = \frac{PTR}{100}$.

Calculate the simple interest due on £650 invested for 3 years at 4% per annum.

3 Find the length of the side marked y in the triangle shown.

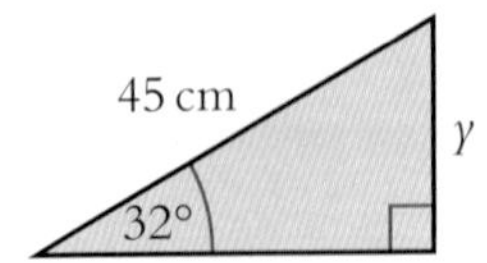

(Give your answer correct to one decimal place.)

4 Calculate the size of the shaded angle, correct to one decimal place.

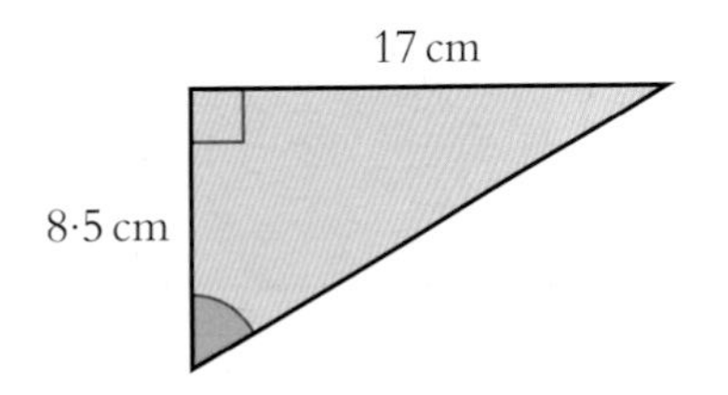

5 Ellen sails 30 kilometres due West from base, then 25 kilometres due North.

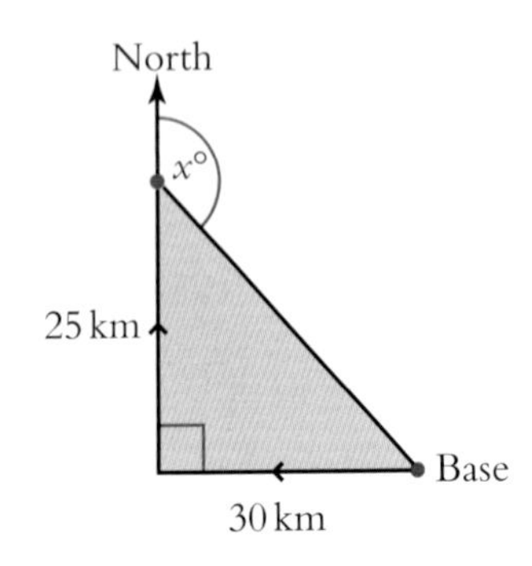

(a) How far is she now from base?

(b) Find the bearing, $x°$, of her return journey to base.

6 Solve the inequality $2x - 7 \leqslant 6$.

7 A multinational company employs 6000 people. The average earning, per employee, is £21 000. Find the total sum which the company pays in earnings. (Give your answer in standard form.)

8 (a) Copy and complete the table for the line whose equation is $y = 2{\cdot}5x + 1$.

x	–4	0	2
y			

(b) Hence draw the line with equation $y = 2{\cdot}5x + 1$ on a piece of squared paper.

9 The length of the orbit path of a satellite is approximately $2{\cdot}41 \times 10^6$ kilometres. How far does the satellite travel in 8 orbits? (Give your answer in scientific notation.)

Test Three (A/B Content)

1 A relationship is expressed by the formula $W = X^2y$. Evaluate W when $X = 2{\cdot}5$ and $y = 8$.

2 The Theorem of Pythagoras is sometimes written in the form $a = \sqrt{c^2 - b^2}$. Calculate a when $c = 10{\cdot}6$ and $b = 5{\cdot}6$.

3 Solve the equations:

(a) $\frac{1}{4}y = 8$ (b) $\frac{1}{7}w = 5$.

4 Solve the equation $p + 12 = 4p - 9$.

5 (a) Write the standard form number $8{\cdot}9 \times 10^{-2}$ as a normal number.

(b) Write the following numbers in order, lowest first:

$8{\cdot}9 \times 10^{-2}$, $\frac{9}{100}$, $0{\cdot}0095$.

6 Triangle ABC is isosceles.

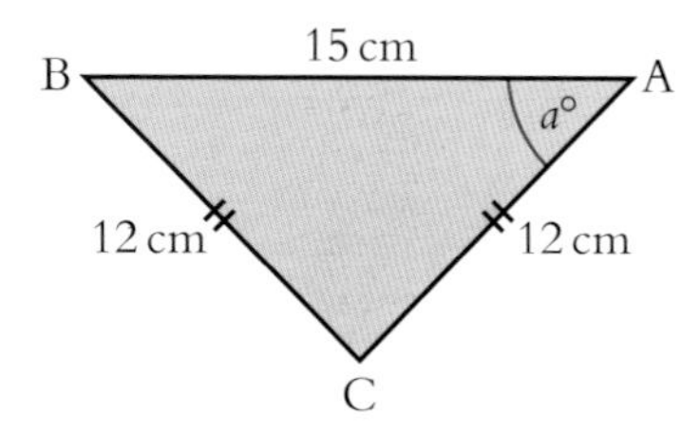

Calculate the size of the angle marked *a*.

7 A banner advertising houses for sale is a composite shape made of of an isosceles triangle and a semi-circle.

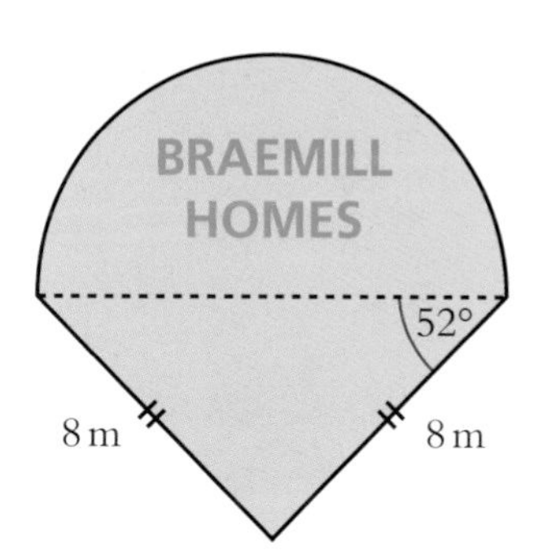

Calculate the perimeter of the banner.

8 The cross-section of the attic in Ed's house is in the shape of an isosceles triangle.

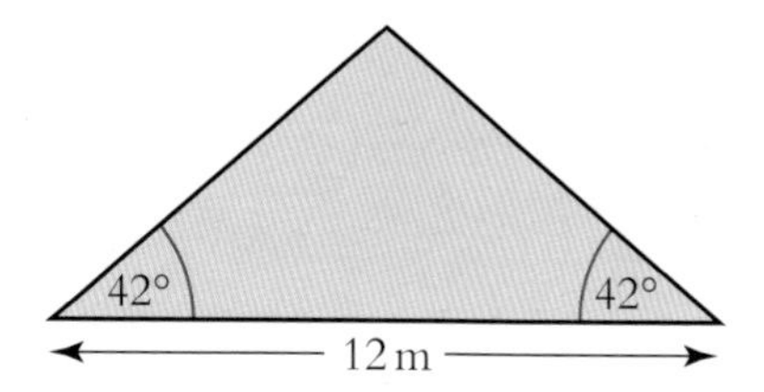

Ed is planning to build a room in the attic which is 7 metres long.

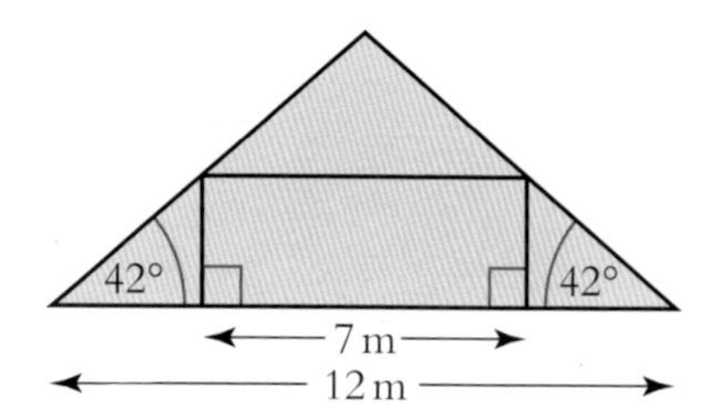

According to regulations, the room must be at least 2·2 metres high before planning permission to build it will be granted.

Will Ed be given permission to build the room?

You must give a reason for your answer.

Applications of Mathematics Unit

Chapter 14 **Social Arithmetic: Earnings and Borrowing**

Chapter 15 **Logic Diagrams**

Chapter 16 **Applications of Geometry**

Chapter 17 **Statistical Assignment**

End of Unit Tests

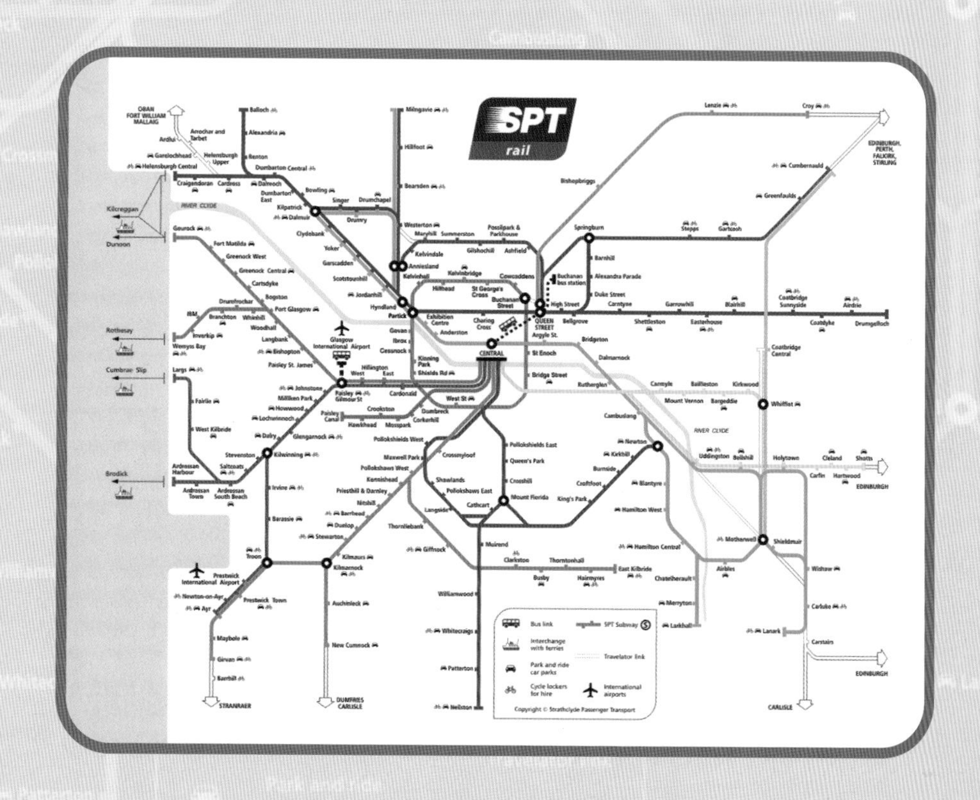

Social Arithmetic: Earnings and Borrowing

The relationship between Income and Expenditure can be a delicate one. Income is determined by salary and deductions. Expenditure involves shopping, monthly bills, savings, and often, repaying loans. Most of us realise that for peace of mind, income should always exceed expenditure. Sadly these days however, with credit so easy to come by, the inequality is in some cases the wrong way round.

In this chapter we look at both Earning and Borrowing.

Earning Money

In this section, you should be able to carry out calculations involving some different ways of earning money.

We studied some aspects of earning money in Chapter 4, where we looked at examples on basic pay, overtime, bonus and annual salary. We begin by revising these topics in the following exercise.

Exercise 1

1 Allan is paid the minimum wage of £5·05 per hour. How much does he earn if he works a 40-hour week?

2 Malcolm has an annual salary of £13 000. How much does he earn each week?

3 Joan earns a basic salary of £1250 per month as a salesperson. In addition she receives 3% commission on all her sales. Calculate her salary for a month in which her sales are £27 000.

Exercise 1 continued

4 Mrs Elder is a senior citizen, and she receives a pension of £84·75 per week. In addition, she receives a Christmas bonus of £200. Find her total income for the year.

5 Tony is a factory worker who is paid at an hourly rate of £7·10. During one week he works 35 hours at the basic rate, plus 6 hours overtime at time and a half. Calculate his total pay for that week.

6 Martine translates documents from French into English. She is paid £27·50 for each page she translates. How much will she earn if she translates a document with 42 pages?

7 Sonia's basic rate of pay is £6·50 per hour. She is paid time and a half for working in the evening, and double time for working at the weekend. Find her wage for a week in which she works 35 hours at the basic rate, 6 hours in the evening, and 4 hours at the weekend.

8 Joshua earns £280 per week. He works 48 weeks in a year, but is paid for 52 weeks.

The extra amount is called **holiday pay**.

Calculate (a) Joshua's total pay for the year.

(b) The amount of this which is holiday pay.

9 Sandra earns £7·25 per hour as a new bus driver. She has been told that she will receive a 12% rise after one year when she is more experienced.

What will be her hourly pay after this rise?

10 Colin is retired but now works part-time. He has three sources of income:

- an annual pension of £14 350
- interest on a sum of £43 050 invested at a rate of 4% per annum
- work for 90 days at a rate of £154 per day.

Calculate Colin's total income for the year.

11 Bill is paid to mark exam scripts. He earns £2·05 for every script he marks plus he is paid a bonus of £1·25 per script for every script he marks above his target of 175 scripts. How much will Bill earn if he marks 234 scripts?

Exercise 1 continued

12 Iain is a printer who earns £6·80 per hour and works a 40-hour week. The company he works for had an urgent order to print new railway timetables. Iain worked 8 hours overtime at time and a half to complete the job within the week. His boss was so pleased with his effort that he gave him a £75 bonus. How much did Iain earn that week?

13 Catherine Steele is a famous author. She receives royalties of 10% on all her book sales. Her latest novel 'Classroom Romance' has sold 135 780 copies at £9·99 each. How much will Catherine receive in royalties?

14 Robin Crooks is a football agent who receives 20% of a client's salary in commission. He negotiates a weekly wage of £28 000 for his star player, Alfredo Biggins. How much will Robin receive in commission in a year?

15 Catriona sells cars. Her basic salary is £800 per month, and she earns 4% commission on all her sales. Calculate her sales in a month in which her total salary was £12 800.

Payslips

In this section, you should be able to:

- carry out calculations involving gross pay, deductions and net pay
- interpret information on a payslip.

Remember

Payslips or wage slips are given to all employees and contain details of earnings and deductions.

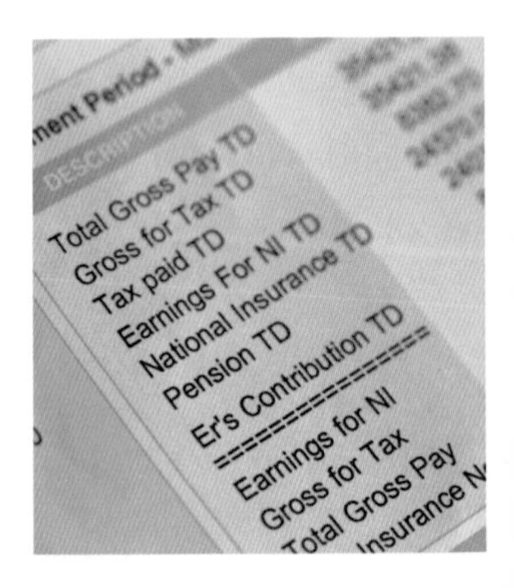

Gross pay is the amount of money earned before any deductions are made. This includes basic pay and possibly overtime, commission or bonuses.

Deductions are the amounts of money taken from your pay. The main deductions are for *Income Tax* (money taken by the government to help run the country) and *National Insurance* (to pay for the National Health Service, state pensions, unemployment and other benefits). Other deductions may include *Superannuation* (extra payments towards pension), union dues and donations to charity.

Net Pay or take-home pay is the pay after deductions have been taken off. That is,

Net pay = Gross pay – Deductions.

Example Darlene Gray works in a bakery. Her payslip for the week ending 23rd March is shown. There are some entries missing however.

Employee: Darlene Gray		Week Ending: 23/03/2007	
Basic Pay £216·00	Overtime £48·60	Bonus £25·00	Gross Pay
Income Tax £37·45	National Insurance £21·85	Superannuation £17·38	Total Deductions
			Net Pay

From the information already on the payslip calculate:

(a) gross pay,
(b) total deductions,
(c) net pay.

Solution

(a) Gross pay = £(216·00 + 48·60 + 25·00) = £289·60.

(b) Total deductions = £(37·45 + 21·85 +17·38) = £76·68.

(c) Net pay = £(289·60 – 76·68) = £212·92.

Exercise 2

1 Vincent works as a painter. His basic pay is £6·80 per hour. During one week he works 40 hours at the basic rate and 5 hours overtime at time and a half. He is also given a £20 bonus for good workmanship. Calculate his gross pay for the week.

2 Tim sees the following deductions on his payslip – 'Income Tax: £43·07', 'National Insurance: £28·21', and 'Superannuation: £18·94'. Calculate his total deductions.

3 Bryan completes his tax calculations for the year. His gross income is £28 356 and his total deductions are £4890·08. Calculate his net earnings.

4 Tallulah repairs computers. The following details appear on her monthly payslip. Basic pay = £1425, Overtime = £85, Bonus = £35.

(a) Find Tallulah's gross pay for the month.

She has deductions of £226·09 for Income Tax, £92·70 for superannuation, and £126·57 for National Insurance.

(b) Calculate her total deductions.

(c) Calculate her net pay.

5 Copy and complete the blank payslip using the information given.

Employee: Andrea Watt		December 2006	
Basic Pay	Overtime	Bonus	Gross Pay
Income Tax	National Insurance	Superannuation	Total Deductions
			Net Pay

Andrea Watt earns a basic pay of £1890, £560 in overtime, and receives a bonus of £65.

She pays £439·50 in Income Tax, £272·80 in National Insurance, and £150·90 in superannuation.

Exercise 2 continued

6 Lynne is a midwife. Her payslip for the month of October is shown.

The deduction for National Insurance and 'total deductions' have still to be entered.

Employee: Lynne Bell		Month Ending: 31/10/06	
Basic Pay £2151·50	Overtime —	Bonus —	Gross Pay £2151·50
Income Tax £359·52	National Insurance	Superannuation £129·09	Total Deductions
			Net Pay £1469·60

Calculate

(a) Lynne's total deductions

(b) her National Insurance contribution.

7 Nicol is a Headteacher. His payslip for March is shown.

The amounts for gross pay, National Insurance and 'total deductions' have still to be entered.

Employee: Nicol Campbell		Month Ending: 31/03/07	
Basic Pay £6207·00	Overtime £40·00	Bonus —	Gross Pay
Income Tax £1809·97	National Insurance	Superannuation £374·82	Total Deductions
			Net Pay £3418·49

Calculate

(a) Nicol's gross pay

(b) his total deductions

(c) his National Insurance contribution.

Exercise 2 continued

8 Emily sells kitchens. Her payslip for April is shown, but the amounts for 'commission', 'gross pay', 'total deductions' and 'net pay' are still to be entered.

Emily earns commission of 3% on her sales each month in addition to her basic pay. In April she sold goods to the value of £85 800.

Employee: Emily Bates		Month Ending: 30/04/07	
Basic Pay £1047·50	Commission	Bonus —	Gross Pay
Income Tax £759·77	National Insurance £354·92	Superannuation £217·29	Total Deductions
			Net Pay

Calculate

(a) Emily's commission

(b) her gross pay

(c) her total deductions

(d) her net pay.

9 Madison worked the following hours during a week in June.

Monday	9 am to 3pm
Tuesday	8 am to 5 pm
Wednesday	9 am to 4 pm
Thursday	7 am to 12 Noon
Friday	10 pm to 12 pm.

She earns £9·50 per hour.

(a) Calculate her gross pay for the week.

(b) If her total deductions are £48·75, calculate her net pay for the week.

Borrowing Money

People sometimes borrow money to pay for things when they cannot afford to pay the full cash price at the time of purchase. Banks, Building Societies, Loan Companies, and some shops too, offer loans to customers. When you borrow money however, you are charged Interest on top of your repayment of the loan. When you borrow money you should be told how much Interest you will be charged. The **APR** (Annual Percentage Rate) tells you the annual rate of interest you are being charged.

Repayments of the loan have to be made, usually monthly. The amount of the repayment depends on how much is borrowed, the APR, and the length of time over which the loan is taken.

Loan repayment tables give information about how much you have to pay when you borrow money.

The **cost of the loan** is the difference between the total repayments and the value of the loan.

In this section, you should be able to:

- **use a loan repayment table to find the monthly repayment for a particular loan.**
- **calculate the total repayments based on a set of conditions**
- **calculate the cost of a loan.**

Example The table shown is used to calculate repayments on a loan of £1000.

(Monthly repayments on a loan of £1000)

APR	12 months	24 months	36 months	48 months
12%	£90·60	£48·83	£34·96	£28·07
14%	£91·44	£49·66	£35·82	£28·95
16%	£92·27	£50·50	£36·67	£29·84
18%	£93·09	£51·32	£37·53	£30·72
20%	£93·90	£52·14	£38·38	£31·61

Paul O'Connor borrows £7500 over 24 months at an annual percentage rate (APR) of 16%. Use the table to calculate the cost of his loan.

Solution

From the table, monthly repayment = £50·50 (24 months, 16%).

This payment is for a loan of £1000.

Hence monthly repayment for a loan of £7500 $= \frac{7500}{1000} \times 50{\cdot}50$
$= 7{\cdot}5 \times 50{\cdot}50$
$=$ £378·75.

Hence total repayment over 24 months $= 24 \times 378{\cdot}75$
$=$ £9090.

The amount borrowed was £7500 and the total repayment was £9090.

Hence cost of the loan = £(9090 – 7500)
= £1590.

Exercise 3

Use the last loan repayment table to answer Questions 1–8

1 Abigail borrows £1000 over 24 months at an APR of 20%.

Calculate (a) her monthly repayment

(b) her total repayment

(c) the cost of the loan.

2 Repeat Question 1 for Umar who borrows £3000 over 48 months at an APR of 12%.

3 Repeat Question 1 for Victoria who borrows £6000 over 24 months at an APR of 18%.

4 Isobel borrows £5000. She repays the loan over 3 years at an APR of 14%.

(a) How many months are in 3 years?

(b) What is Isobel's monthly repayment?

(c) Calculate her total repayment.

(d) Calculate the cost of the loan.

Exercise 3 continued

Use the last loan repayment table to answer Questions 1–8

5 Matthew borrows £3500 over 4 years at an APR of 18%.

Calculate (a) his monthly repayment

(b) the cost of the loan.

6 Ashley borrows £6500 over 1 year at an APR of 12%. Find the cost of her loan.

7 Caroline borrows £25 000 over 48 months at an APR of 12%. Calculate the cost of her loan.

8 Gordon borrows £8500 to set up his own business. The APR is 14% and the loan is to be paid back over 36 months.

(a) Find the cost of the loan.

(b) How much would Gordon have saved if he had taken the loan over 24 months?

9 The table shows the monthly repayments to be made when £1000 is borrowed from the 'Money for Old Rope' Loan Company.

(Monthly repayments on a loan of £1000)

APR	12 months	24 months	36 months
12%	£93·80	£52·03	£38·16
14%	£94·64	£52·86	£39·02
16%	£95·57	£53·70	£39·87
18%	£96·29	£54·52	£40·73

Meg takes out a loan for £6500 for a car, over 24 months at an APR of 16%. Find the cost of her loan.

10 Anna borrows £4500 from 'Money for Old Rope' to pay for an extension to her house. She will repay the loan over 36 months at an APR of 18%.

(a) Calculate the cost of her loan.

(b) How much could Anna save if she were to repay the loan over 24 months?

Loan Protection

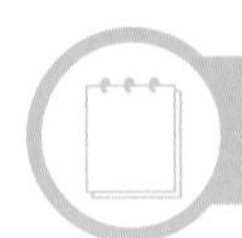

Main points

In practice, when you borrow money, you can do so **with or without loan protection**. Sometimes this is described as 'with or without **payment** protection'. Any loan will cost more if taken out with loan protection, but there are advantages in having loan protection. Loan protection means that should your circumstances change through illness, accident, or financial difficulty, the payments will be made for you.

Example The table shows the monthly repayments to be made when money is borrowed from the Albion Bank. (Repayments can be made with or without loan protection.)

(Monthly repayments: Albion Bank)

	24 months		36 months		48 months	
Loan amount	With loan protection	Without loan protection	With loan protection	Without loan protection	With loan protection	Without loan protection
£10 000	£492	£444	£342	£304	£276	£236
£8000	£394	£355	£274	£243	£221	£189
£5000	£246	£222	£171	£152	£138	£118
£4000	£197	£178	£137	£122	£110	£94

(a) Valdas borrows £8000 over 3 years **with** loan protection. How much is his monthly repayment?
(b) Calculate the cost of his loan.
(c) How much would he have saved if he had borrowed the money without loan protection?

Solution

(a) Period = 3 years = 3 × 12 = 36 months.

Hence monthly repayment (£8000, 36 months, with LP) = £274.

(b) Total repayment over 36 months = 36×274 = £9864, so cost of loan = £(9864 – 8000) = £1864.

(c) Monthly repayment (£8000, 36 months, without LP) = £243.

Total repayments over 36 months = 36×243 = £8748.

Hence amount saved without LP = £(9864 – 8748) = £1116.

Exercise 4

Use the last loan repayment table to answer Questions 1–8

1 Sven borrows £5000 over 24 months **without** loan protection.

How much is his monthly repayment?

2 Clarissa borrows £10 000 over 48 months **without** loan protection.

(a) How much is her monthly repayment?

(b) How much will she pay altogether?

3 Adrian takes out a loan of £4000 over 24 months **with** loan protection.

(a) How much is his monthly repayment?

(b) How much will he pay altogether?

4 Basharat borrows £8000 over 36 months **without** loan protection.

(a) How much is his monthly repayment?

(b) Over the 36 months, how much extra would Basharat pay in total for loan protection on his loan of £8000?

5 (a) Danny borrows £5000 to buy a new car. He will repay the loan over 36 months with payment protection. How much is his monthly repayment?

(b) After 30 months, Danny is made redundant. The bank makes the remainder of the payments for him. How much does the bank pay?

6 Angela borrows £10 000 over 48 months with loan protection.

(a) How much is her monthly repayment?

(b) Due to an illness after 40 months, she is unable to pay the remainder. How much will it cost the bank to make up the remaining payment?

Exercise 4 continued

Use the last loan repayment table to answer Questions 1–8

7 Jill borrows £8000 over 24 months without loan protection.

(a) How much is her monthly repayment?

(b) Calculate the cost of her loan.

8 Gloria borrows £5000 over 48 months without payment protection.

(a) Calculate the cost of her loan.

(b) How much more would the loan have cost if it has been taken with loan protection?

Exercise 5 – Revision of Chapter 14

1 Leanne earns £5·90 per hour as a shop assistant.

(a) How much will she earn weekly if she works 40 hours?

(b) How much will she earn in a year if she is paid for 52 weeks including holiday pay?

2 Katie is paid £6·40 per hour for a 35-hour week. During one week she works a total of 41 hours. Overtime is paid at time and a half. Calculate Katie's gross pay for that week.

3 Barry works in a supermarket and he is paid £5·25 per hour. He works 38 hours each week. His payslip, partly completed, is shown.

Calculate

(a) Barry's gross pay

(b) his total deductions

(c) his net pay.

Name: Barry Oldman	Week ending: 8/6/07	
Hours Worked 38	Hourly Rate £5·25	Gross Pay
Income Tax £17·63	National Insurance £11·94	Total Deductions
		Net Pay

Exercise 5 – Revision of Chapter 14 continued

4 Holly is a trainee electrician. Her basic rate of pay is £5·05 per hour for a 36-hour week. Her overtime rate of pay is double time. Her payslip is as shown, but some entries are missing.

Earnings				Deductions	
	Hours	Rate	Amount		Amount
Basic	36	£5·05	£181·80	Income Tax	£22·13
Overtime	4			National Insurance	£14·43
		Gross Pay		Total Deductions	
				Net Pay	

Copy and complete her payslip.

5 The following tables show the monthly repayments to be made, with and without payment protection, when £1000 is borrowed from a Finance Company.

(With Payment Protection)

APR	12 months	24 months	36 months	48 months
10%	£93·67	£50·35	£36·04	£28·74
12%	£94·57	£51·28	£37·01	£29·75
14%	£95·47	£52·20	£37·98	£30·76
16%	£96·35	£53·11	£38·95	£31·78

(Without Payment Protection)

APR	12 months	24 months	36 months	48 months
10%	£87·53	£45·75	£31·88	£24·97
12%	£88·37	£46·70	£32·73	£25·84
14%	£89·21	£47·43	£33·59	£26·72
16%	£90·04	£48·27	£34·44	£27·61

Ronald wishes to borrow £5000 and to make repayments over 36 months at an APR of 12%, without payment protection.

(a) How much will Ronald's monthly repayment be?

(b) How much more would the loan cost if it is taken with payment protection?

Exercise 5 – Revision of Chapter 14 continued

6 The table shows the monthly repayments to be made when money is borrowed from the Angus Building Society.

(Angus Building Society)

	Monthly Repayments		
Loan \ Term	20 years	25 years	30 years
£30 000	£241·65	£219·88	£208·18
£40 000	£322·20	£293·17	£275·57
£50 000	£402·75	£366·46	£346·97

Elle wishes to borrow £50 000 to buy a house.

(a) She decides to repay the loan over 30 years. How much will she pay each month?

(b) How much will she pay altogether?

(c) How much would she have saved if she had borrowed the same amount over 25 years?

7 Nicola reads the following advert for a dental assistant in the local job centre.

DENTAL ASSISTANT

Hours:	Mon – Wed	0830–1230 and 1300–1700
	Thur – Fri	0830 –1230
	Sat	0900–1200
Pay:	Mon – Fri	£6·90 per hour
	Sat	time and a half

How much would Nicola earn each week if she got this job?

Summary

(The calculations in this Chapter are quite straightforward. You should now be clear in your own mind about the meanings of the following terms.)

1 *Overtime*

The most common rates of overtime are *time and a half* (1·5 × basic rate) and *double time* (2 × basic rate).

2 *Payslips*

Payslips give details of:

(a) Gross Pay (basic pay, overtime, commission, bonuses).

(b) Deductions (Income Tax, National Insurance, Superannuation).

(c) Net Pay (pay after deductions), also called take-home pay.

For example.

Name H. Worker	Employee No. 001	Tax Code 404L	Month February
Basic Salary £1200·00	Bonus £64·00	Overtime £42·00	Gross Salary £1306·00
Nat. Insurance £100·21	Income Tax £171·40	Superannuation £78·36	Total Deductions £349·97
			Net Salary £956·03

3 *Borrowing Money*

When money is borrowed, details of *monthly repayments* are given in *loan repayment tables*. The tables give details of how much is borrowed, the APR (Annual Percentage Rate), and the term of the loan. The tables usually detail repayment amounts *with* and *without* payment protection. The *cost of the loan* is the difference between the total repayments and the loan.

Logic Diagrams

A logic diagram gives us a means of setting out information in a way which is easy to understand. Without such diagrams, information can easily tie us in mental knots. In this chapter we look at network diagrams (including tree and activity diagram problems), flowcharts, and spreadsheets.

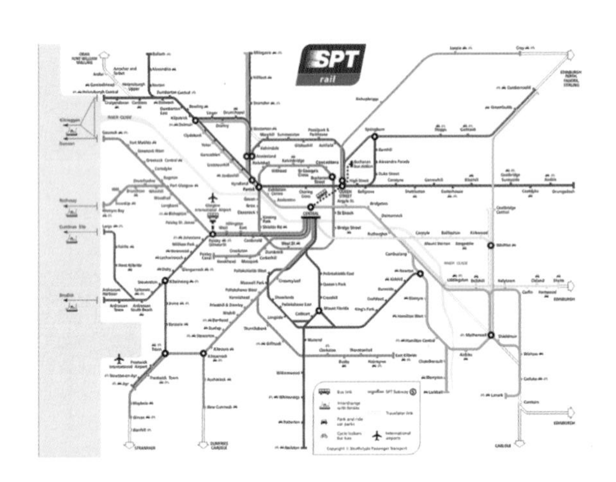

Network Diagrams

In this section, you should be able to:

- interpret the terms vertex, node, order of a node, and arc in a network diagram
- interpret a network diagram.

Consider the map which shows some towns and the roads connecting them.

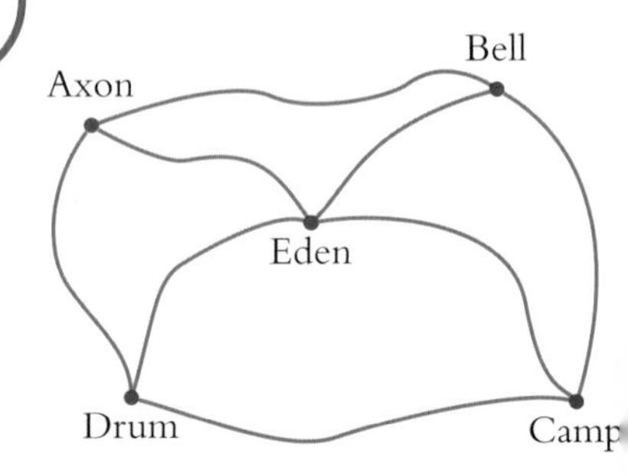

Key words and definitions

This may be drawn in a simpler form, called a **network diagram**.

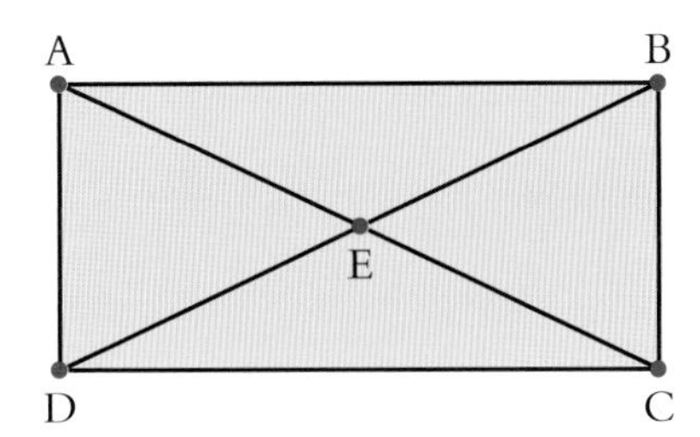

A network diagram consists of vertices (or nodes) connected by arcs (or edges).

The network shown has 5 vertices (or nodes). These are represented by the dots at A, B, C, D, and E. Note that *vertices* is the plural of the word *vertex*.

This network has 8 arcs (or edges) joining the vertices.

The order (or degree) of a vertex is the number of arcs meeting there. The order of each vertex in the diagram is:

A – 3, B – 3, C – 3, D – 3, E – 4.

A, B, C, and D are *odd* nodes, while E is an *even* node.

Example The network diagram shows the distances (in metres) between the Station and some other buildings in a town.

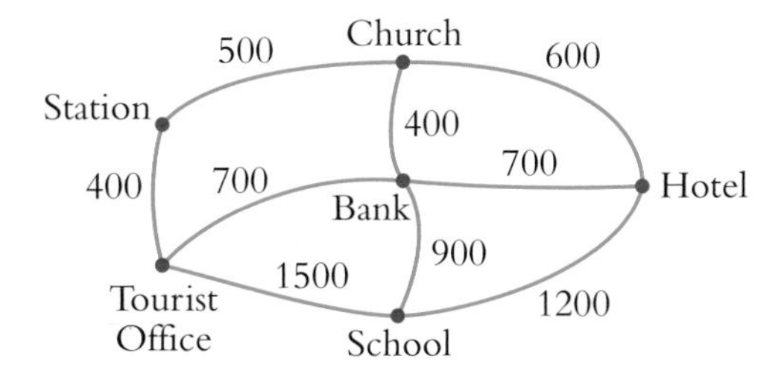

(a) State the order of the node at the Bank.

(a) What is the shortest distance from the Station to the School?

Solution

(a) 4 arcs meet at the bank, so the order is 4.

(b) Try different routes, such as:

Station → Tourist Office → School = 400 + 1500 = 1900

Station → Tourist Office → Bank → School
= 400 + 700 + 900 = 2000

Station → Church → Bank → School
= 500 + 400 + 900 = 1800

Hence the shortest distance is 1800 metres.

Exercise 1

1 Look at the network diagram.

(a) How many nodes are there?

(b) How many arcs are there?

(c) Write down the order of each node.

(d) Which nodes are even?

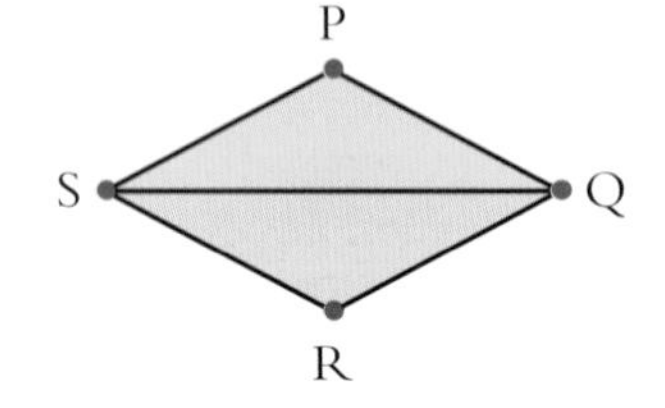

2 Repeat Question 1 for the network diagram shown here.

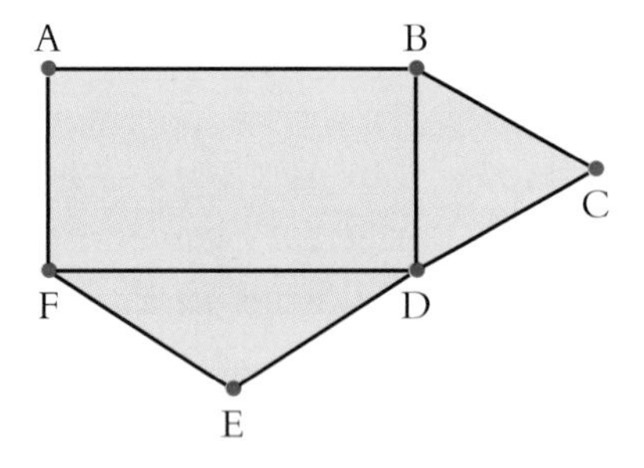

3 Copy and complete the table for the network diagram shown.

Node	Order	Even/Odd
U	2	Even
V		
W		
X		
Y		
Z		

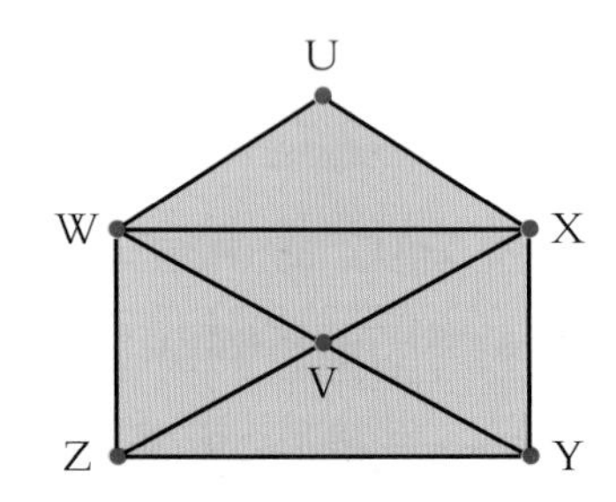

Exercise 1 continued

4 The network diagram shows the streets to be visited by a postman.

(a) How many arcs are there in the network diagram?

(b) The postman is planning his route. He starts at the Sorting Office and he goes along every street once. He does not need to finish at the Sorting Office. List the streets in order for one possible route.

Sorting Office
High Street
Cart Road
Millbrae Street
Airth Road
Battle Avenue

5 The network diagram shows the distances in miles between six towns.

(a) State the order of the node at Strathvale.

(b) What is the length of the shortest route from Ingleton to Greenview?

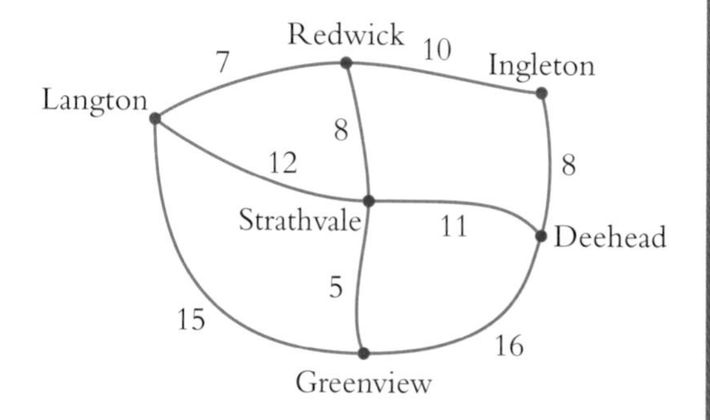

6 Here is a simple network diagram.

(a) How many arcs are there?

(b) Write down the even nodes.

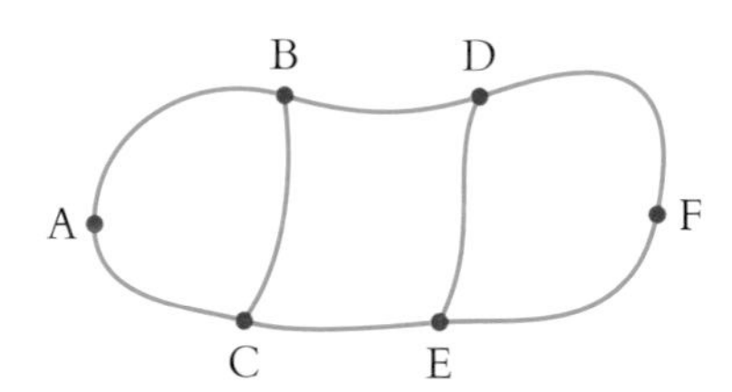

7 Jenson has measured his journey times in minutes along the streets in his neighbourhood.

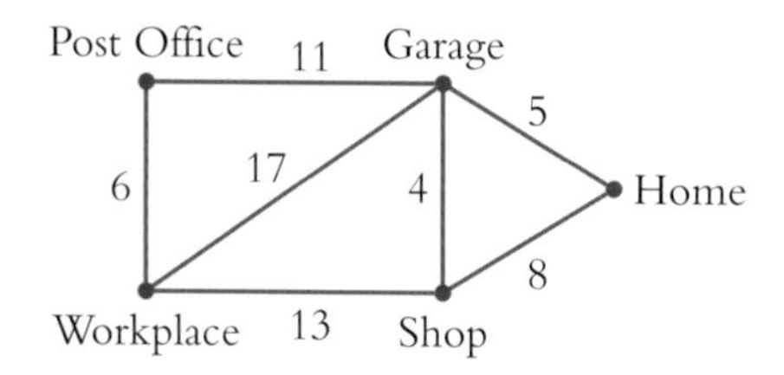

(a) State the degree of the node at his home.

(b) What is the shortest journey time from his home to his workplace?

Exercise 1 continued

8 The network diagram shows the streets to be visited by a bus inspector.

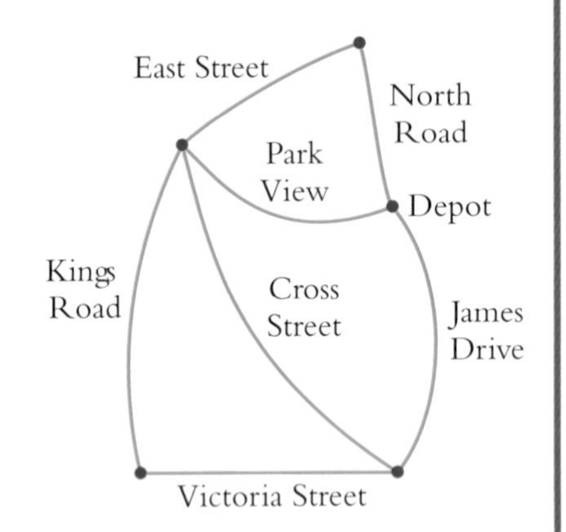

(a) How many arcs are there in this network diagram?

(b) The inspector is planning his route. He starts at the depot and goes along every street once. He does not need to finish at the depot. List the streets in order for one possible route.

9 Study this network diagram.

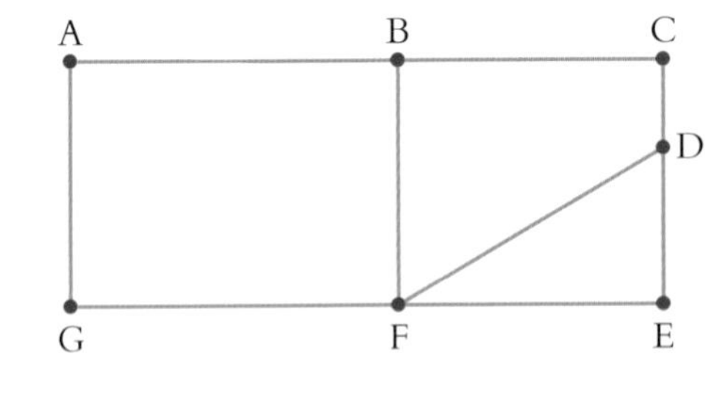

(a) How many odd nodes are there in the diagram?

(b) Travers has been asked to draw a pathway through the network by going over every line once and only once without lifting his pencil.

He starts D → E → …

Copy and complete a possible pathway for Travers to achieve this.

10 The network diagram shows the distances in miles between Scotland's four largest cities.

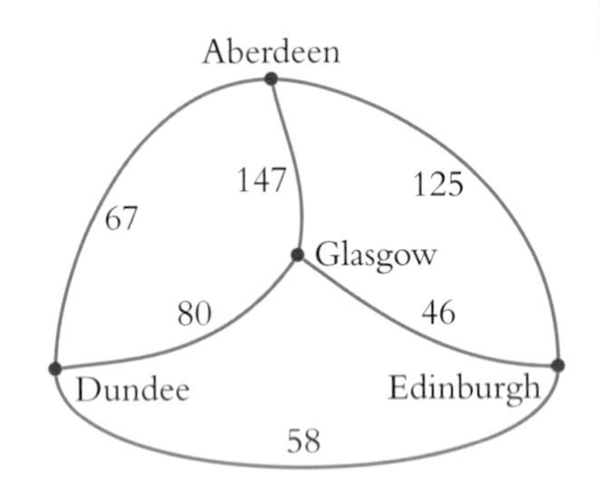

A courier leaves Glasgow. He has to deliver packages to Aberdeen, Dundee and Edinburgh.
He cannot go through any city more than once and he does not have to return to Glasgow.

His possible routes may be shown on a **tree diagram**.

(a) Copy and complete this tree diagram.

(b) Which is the shortest route?

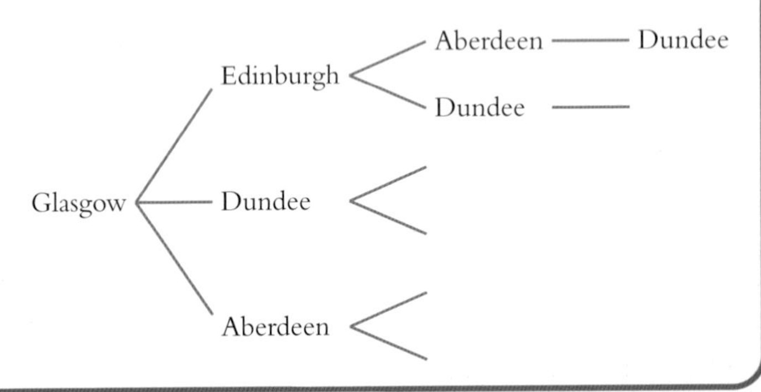

Exercise 1 continued

(A special type of network in which vertices represent the 'starts' and 'finishes' of stages in a job is called an *activity network*. The 'length' of the longest path is the minimum time it will take to complete the job. The longest path is called the critical path.)

11 Anthony wishes to cook a meal of haddock with boiled potatoes and peas. He has four rings on his cooker, so that the different dishes may be cooked at the same time. The network diagram shows how this may be done.

(All times are in minutes.)

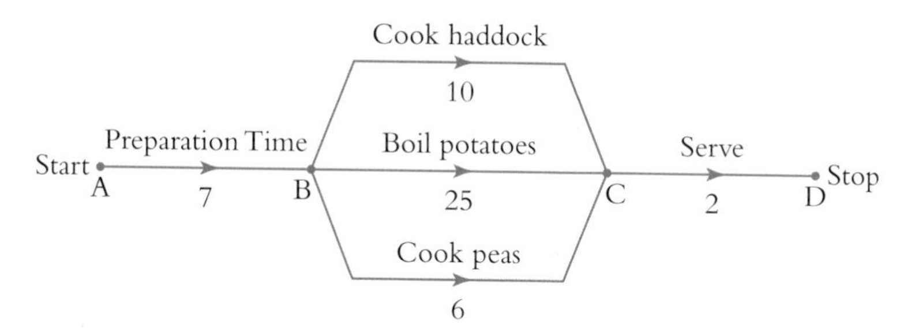

By considering the critical path, find the minimum time needed to complete the meal.

12 The network diagram shows how three people tidy a garden.

All three work at the same time and all times are in minutes.

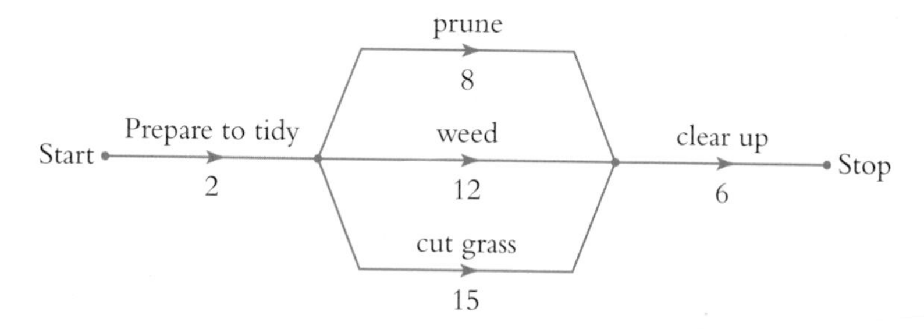

By considering the critical path, find the minimum time required to tidy the garden.

Flowcharts

A flowchart is a useful way of organising a set of instructions so that we may be guided along any of the possible 'routes' of a process. The technique of flowcharting is an essential one in computer programming.

In this section you should be able to use and interpret simple flowcharts.

Main points

A flowchart is a logic diagram with a set of *instructions* which must be followed in order. It contains:

- start and stop boxes
- statement boxes (sometimes in the form of instructions) which are rectangular
  (and possibly)
- decision boxes (with questions to be answered yes or no) which are diamond-shaped.

Example Kathleen sells encyclopaedias. The flowchart shown is used to calculate her monthly salary in pounds.

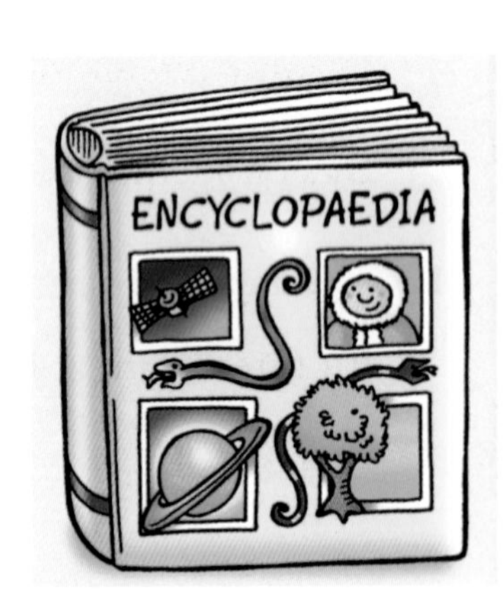

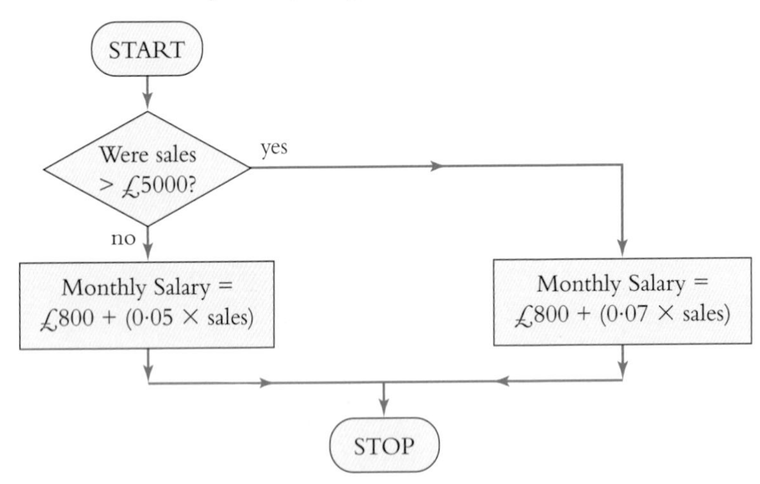

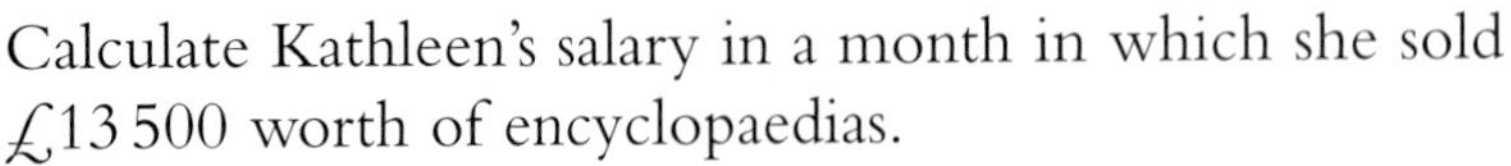

Calculate Kathleen's salary in a month in which she sold £13 500 worth of encyclopaedias.

Solution (Decision Box) 'Were sales > £5000?' Answer: YES.

Hence 'Monthly Salary = £800 + (0·07 × sales)'

Hence monthly salary = £800 + (0·07 × 13 500)
= £800 + (945)
= £1745.

Be careful in flowchart calculations that you perform the operations in the correct order. Note that the multiplication was done before the addition. (Remember 'BOMDAS'!)

Exercise 2

1 Stilian is a salesman. The flowchart shows how his monthly salary may be calculated.

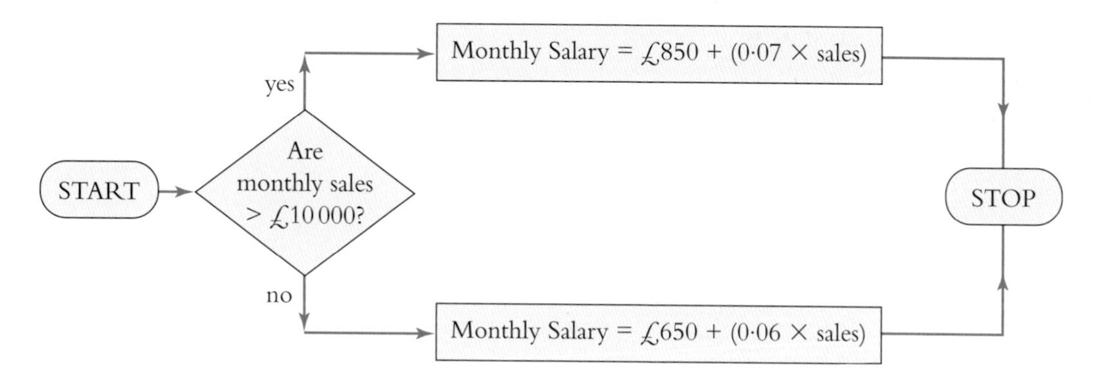

In April, Stilian sold £9000 worth of goods. Calculate his salary for April.

2 Tessa sells cars. The flowchart shows how her monthly salary may be calculated.

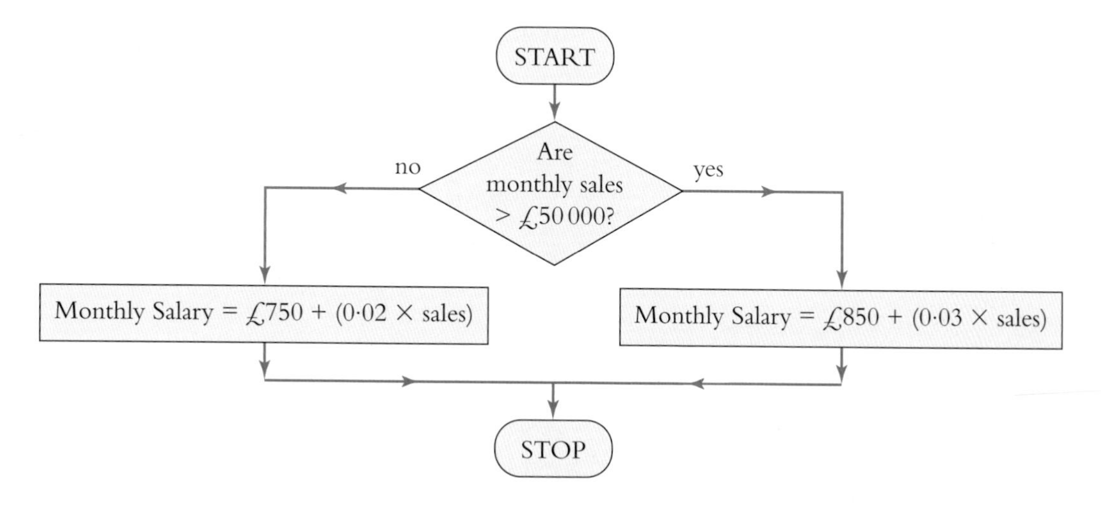

In May, Tessa sold £75 000 worth of cars. Calculate her salary for that month.

Exercise 2 continued

3 A publisher offers discounts to schools on large orders of educational books.

The flowchart shown may be used to calculate the discount in pounds.

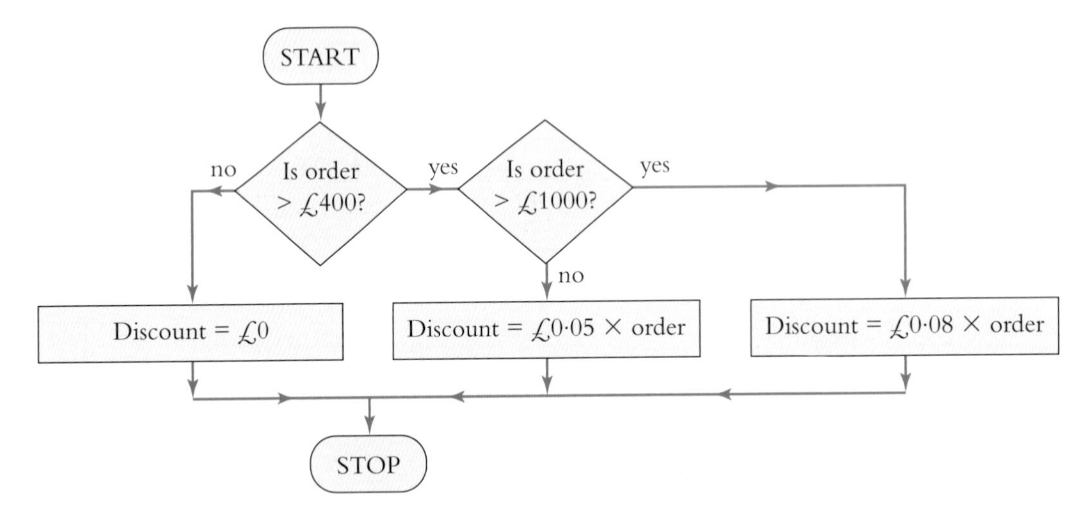

(a) A school orders £800 of educational books from this publisher.

How much will the school pay?

(b) The following month, this school orders a further £600 of books.

How much extra would the school have saved by placing both orders at the same time?

4 A farmer gives a discount to a supermarket on large orders of produce.

The flowchart shown is used to calculate the discount in pounds.

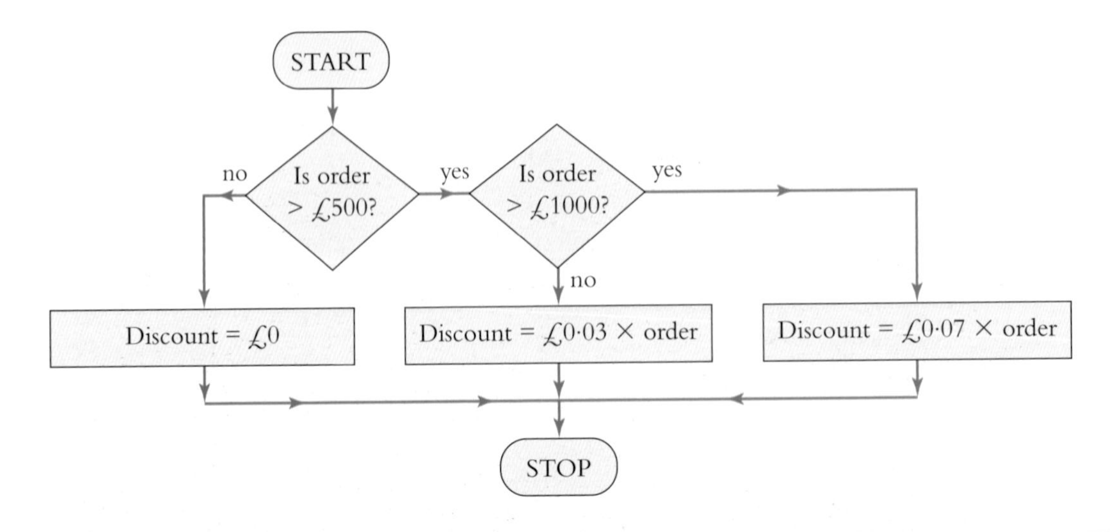

Exercise 2 continued

(a) The supermarket orders £900 of produce from the farmer every week. How much does the supermarket pay each week for this order?

(c) The supermarket then decides to order £1800 of produce once a fortnight instead of placing a £900 order each week. How much does the supermarket save by doing this?

5 This flowchart may be used to find the cost of tickets for a concert. Adults pay full price for their tickets. Reduced prices apply to concession and child tickets.

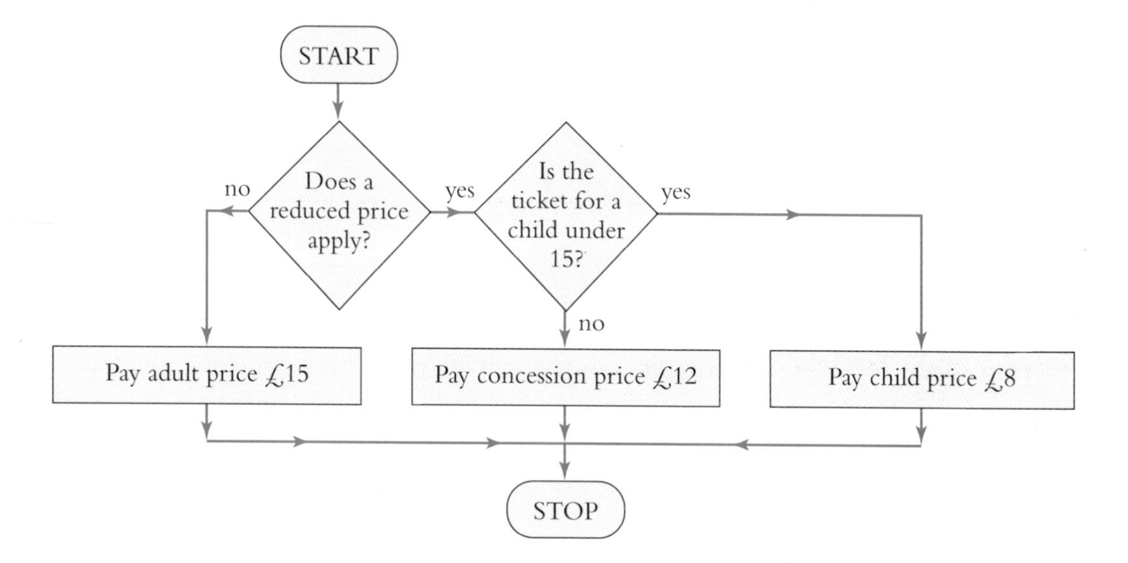

The Britten family buys 3 adult tickets, 2 concession tickets and 1 ticket for a child who is under 15. Find the total cost of the tickets.

Exercise 2 continued

6 Brian works in a woollen mill making jumpers.

The flowchart shown is used to calculate his weekly bonus.

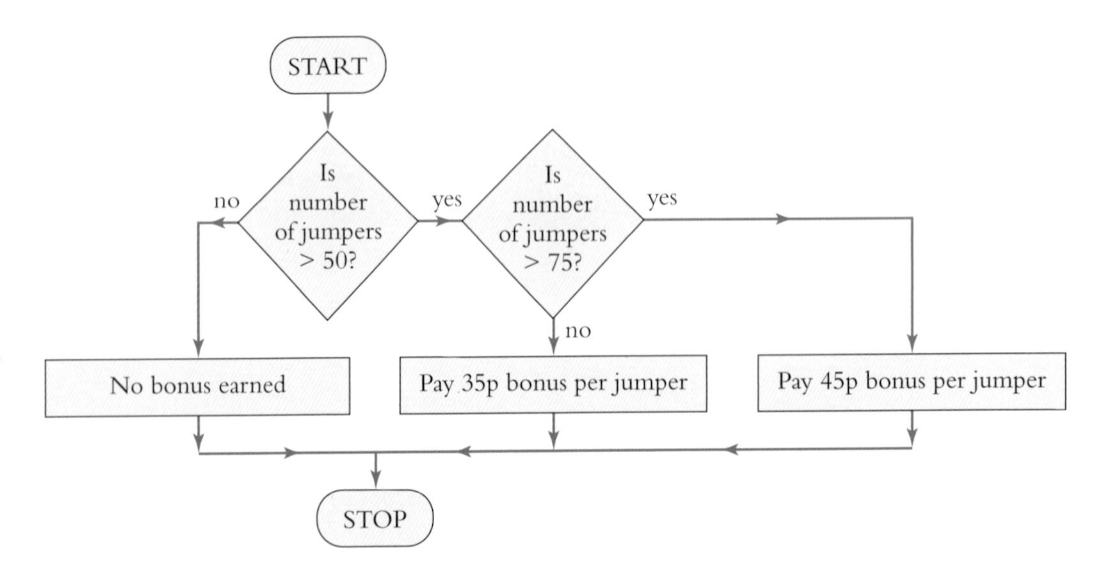

(a) Brian made 69 jumpers during one week. Calculate his bonus for that week.

(b) Brian hopes to make a bonus of £40 one week. Is it possible to make **exactly** £40 in a week? Explain your answer.

7 A *decision tree diagram* displays information and can be used to *sort items* into *different categories*. The decision tree diagram shown on the opposite page describes the life forms on the planet Zob.

(a) What life form on the planet Zob has five legs, two mouths and hair?

(b) What life form has three legs, five eyes and is purple?

(c) Describe a poppet.

Exercise 2 continued

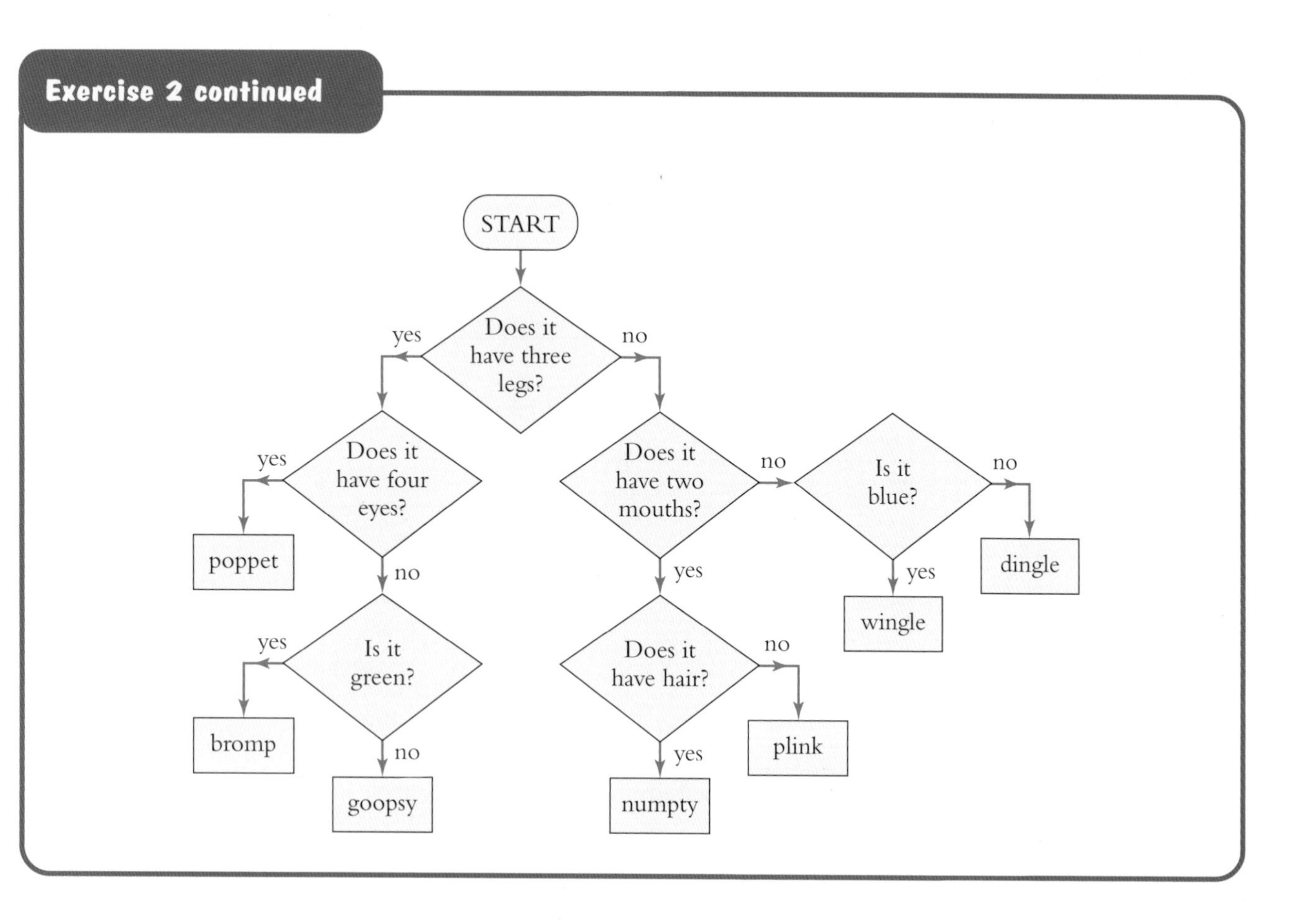

Spreadsheets

A spreadsheet is not so much a logic diagram as a grid of cells. The data entered on a spreadsheet may be analysed and manipulated by a computer however, and diagrams and graphs may be created in turn. Computers perform calculations extremely quickly on information in spreadsheets.

In this section you should be able to:

- enter given data into a spreadsheet
- enter simple formulae into a spreadsheet
- use the SUM function in a spreadsheet
- use the AVERAGE function in a spreadsheet.

Main points

The spreadsheet shown may be used to calculate the gross wage of each worker in a warehouse.

	A	B	C	D	E
1					
2	**First Name**	**Second Name**	**Basic hourly rate**	**Hours worked at basic rate**	**Gross wage**
3	John	O'Connor	£7·00	40	£280·00
4	Mary	Clarke	£7·40	40	£296·00
5	Desmond	Reilly	£8·50	35	£297·50
6	Paul	Curran	£7·00	40	£280·00
7	Rowan	Docherty	£9·60	35	£336·00
8	Connor	Regan	£6·50	30	£195·00
9				**Total gross wages**	£1684·50
10					
11				**Mean (average) gross wage per worker**	£280·75

1 The *rows* of the spreadsheet are identified by *numbers*.

2 The *columns* are identified by letters.

3 Each box in the spreadsheet is called a **cell**.

4 In the spreadsheet shown, the word 'Clarke' is in cell B4 and the number '30' is in cell D8.

5 A calculation may be performed by entering a *formula* into the cell in which you wish an answer to appear.

6 To calculate John O'Connor's gross wage and show the answer in cell E3, the formula **=C3*D3** is entered in cell E3. (This multiplies the number in cell C3 by the number in cell D3 and shows the answer in cell E3.)

(In this work we use the symbol '*' instead of '×' to indicate 'multiply'.)

7 To calculate the total of the gross wages of the six workers and show the result in cell E9, we would enter the formula **=E3+E4+E5+E6+E7+E8** in cell E9. Another, but more concise formula, which carries out the same calculation is **=SUM(E3..E8)**.

8 To calculate the mean (average) gross wage per worker and show the result in cell E11, we would enter the formula **=E9/6** in cell E11. (We use the symbol '/' instead of '÷' to indicate 'divide'.)

Another formula which carries out the same calculation is **=AVERAGE(E3..E8)**.

9 **All** formulae **must** start with an equal sign '='.

Example Mr Clark enters the results of three tests taken by class 5G in a spreadsheet.

	A	B	C	D	E
1	**Class 5G**	**Test results**	**in Maths**		
2					
3		Test 1	Test 2	Test 3	Total
4	Anna	46	47	55	
5	Barbara	53	55	50	
6	Cara	47	55	61	
7	David	65	68	71	
8	Edna	75	70	76	
9	George	60	70	76	
10	Mary	80	83	85	
11	Shaheen	56	65	69	
12	Tariq	56	46	56	
13	Umar	76	83	87	
14	Walter	46	56	66	
15					
16	Average				

(a) The formula = SUM (B4..D4) is entered into cell E4. What value will appear in cell E4?

(b) Mr Clark wants the mean mark for Test 1 to be shown in cell B16. Write down the formula he should enter in cell B16.

Solution (a) = SUM (B4..D4) becomes =B4 + C4 + D4
= 46 + 47 + 55 = 148.

(b) = AVERAGE (B4..B14).

Exercise 3

(Do not use a calculator in questions 1 to 5) 

1 The spreadsheet shows the marks of a group of students in four tests.

	A	B	C	D	E	F
1		Test 1	Test 2	Test 3	Test 4	Total
2	Andrews	57	48	57	50	
3	Brown	62	54	67	65	
4	Caldwell	47	54	56	60	
5	Fletcher	57	67	68	69	
6	Gordon	49	50	58	62	
7	McManus	67	72	75	70	
8	Martin	35	37	45	56	
9	Neilson	61	67	72	65	
10						
11	Average					

(a) What mark appears in cell D7?

(b) In what cell does a mark of 35 appear?

(c) The result of the formula = SUM (B8..E8) is to appear in cell F8. What value will appear in cell F8?

(d) Write down a formula to enter in cell C11 to show the average mark for test 2.

2 The manager of a small firm uses a spreadsheet to calculate the gross wage of each worker.

	A	B	C	D	E
1	**First Name**	**Second Name**	**Basic hourly rate**	**Hours worked at basic rate**	**Gross wage**
2	Jim	Barr	£5·05	40	£202·00
3	Mary	Newman	£6·25	40	
4	Pat	McIntyre	£6·80	35	
5	Dorothy	Shaw	£7·00	36	
6	Jack	Jones	£7·50	40	
7	Martha	Jackson	£10·00	38	
8					
9		**Total gross wage bill for the week =**			

Exercise 3 continued

(Do not use a calculator in questions 1 to 5) 

(a) What appears in cell D7?

(b) In which cell does £6·80 appear?

(c) The result of the formula =C6*D6 is to appear in cell E6. What value will appear in cell E6?

(d) Write down a formula to enter into cell E9 to show the total gross wage bill for the week.

3 A spreadsheet is used to process data for a school tuck shop.

	A	B	C	D	E	F	G
1		**SALES OF GOODS IN TUCK SHOP**					
2							
3		**Crisps**	**Juice**	**Fruit**	**Ices**	**Chocolate**	**Salad**
4	Monday	80	35	25	40	58	13
5	Tuesday	96	37	35	34	50	16
6	Wednesday	76	34	37	45	47	18
7	Thursday	70	40	38	47	56	20
8	Friday	76	43	46	47	53	21
9							
10	Totals						
11	Averages						

(a) On which day was most fruit sold?

(b) In which cell does the number 96 appear?

(c) The result of the formula =SUM(E4..E8) is to appear in cell E10. What value will appear in cell E10?

(d) What formula would be used to enter the average daily sale of ices in cell E11?

Exercise 3 continued

4 A shopkeeper uses a spreadsheet to calculate her daily sales.

(a) The result of the formula =B6*C6 is to appear in cell D6. What value will appear in cell D6?

(b) What formula would be used to enter the total takings for the day in cell D11?

	A	B	C	D
1	**Item**	**Number Sold**	**Cost per item**	**Total Cost**
2				
3	Shirts	25	£30·00	£750·00
4	Trousers	12	£40·00	
5	Jackets	7	£45·00	
6	Ties	26	£11·00	
7	Shoes	14	£50·00	
8	Socks	15	£8·00	
9	Coats	10	£75·00	
10				
11	**Total takings for the day =**			

5 A spreadsheet is used to show the daily takings in a fishmonger's shop.

	A	B	C	D	E	F	G	H
1		**Haddock**	**Salmon**	**Cod**	**Shellfish**	**Poultry**	**Assorted**	**Totals**
2								
3	Tuesday	£78·55	£55·24	£27·45	£18·96	£75·89	£23·56	£279·65
4	Wednesday	£115·79	£46·80	£37·55	£67·90	£117·89	£47·38	
5	Thursday	£120·35	£75·00	£46·00	£65·78	£69·01	£13·34	
6	Friday	£99·67	£57·50	£58·00	£32·21	£123·45	£25·56	
7	Saturday	£87·65	£101·20	£50·00	£39·20	£150·99	£56·75	
8								
9	**Totals**	£502·01						

(a) What is the amount in Cell B4?

(b) The result of the formula =SUM (B3..G3) is shown in cell H3. What formula should be entered in cell H4?

(c) The result of the formula =SUM(D3..D7) is to appear in cell D9. What value will appear in cell D9?

Exercise 3 continued

6 Gordon uses a spreadsheet to record his transactions at the bank.

	A	B	C	D	E
1			Deposit	Withdrawal	Balance
2		**Previous Balance**			£1156·89
3	5.1.07	Telephone		£32·50	£1124·39
4	9.1.07	Gas/Electricity		£72·50	£1051·89
5	17.1.07	Credit card		£258·94	£792·95
6	23.1.07	Cash withdrawal		£250·00	£542·95
7	26.1.07	Salary	£1506·23		
8	29.1.07	Insurance		£42·25	
9					
10	Totals for month				

(a) The formula =E6 + C7 – D8 is entered in cell E8. What value will appear in cell E8?

(b) What formula would be used to enter the total withdrawals for the month in cell D10?

7 A company uses a spreadsheet to calculate its weekly wage bill.

(a) What formula would be used to enter the total weekly wage for grade 9 in cell D12?

(b) What formula would be used to enter the total weekly wage bill in cell D14?

	A	B	C	D
1	Employee	Weekly	Number of	Total Weekly
2	Grade	Wage in £	Employees	Wage in £
3				
4	1	210	7	1470
5	2	225	10	2250
6	3	240	12	2880
7	4	260	13	3380
8	5	285	25	7125
9	6	315	20	6300
10	7	350	14	4900
11	8	390	8	3120
12	9	450	6	
13				
14		Total Weekly Wage Bill		

Exercise 3 continued

8 A football club uses a spreadsheet to calculate its weekly wage bill.

	A	B	C	D	E	F	G
1			**Strathclyde United FC**				
2							
3	Employee		Weekly		Number of		Total Weekly
4	Salary Scale		Wage in £		Employees		Wage in £
5							
6	1		30 000·00		2		60 000·00
7	2		25 000·00		3		75 000·00
8	3		20 000·00		3		60 000·00
9	4		15 000·00		4		60 000·00
10	5		10 000·00		5		50 000·00
11	6		6000·00		7		42 000·00
12	7		4000·00		5		20 000·00
13	8		2000·00		6		12 000·00
14	9		1000·00		5		5000·00
15	10		500·00		13		
16							
17				**Total Weekly Wage Bill**			

(a) The result of the formula =C15*E15 is to appear in cell G15. What value will appear in cell G15?

(b) What formula would be used to enter the total weekly wage bill in cell G17?

(c) What will appear in cell G17?

Exercise 3 continued

9 Arthur sells antiques. He keeps a record of his profits on a spreadsheet.

	A	B	C	D	E
1					
2	**Item**	**Cost Price**	**Selling Price**	**Profit**	**Percentage Profit**
3					
4	Vase	£70·00	£77·00	£7·00	10%
5	Clock	£40·00	£50·00	£10·00	
6	Painting	£300·00	£345·00	£45·00	
7	Cutlery	£75·00	£90·00		
8	Chair	£80·00	£112·00		
9	Table	£250·00	£290·00		
10	China	£150·00	£168·00		

(a) The result of the formula =C8 – B8 is to appear in cell D8. What value will appear in cell D8?

(b) The result of the formula =D5/B5*100 is to appear in cell E5. What value will appear in cell E5?

(c) The result of the formula =C9 – B9 is to appear in cell D9. What value will appear in cell D9?

(d) The result of the formula =D9/B9*100 is to appear in cell E9. What value will appear in cell E9?

Exercise 3 continued

10 Mr McCallum records the results of three class tests on a spreadsheet.

	A	B	C	D	E	F
1	**Class 4H - Test Results**		**Teacher: Mr McCallum**			
2						
3		Test 1	Test 2	Test 3	Total	Percentage
4	Possible Mark	50	65	85		
5	Student					
6						
7	Clarke	38	50	66		
8	Di Marco	30	41	53		
9	Donald	35	46	57		
10	Henry	37	48	62		
11	Howell	29	35	60		
12	Johnson	26	32	44		
13	Montgomery	43	56	71		
14	Taylor	24	33	46		
15	Woods	32	43	59		
16						
17	Average					

(a) What formula should he enter in cell B17 to calculate the average for test 1?

(b) What formula should he enter in cell E4 to calculate the total possible mark for all three tests?

(c) What will appear in cell E4?

(d) The result of the formula =SUM(B7..D7) is to appear in cell E7. What value will appear in cell E7?

(f) The result of the formula =E7/2 is to appear in cell F7. What value will appear in cell F7?

Exercise 4 – Revision of Chapter 15

1 Study the network diagram shown.

(a) How many nodes are there?

(b) How many arcs are there?

(c) Write down the degree of each node.

(d) Which nodes are odd?

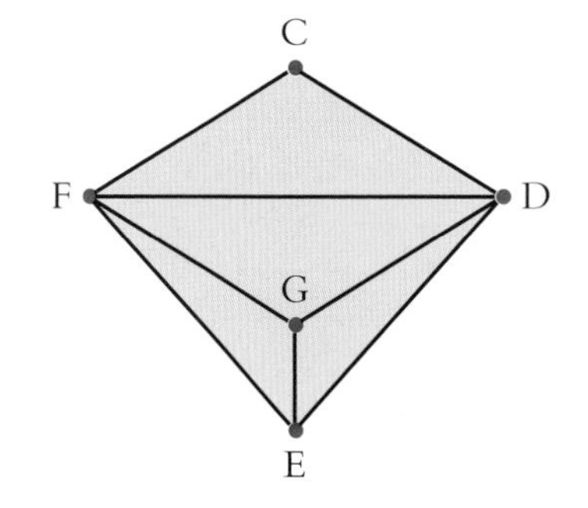

2 This network diagram shows the distance, in kilometres, between several towns.

(a) State the order of the node at Garrison.

(b) What is the length of the shortest route from Forthdee to Bankside?

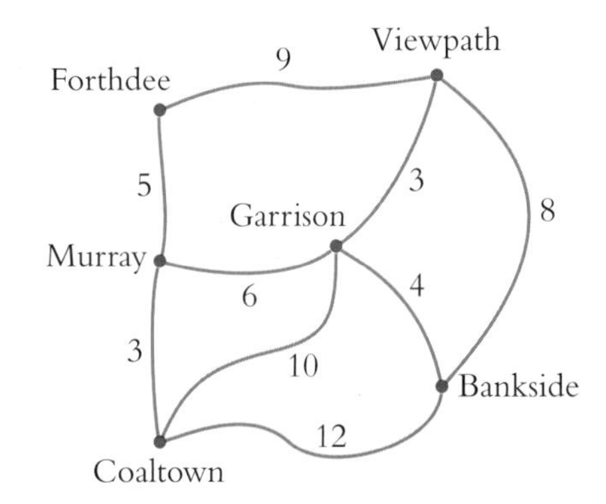

3 James works for an insurance company.

The flowchart shown is used to work out his monthly salary in pounds.

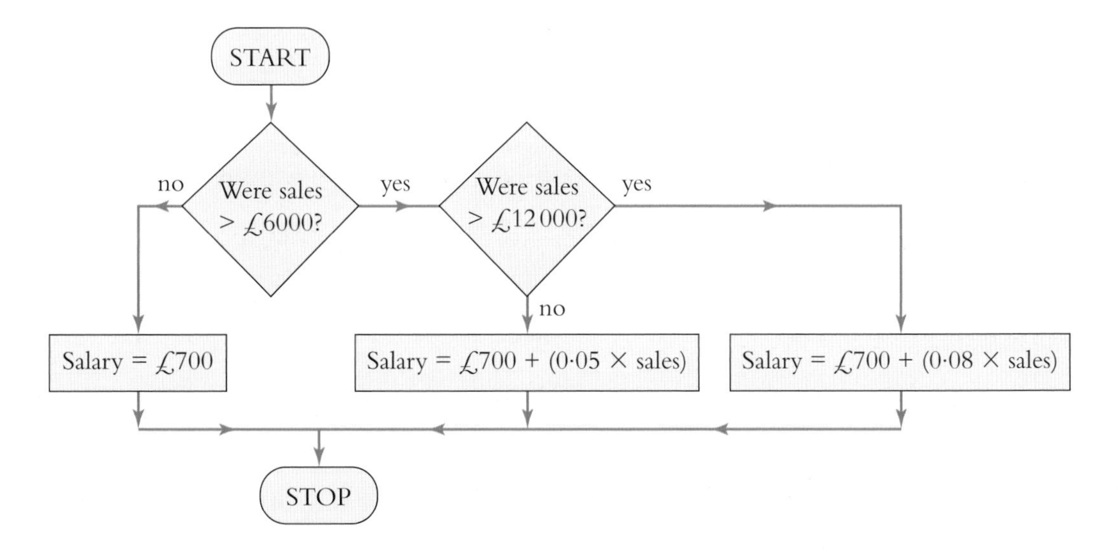

James sells insurance to the value of £15 000 in March. Calculate his salary for the month.

Exercise 4 – Revision of Chapter 15 continued

4 The flowchart shows how to calculate interest when a sum of money is deposited with the Bank of Alba for one year.

START → Is the amount ≤ £1000? — yes → Interest = 4·2% of amount → STOP

no → Is the amount ≤ £15 000? — yes → Interest = 4·4% of amount → STOP

no → Interest = 4·7% of amount → STOP

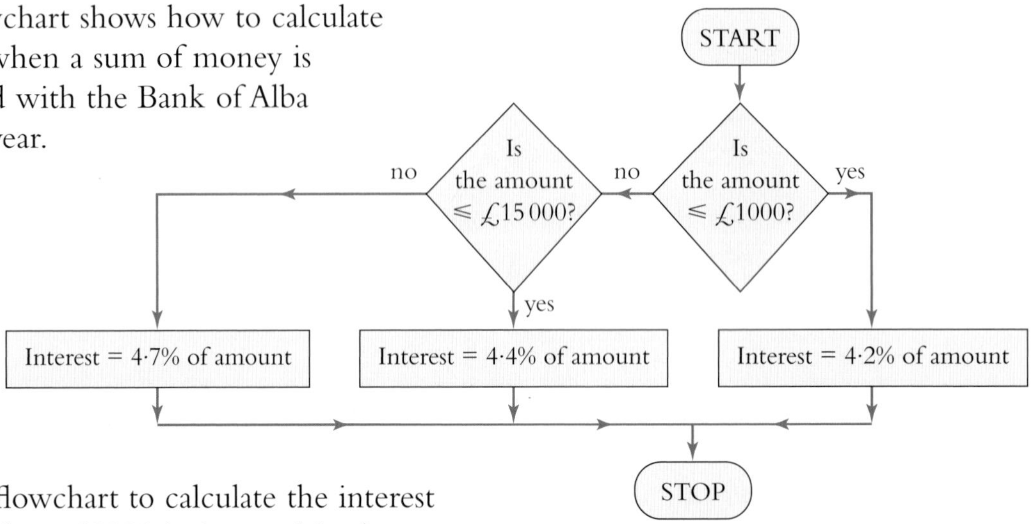

Use the flowchart to calculate the interest earned when £8500 in invested in the Bank of Alba for one year.

5 Marjorie uses a spreadsheet to keep track of her standing orders.

	A	B	C	D	E	F
1		**Standing Orders: payments from my bank account**				
2						
3						
4			**Amount of payment**	**Number of payments per year**	**Total amount paid per year**	
5						
6	**Electricity**		£32·00	12	£384·00	
7	**Gas**		£15·50	12	£186·00	
8	**Telephone**		£32·50	12		
9	**Insurance**		£176·08	1	£176·08	
10	**TV licence**		£126·50	1	£126·50	
11	**Magazine Subscription**		£29·50	1	£29·50	
12	**Council Tax**		£191·00	10	£1910·00	
13						
14		**Total payments per year**				

Exercise 4 – Revision of Chapter 15 continued

(a) The result of the formula =C8*D8 is to appear in cell E8. What value will appear in cell E8?

(b) Write down the formula to be entered in cell E14 to find the total payments for the year.

(c) What value will appear in cell E14?

6 The numbers of matches in a sample of matchboxes are counted.

The results are entered in a spreadsheet.

(a) The result of the formula =SUM(B5..B11) is to appear in cell B13. What value will appear in cell B13?

(b) The result of the formula =SUM(C5..C11) is to appear in cell C13. What value will appear in cell C13?

(c) The formula =C13/B13 is used to show the mean number of matches per box in cell C15. What value will appear in cell C15?

	A	B	C	D
1	**The numbers of matches in a sample of matchboxes**			
2				
3	**number of matches**	**frequency**	**number of matches × frequency**	
4				
5	45	1	45	
6	46	9	414	
7	47	21	987	
8	48	35	1680	
9	49	22	1078	
10	50	10	500	
11	51	2	102	
12				
13	**Totals**			
14				
15	**Mean number of matches per box**			

Summary

(Before leaving this chapter look at this summary of items covered in the chapter. They are relatively straightforward, and should present little difficulty, provided, as usual, you *think* what you are doing.)

1 ***Network Diagrams***

A network diagram consists of nodes (or vertices) joined by arcs (or edges).

The diagram has 5 nodes (at A, B, C, D and E). They are connected by 8 arcs. The nodes at A, B, C, and D are *odd* nodes (of degree 3). The node at E is an *even* node (of degree 4).

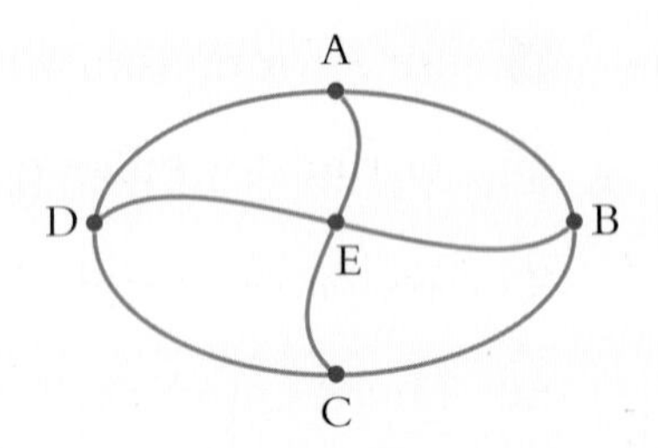

2 ***Flowcharts***

Flowcharts consist of instruction boxes, statement boxes and decision boxes.

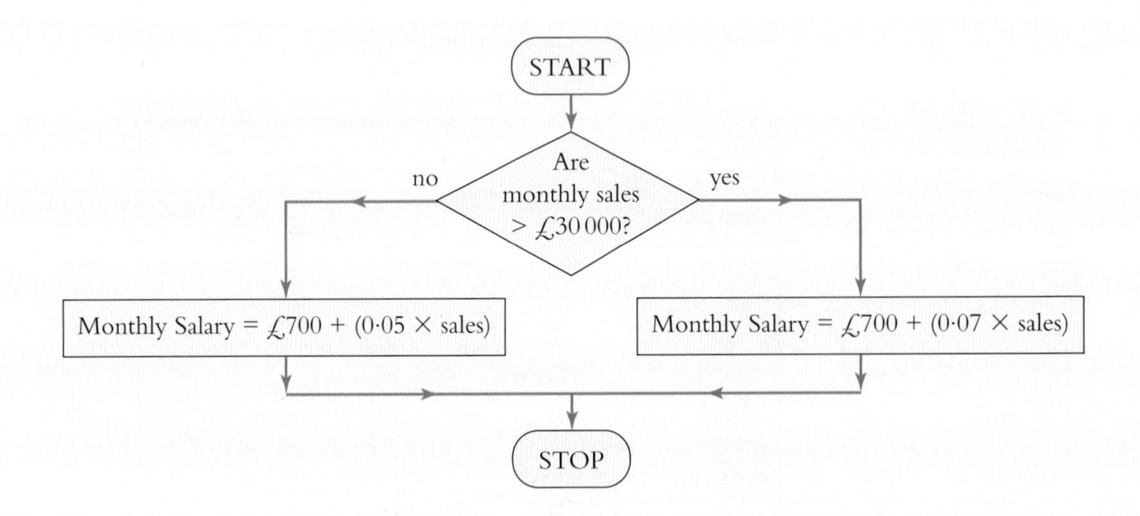

Using this flowchart you should be able to show that if Danny sold £38 000 worth of cars in a month, then his monthly salary would be £3360.

Summary continued

3 *Spreadsheets*

Spreadsheets are used to display, organise, and analyse data.

In a spreadsheet:

- use * for multiplication
- use / for division
- start **all** formulae with an equal sign (=)
- use SUM to add several cells, such as = SUM (A2..A5)
- use AVERAGE to find the mean of several cells such as = AVERAGE (A2..A5).

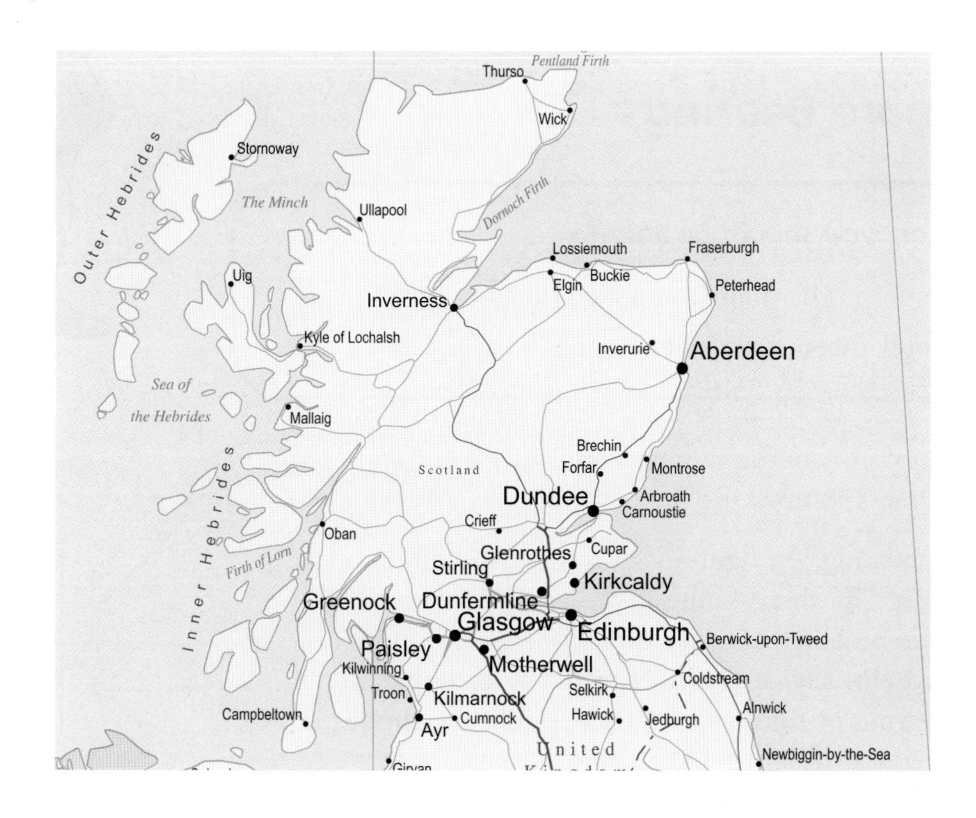

16 Applications of Geometry

In Chapter 2 ('Basic Geometry') we looked at *plane geometry* and calculated areas of rectangles, squares, triangles, and circles. In this chapter we move into *solid geometry*, and calculate the surface areas of some three-dimensional shapes. We also look at the geometry of maps and charts by considering three-figure bearings and scale drawings.

Three-Figure Bearings

In this section, you should be able to

- **identify the eight main points of the compass**
- **understand three-figure bearings.**

The eight main points of the compass are shown in the diagram, each one 45° from the next.

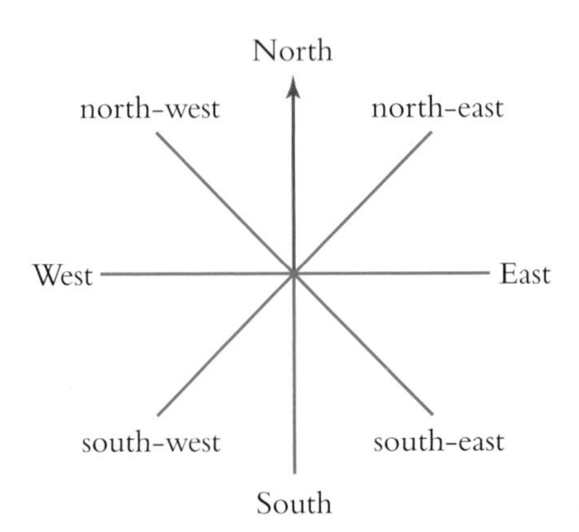

A three-figure bearing is a figure used to describe direction relative to North. The three-figure bearing of North is 000°. The three-figure bearings of other directions are given by angles measured **clockwise from North**. For example, the three-figure bearing of north-east is 045°, while the three-figure bearing of south-east is 135°. (Note the use of the zero in the bearing of north-east).

Example What is the three-figure bearing of south-west?

Solution If you turn in a clockwise direction from North, the angle to south-west is 225° (= 180 + 45). Therefore the three-figure bearing of south-west is 225°.

Example Using a protractor, find the bearing of each of the directions indicated in the diagram.

(a)

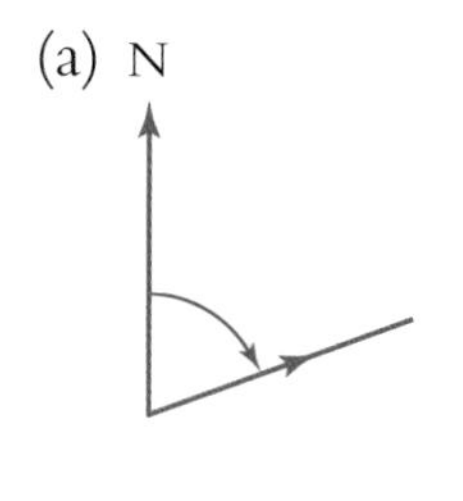

(b)

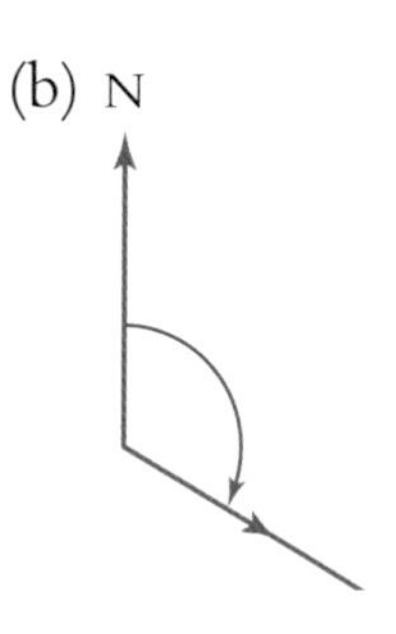

(c)

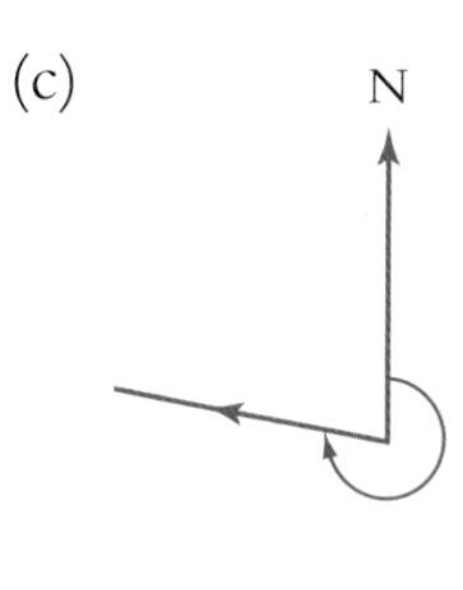

Solution

(a) 070°

(b) 120°

(c) 280°.

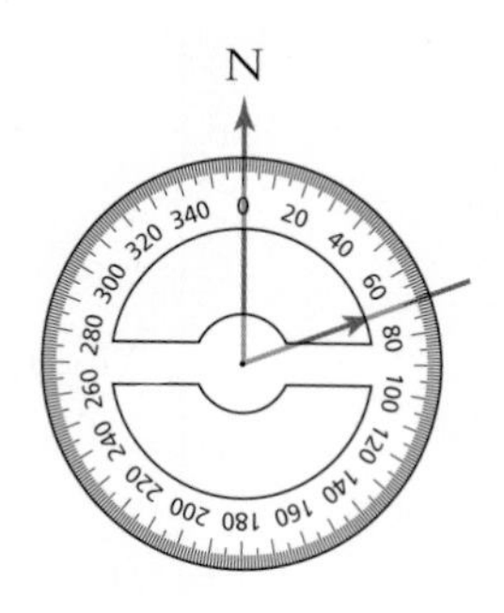

Example Indicate in a diagram a direction with a bearing of 200°.

Solution Choose a point P on the page. Draw a line North, and then measure 200° clockwise from it. Clockwise from north, 180° takes us to South, then a further 20° (clockwise) to the bearing of 200°.

(Also, since 200° = (360 – 160)°, we could measure 160° *anticlockwise* from North to show the correct bearing.)

(i)

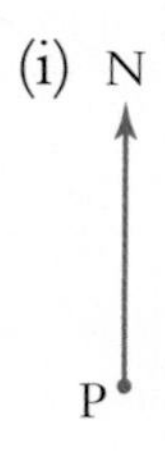

(ii)

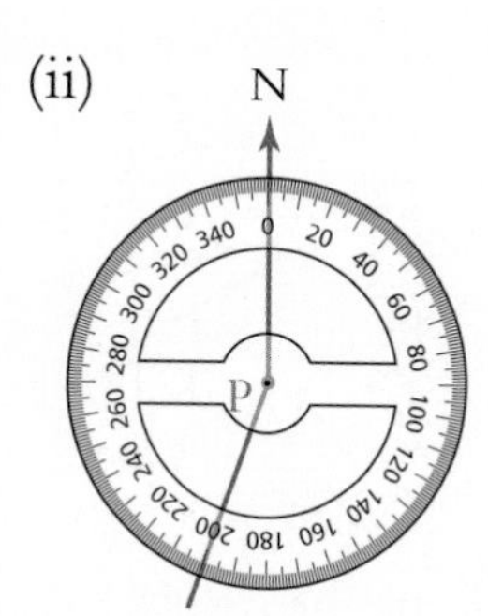

(iii)

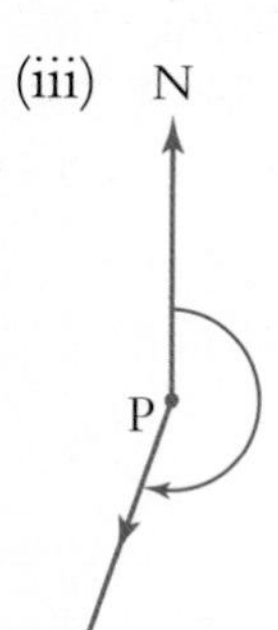

Exercise 1

1 Suppose I am facing north-east and I make a half turn. What direction am I facing now?

2 Suppose I am facing West and I turn 45° clockwise. What direction am I facing now?

3 Suppose I am facing south-east and I turn 135° anticlockwise. What direction am I facing now?

4 A map of Buccaneer Island is shown in the diagram. Some landmarks are shown on the island. The cove is at position (3, 2).

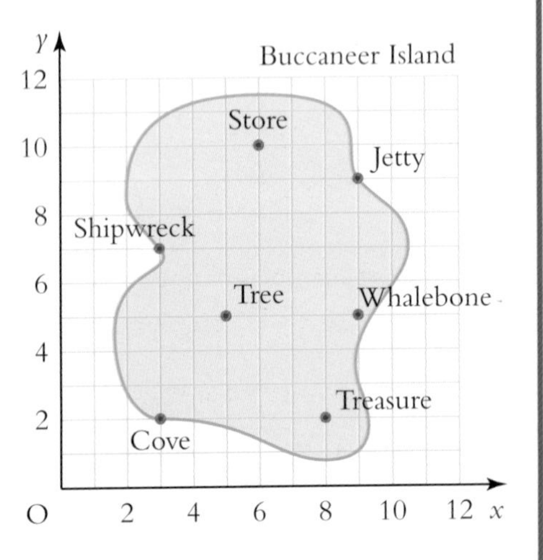

(a) What landmark is South of the jetty?

(b) What landmarks are south-east of the shipwreck?

(c) Captain Pugh walks south-west from the jetty. Where will he come to?

(d) In what direction must you walk from the cove to reach the shipwreck?

5 North has a three-figure bearing of 000°. Write down the three-figure bearings of:

(a) East (b) south-east (c) South

(d) West (e) north-west.

6 Using a protractor, measure the three-figure bearing of each direction from point P.

(a) N P

(b)

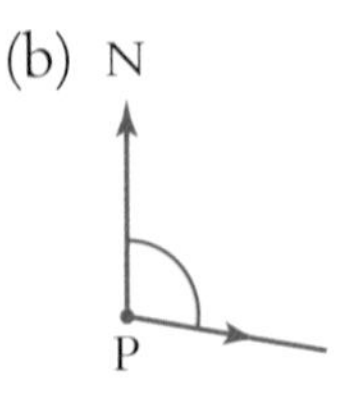

(c)

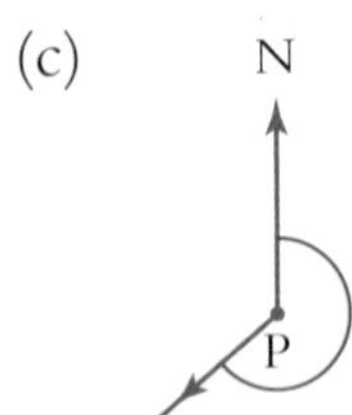

(d)

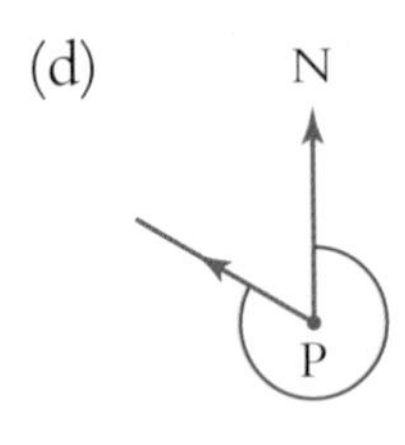

7 In separate diagrams show the following three-figure bearings:

(a) 050° (b) 110° (c) 240° (d) 330°.

Exercise 1 continued

8 Rachel looks at the radar screen on a warship.

The centre of the screen represents the position of the warship.

P and Q show the positions of two submarines.

Submarine P is on a bearing of 120° from the warship.

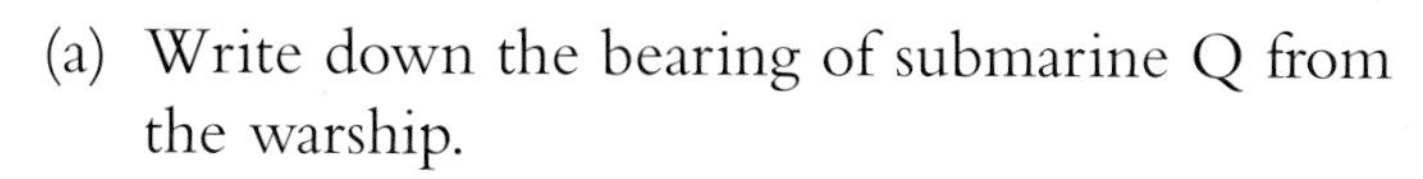

(a) Write down the bearing of submarine Q from the warship.

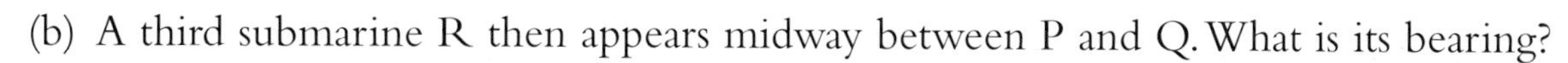

(b) A third submarine R then appears midway between P and Q. What is its bearing?

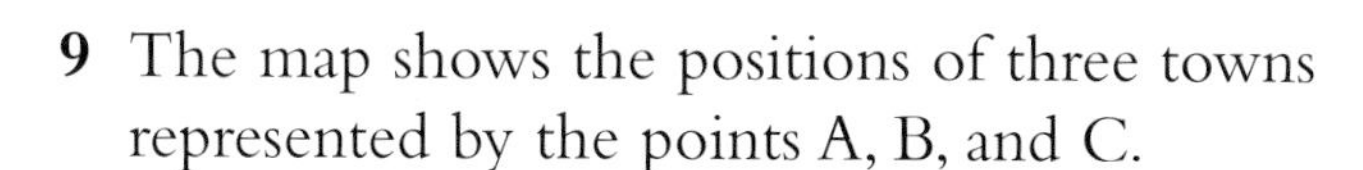

9 The map shows the positions of three towns represented by the points A, B, and C.

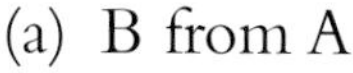

Measure the bearing of (a) B from A

(b) C from B

(c) C from A.

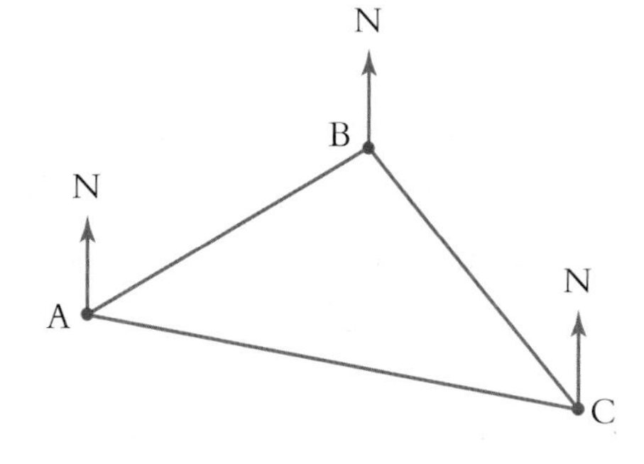

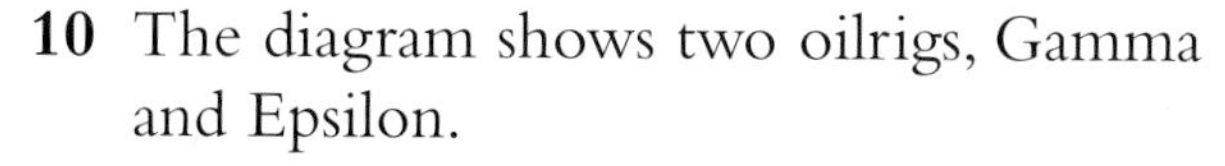

10 The diagram shows two oilrigs, Gamma and Epsilon.

The bearing of Epsilon from Gamma is 130° as shown.

Find the bearing of Gamma from Epsilon.

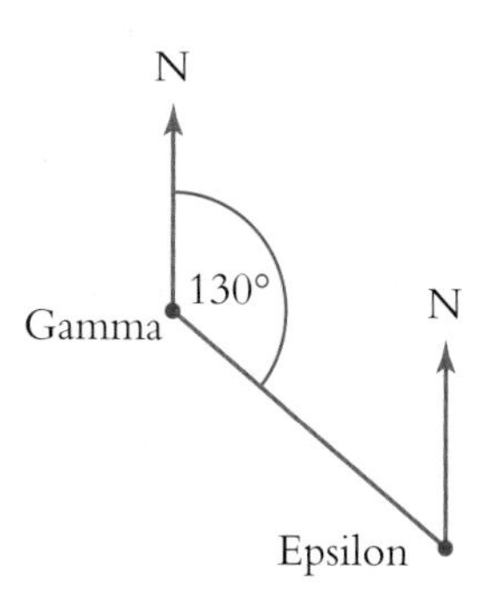

Scale Drawings

Being able to interpret a scale drawing is a valuable skill which you may have to use throughout life, as well as in this section of the course.

Understanding Scale

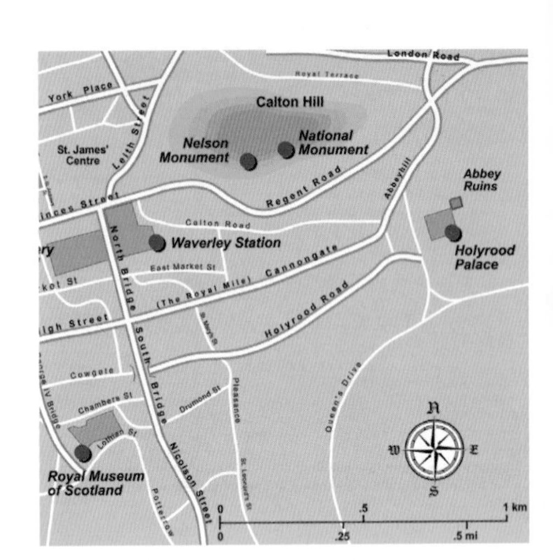

When a map is published, a **scale** is always included so that anyone using the map can find the **actual** distance between two places shown on the map. A scale is also shown in architects' plans. For example, if the scale of a map is '1 centimetre represents 100 metres', then every centimetre measured on the map will equal 100 metres of 'ground', so the actual distance between two points 5 centimetres apart on the map will be 500 m. (We multiply the map distance by the scale factor.)

The scale '1 centimetre represents 100 metres' can be written as '1 cm to 100 m'.

Suppose two towns which are 80 kilometres apart appear 8 centimetres apart on a map. Then the scale would be '1 centimetre represents 10 km' or '1 cm to 10 km.' (We divide the actual distance by the map distance.)

The scaled ruler

Consider the scale '1 centimetre represents 10 kilometres'.

This is sometimes shown on a map or plan as part of a 'ruler'.

0 10 20 30 40 50 km

On this 'ruler', 1 cm = 10 km, 2 cm = 20 km, and so on.

Example The scale drawing shows the positions of two ships represented by points A and B.

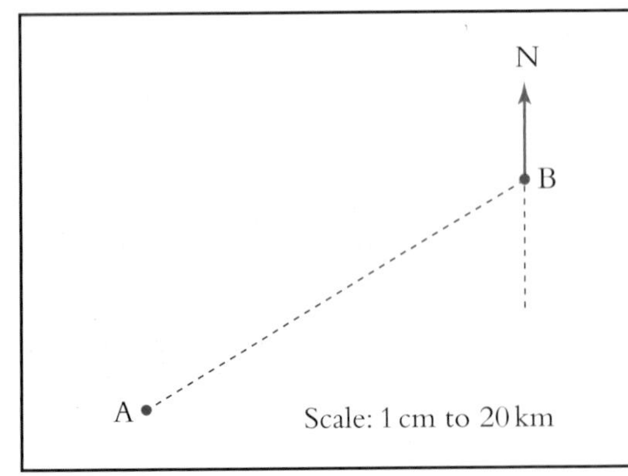

Use the scale drawing to find the distance and bearing of A from B.

Solution By measurement, AB = 5 cm.

Since 1 cm = 20 km, then 5 cm = 5 × 20 = 100 km.

Hence the actual distance AB = 100 km.

By measurement, the bearing of A from B = (180 + 60)° = 240°.

Exercise 2

1 For a scale of '1 cm represents 100 m', what actual distances do the following measurements on a map represent?

(a) 4 cm (b) 6 cm (c) 7·5 cm (d) 2·8 cm.

2 The scale of a map is '1 cm represents 30 km'. On the map two towns are measured to be 8 cm apart. What is the actual distance between the two towns?

3 The lines shown represent distances on a plan. What actual distance does each line represent if the plan has been drawn to a scale of '1 cm to 50 metres'?

(a) ____________

(b) _____________________

(c) ________

4 Two cities are 100 kilometres apart. On a map, the distance between them is 5 cm. Find the scale of the map.

5 The distance from Glasgow to Edinburgh is 70 km. On one map, the two cities appear 7 cm apart. Find the scale of this map.

6 This is a scale drawing of a sideboard.

(a) Measure the width of the sideboard in the drawing.

(b) The scale of the drawing is '1 cm represents 20 cm'. Hence find the actual width of the sideboard.

Exercise 2 continued

7 This map shows the positions of some landmarks in a town.

(a) What is the direction of the Church from the Library?

(b) Measure the distance from the Town Hall to the Cinema on the map.

(c) The scale of the map is '1 centimetre represents 100 metres'. Hence calculate the actual distance from the Town Hall to the Cinema.

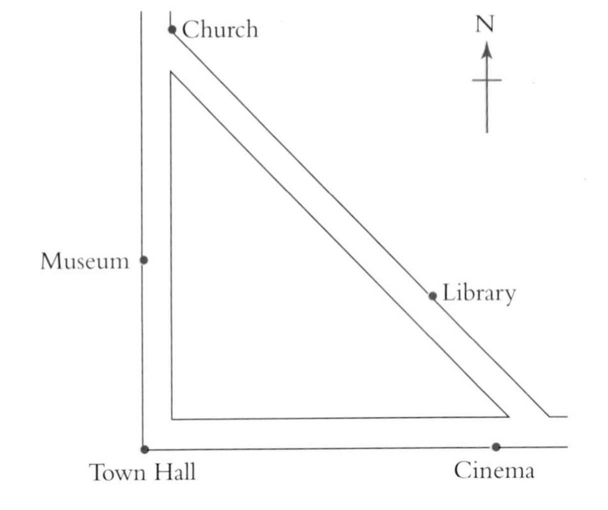

8 The scale drawing shows the positions of two boats represented by points P and Q.

Use this scale drawing to find the distance and bearing of Q from P.

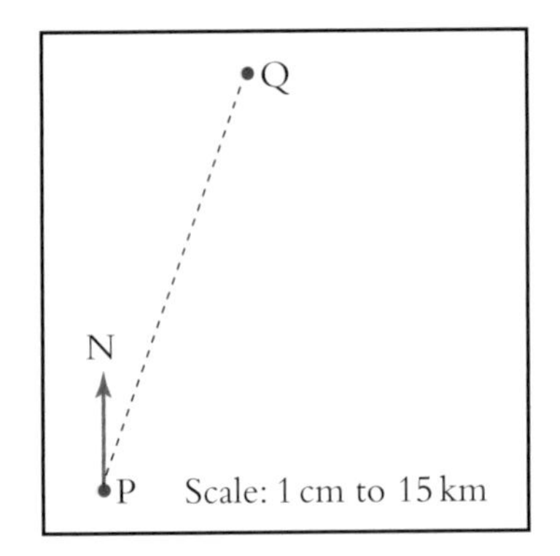

9 The scale drawing shows the positions of two towns, X and Y.

Use this scale drawing to find the distance and bearing of X from Y.

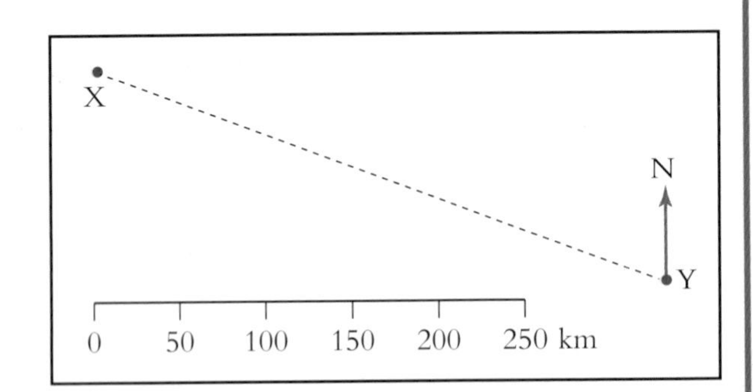

10 On a map two cities are represented by points D and E.

The actual distance between D and E is 250 km. Find the scale of the map.

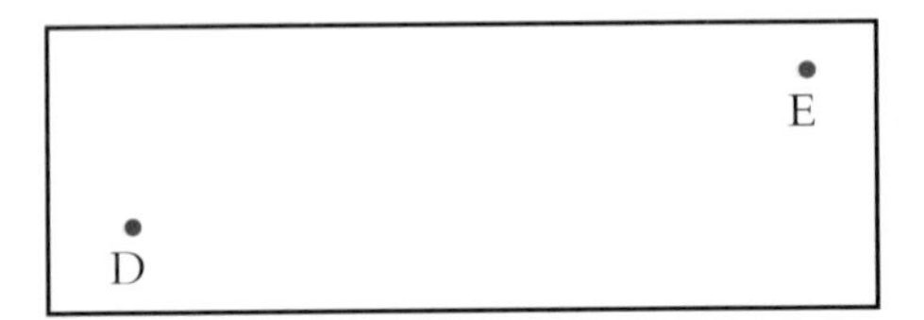

Expressing Scale by a ratio

A scale may be written as a ratio, as well as in words and numbers. For example:

'1 cm represents 50 cm' may be written as '1 : 50'.

However, '1 cm represents 5 m' must first be changed to '1 cm represents 500 cm' before it can be written as '1 : 500'.

A scale of 1 : 500 means that the actual distances are 500 times larger than those shown on the map.

Example The scale of a map is 1 : 2000. On this map, two buildings are 7 cm apart. What is the actual distance between the buildings in metres?

Solution Actual distance $= 7\,\text{cm} \times 2000 = 14\,000\,\text{cm}$
$= (14\,000 \div 100)\text{m} = 140\,\text{m}.$

Exercise 3

1 The scale of a map is 1 : 1000. On this map, two churches are 6 cm apart. What is the actual distance between the churches? (Give your answer in metres.)

2 A map has a scale of 1 : 1 000 000. Two cities are shown 8 cm apart on this map. What is the actual distance between them? (Give your answer in kilometres.)

3 A map has a scale of 1 : 2000. On this map 2 towns appear 6·5 cm apart. What is the actual distance between the towns? (Give your answer in metres.)

4 The scale of a map is 1 : 500. On this map two landmarks are measured to be 12 cm apart. How far apart are the landmarks really? (Give your answer in metres.)

5 A map has a scale of 1 : 500 000. Two cities are shown 4·8 cm apart on this map. What is the actual distance between the two cities? (Give your answer in kilometres.)

Constructing Scale Drawings

As well as being able to interpret scale drawings, you also have to be able to construct them. This too is a valuable skill for life!

Key words and definitions

In the construction of some scale drawings, the terms 'angle of elevation' and 'angle of depression' may have to be taken into account.

The **angle of elevation** is the angle measured above the horizontal that an observer must look **up** to see an object.

The **angle of depression** is the angle measured below the horizontal that an observer must look **down** to see an object.

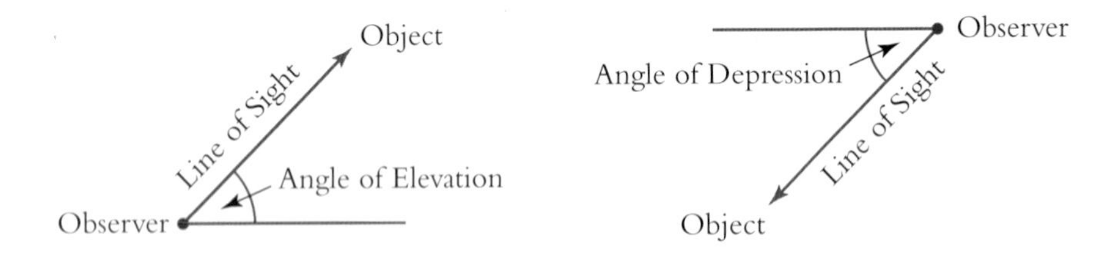

Example Two observers at positions A and B are watching a weather balloon.

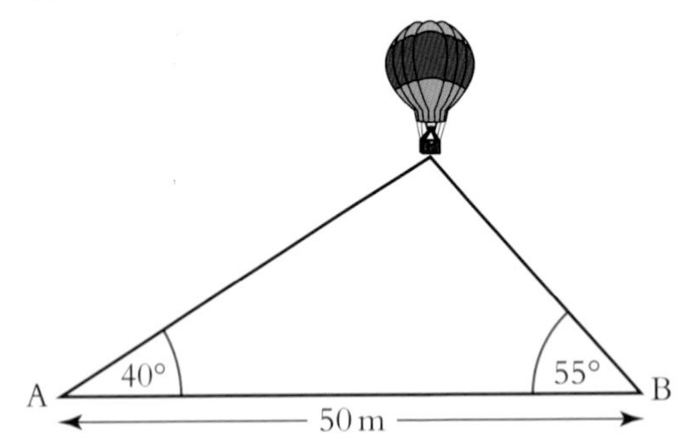

The distance between A and B is 50 metres. The angle of elevation of the balloon is 40° from A and 55° from B.

(a) Make a scale drawing to show the position of the balloon. (Use a scale of 1 cm to 5 m.)

(b) Use your scale drawing to find the actual height of the balloon.

Solution

(a) Since: 5 m = 1 cm, then 50 m = 50 ÷ 5 cm = 10 cm.

1 First we draw a base line of 10 cm to represent 50 m.

2 Then we use a protractor to measure the angles of 40° and 55°.

3 Next we draw the 'lines of sight'.

4 We extend these lines until they meet. This is position of balloon.

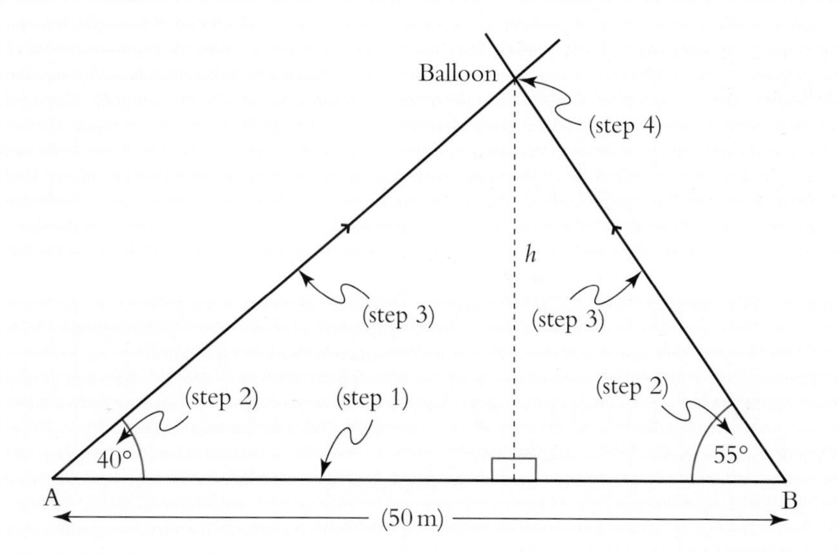

(b) The dotted line shows the height, h m, of the balloon.

On the drawing, $h = 5{\cdot}3$ cm, hence the

actual height of balloon = $(5{\cdot}3 \times 5)$m = 26·5 m.

Example (a) A ship sails from port P for 80 kilometres and on a bearing of 070° to port Q.
Make a scale drawing to illustrate this using a scale of 1 cm to 10 km.

(b) A second ship S is on a bearing of 045° from P and on a bearing of 310° from Q.

Complete the scale drawing to show the position of S.

Solution

(a) Since 10 km = 1 cm, then 80 km = 80 ÷ 10 cm = 8 cm.

First we mark a point for P, draw a north line, and measure 70° clockwise.

Then we draw a line 8 cm long in this direction and mark point Q.

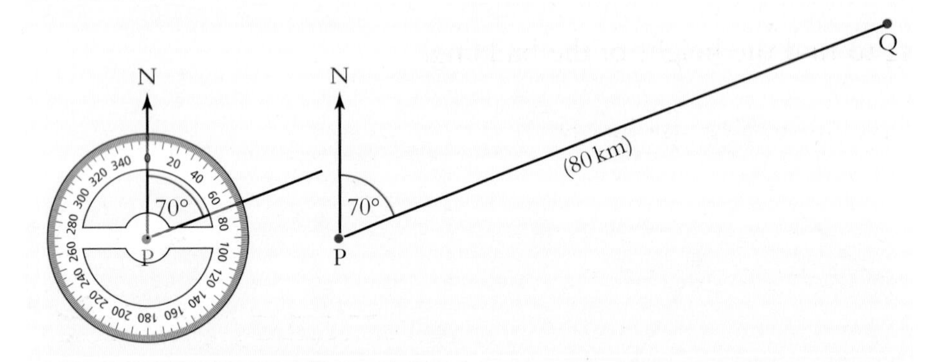

(b) **1** First we measure clockwise 45° from north at P and extend the line.

2 Next we measure clockwise 310° from north at Q and extend the line. (We actually measure 50° anticlockwise from north at Q.)

3 Where the two lines cross then indicates the position of the second ship S.

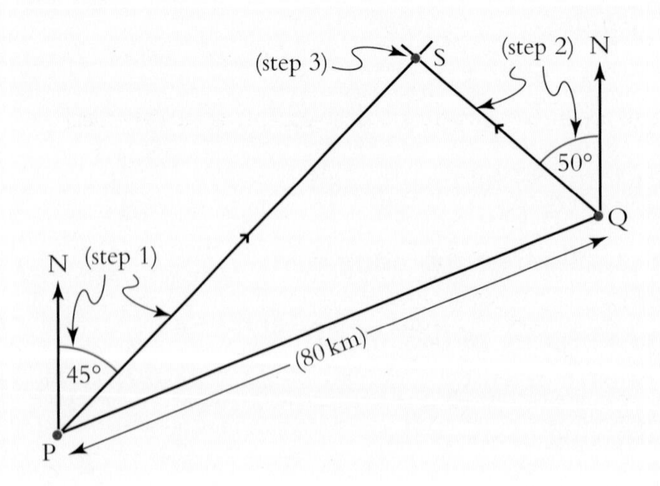

Exercise 4

1 A surveyor stands 80 metres from the foot of a building and measures an angle of elevation to the top of the building as 25°.

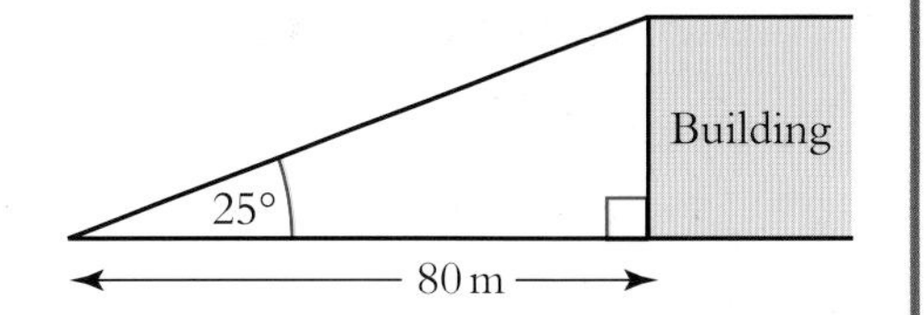

Use a scale of 1 cm to 10 m, and make a scale drawing.

Use your scale drawing to find the height of the building.

2 Alana is using a scale drawing to find the height of a tree.

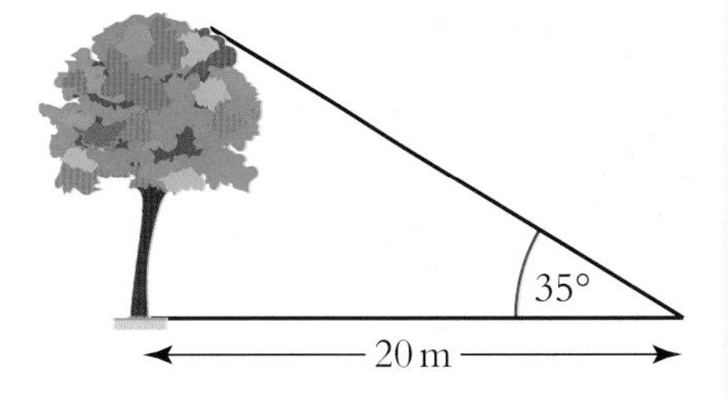

Use a scale of 1 cm to 2 m to produce an accurate drawing and hence find the height of the tree.

3 A ship leaves Aberdeen and sails 50 kilometres East. It then sails 35 kilometres South, and finally 15 kilometres West to a point P.

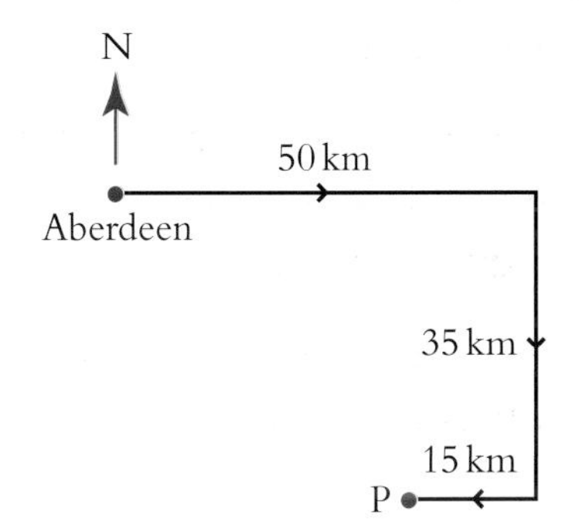

(a) Make a scale drawing to show the ship's journey. (Use a scale of 1 cm to 5 km.)

(b) The ship now sails back to Aberdeen from P. Find the bearing of this return journey.

(c) Calculate the total length of the ship's journey.

4 From the top of a 50 metre high cliff, the angle of depression to a moored buoy is 35°.

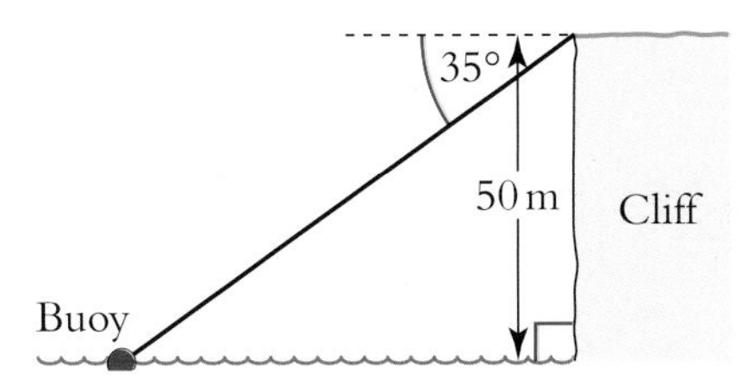

Using a scale of 1 cm to 5 m, draw an accurate diagram and hence find how far the buoy is from the foot of the cliff.

Exercise 4 continued

5 A flagpole is supported from the ground by wires at X and Y.

The distance from X to Y is 40 metres.

The angle of elevation of the top of the flagpole is 50° from X and 40° from Y.

Use a scale of 1 cm = 5 m to draw a diagram and hence find the height of the flagpole.

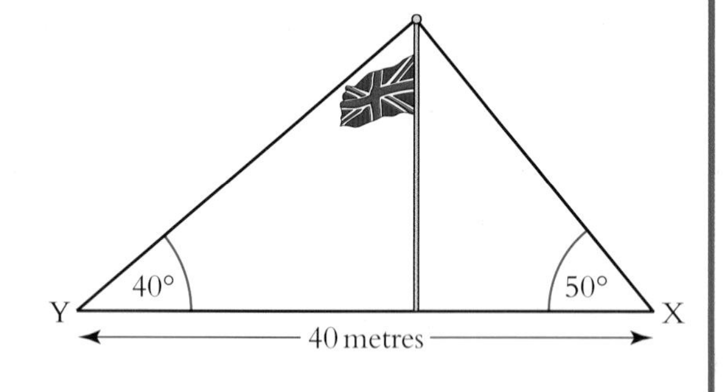

6 A triangular sail on a yacht is as shown.

(a) Make a scale drawing of this sail using a scale of 1 cm to 50 cm.

(b) Hence find the length of the lower edge of the sail PQ.

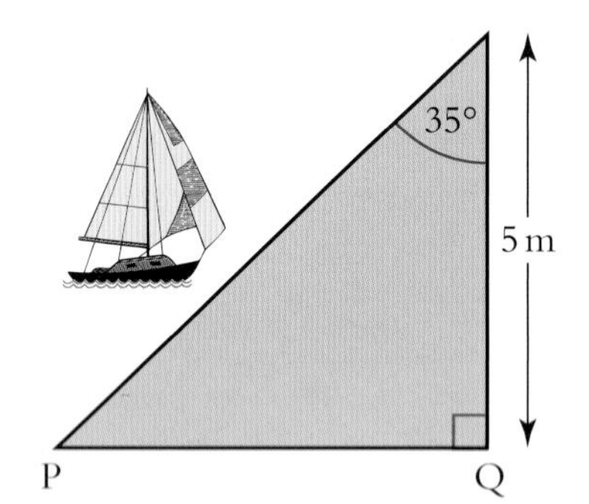

7 A sketch of a piece of metal is as shown.

Using a scale of 1 cm to 2 cm, draw an accurate diagram and hence find the length of the edge AB.

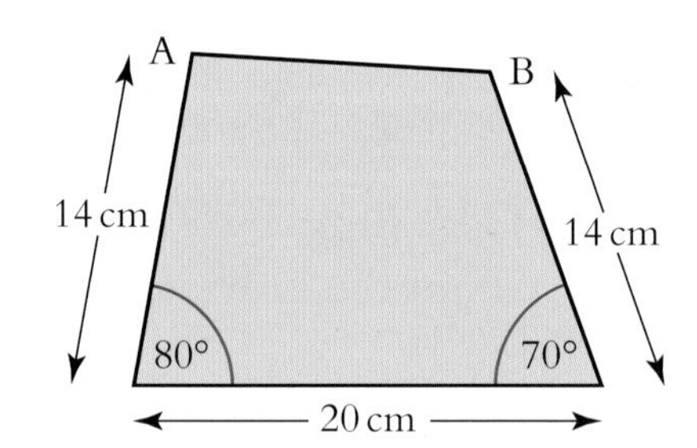

8 A ship leaves port P and sails for 50 km on a bearing 030° to port Q.

From Q, it then sails 40 km on a bearing of 100° to port R.

(a) Show the ship's journey on a scale diagram using a scale of 1 cm to 10 km.

(b) Hence find the distance from R back to P.

9 The town of Knox is 60 km North of the town of Laird.

(a) Show this on a scale drawing using a scale of 1 cm = 10 km.

(b) The town of Middleton has a bearing of 120° from Knox, and 035° from Laird. Complete your scale drawing to show the position of Middleton.

Exercise 4 continued

10 Station S is 50 km West of station T.

(a) Show this on a scale drawing using a scale of 1 cm to 5 km.

(b) Station R has a bearing of 150° from S and 230° from T. Complete your scale drawing to show the position of station R.

11 (a) An oil rig A is 80 km north-east of a second oil rig B. Show this on a scale drawing. Choose a suitable scale.

(b) A supply ship is on a bearing of 030° from oil rig B and on a bearing of 320° from oil rig A. Complete your scale drawing to show the position of the ship.

The Surface Areas of Prisms

A prism is a solid geometric figure whose two ends are similar, equal, and parallel. The prism might be triangular, cylindrical, or square, or any of several other shapes. For the remainder of this chapter, we look at the triangular prism and the cylinder, and how to calculate their surface areas.

The surface area of a triangular prism

In this section, you should be able to:

- recognise triangular prisms from their nets
- calculate the surface area of a triangular prism.

A triangular prism is a solid figure in which the two ends are matching triangles and the edges joining them are parallel. The diagram shows two examples.

Every triangular prism has **five** faces. Two of these are triangular, and the other three are rectangular.

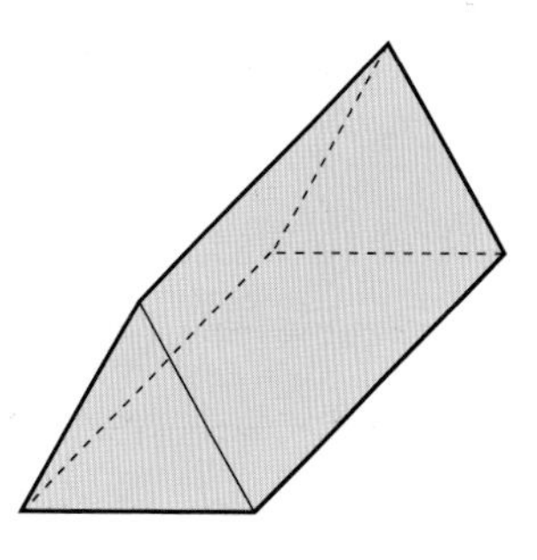

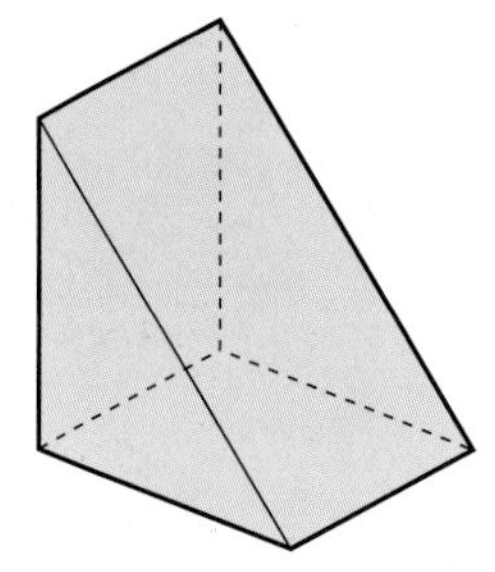

The **net** of a triangular prism is a flat two-dimensional shape which can be folded up to make the triangular prism. The diagram shows two examples.

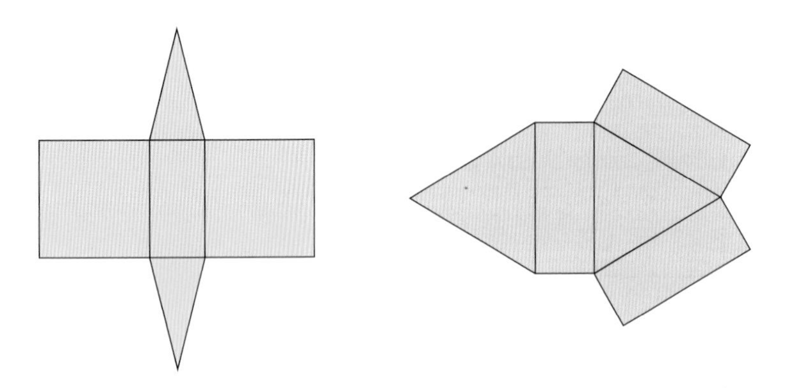

You may be asked to calculate the **surface area** of a triangular prism. To do this, you must calculate the area of each of the five faces and add these areas together.

Remember

Remember the formulae:

Area of a rectangle: $A = lb$

Area of a triangle: $A = \frac{1}{2}bh$.

Example The diagram below shows a triangular prism and its net.

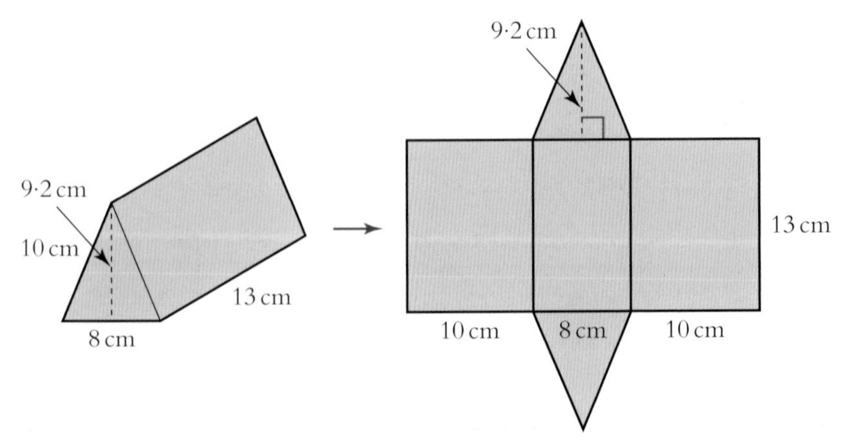

Calculate the total surface area of the triangular prism.

Solution We re-draw the net and letter each face A → E.

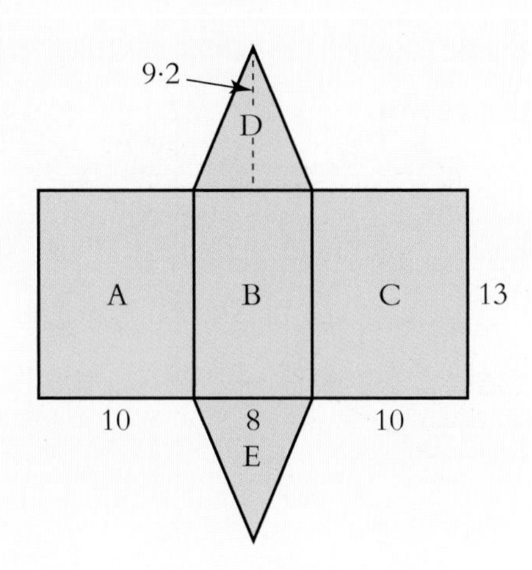

The areas of the faces may then be calculated as follows.

A: $A = lb = 10 \times 13 = 130\,\text{cm}^2$.

B: $A = lb = 8 \times 13 = 104\,\text{cm}^2$.

C: $A = lb = 10 \times 13 = 130\,\text{cm}^2$.

D: $A = \frac{1}{2}bh = 0{\cdot}5 \times 8 \times 9{\cdot}2 = 36{\cdot}8\,\text{cm}^2$.

E: $A = \frac{1}{2}bh = 0{\cdot}5 \times 8 \times 9{\cdot}2 = 36{\cdot}8\,\text{cm}^2$.

Hence the total surface area of the prism
$= (130 + 104 + 130 + 36{\cdot}8 + 36{\cdot}8) = 437{\cdot}6\,\text{cm}^2$.

Exercise 5

1 Which **one** of the following shapes is the net of a triangular prism?

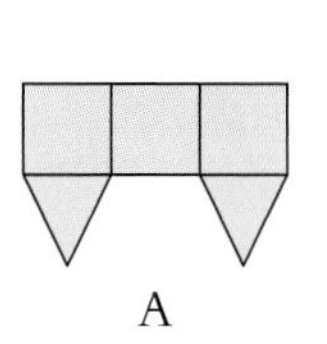

A

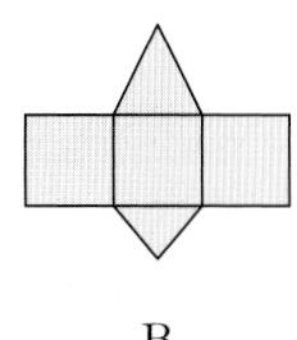

B

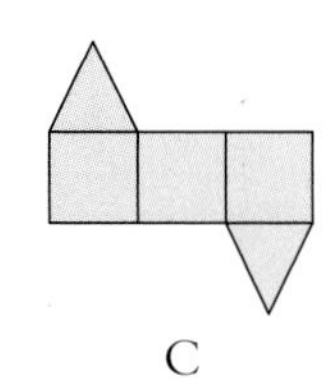

C

Exercise 5 continued

2 The diagram shows the net of a triangular prism.

Calculate the surface area of the prism.

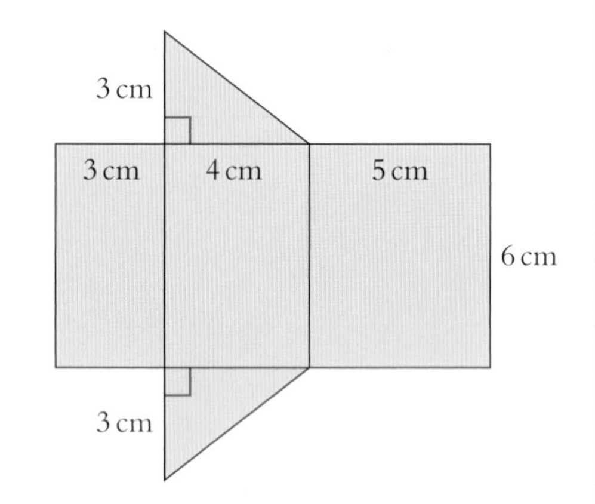

3 The diagram shows the net of another triangular prism.

Calculate the surface area of this prism.

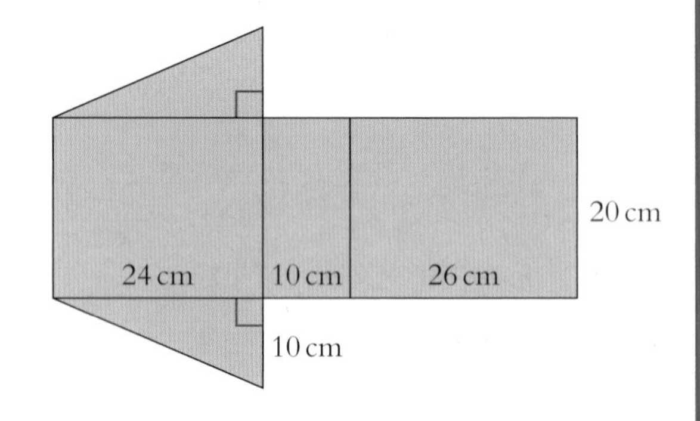

4 The diagram shows the net of a solid shape.

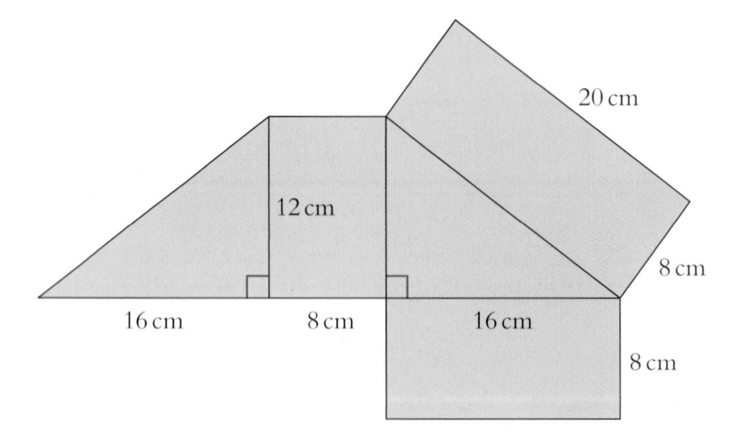

(a) Name the solid shape formed from this net.

(b) Calculate the surface area of the solid shape.

Exercise 5 continued

5 The diagram shows a triangular prism and its net.

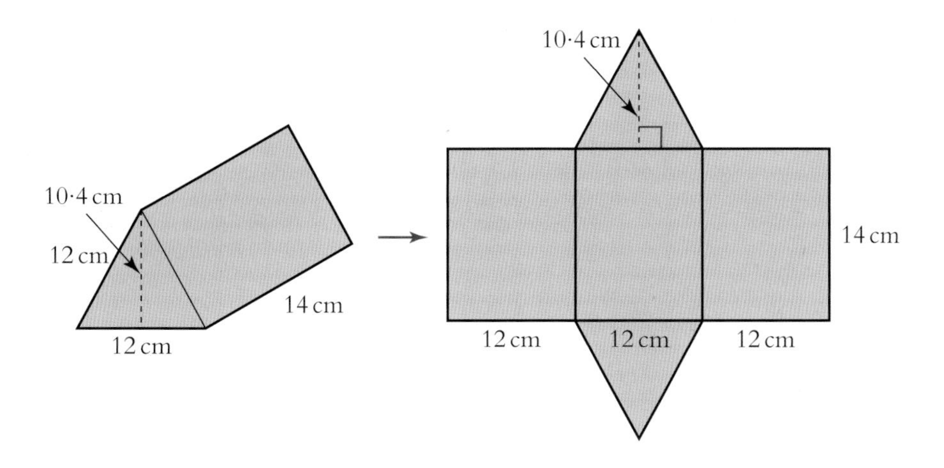

Calculate the surface area of this triangular prism to the nearest cm^2.

6 This diagram shows the net of a triangular prism.

Find the total surface area of this triangular prism.

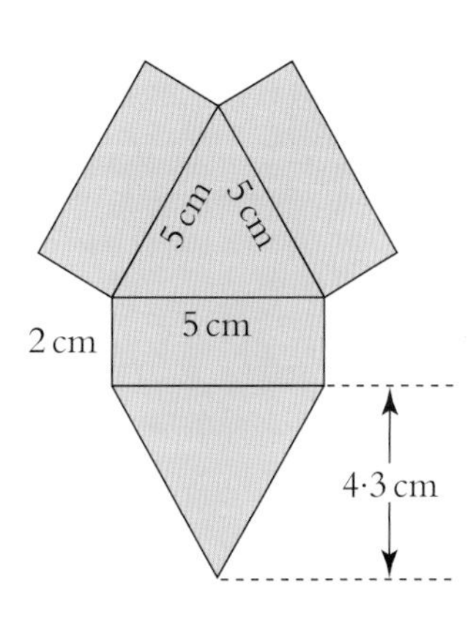

The Surface Area of a Cylinder

In this section, you should be able to:

- recognise cylinders from their nets
- calculate the surface area of a cylinder.

A cylinder is a prism whose two ends are matching circles.
The diagram shows two examples.

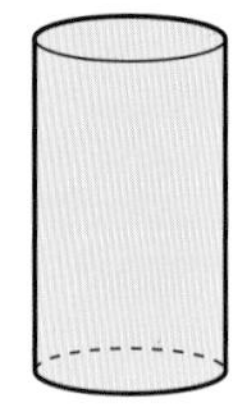
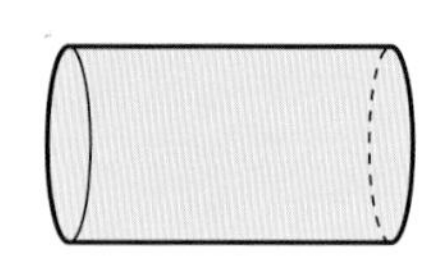

A cylinder has **three** faces. Two of them are circular and one is curved.

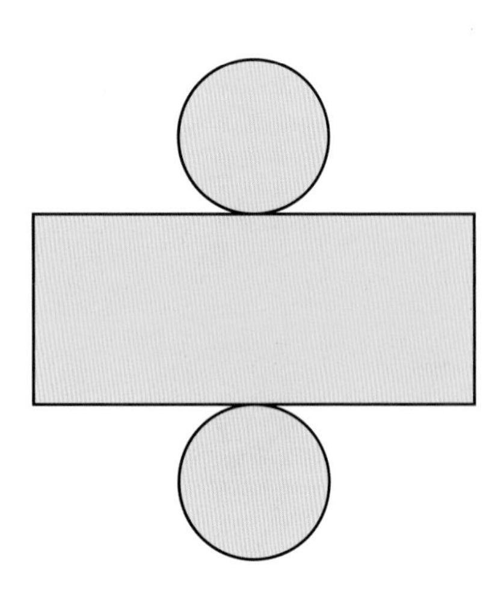

The **net** of a cylinder is a flat two-dimensional shape which can be folded up to make the cylinder. The diagram shows how this is achieved.

The length of the rectangle in the net must equal the circumference of the circle.

The breadth of the rectangle in the net must equal the height of the cylinder.

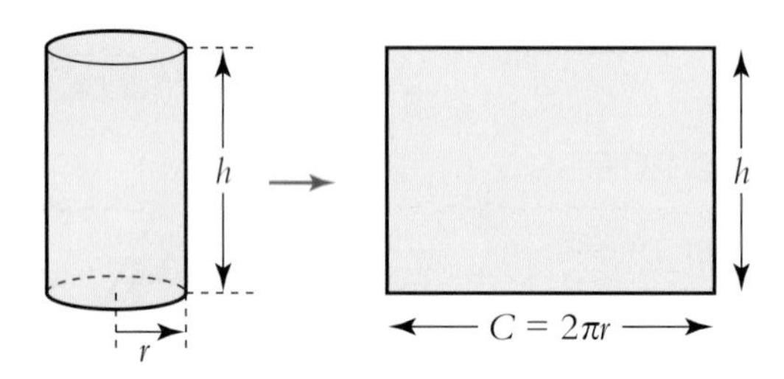

Hence the curved surface area of a cylinder is the area of a rectangle.

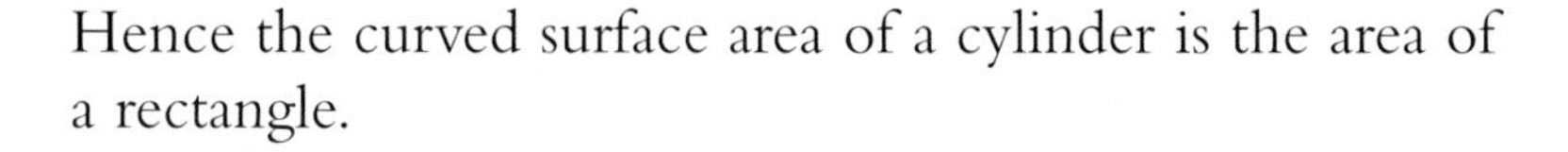

Technique

The area of the curved surface of the cylinder may be calculated as follows:

area of rectangle $= \text{length} \times \text{breadth}$
$= (\text{circumference of circle}) \times (\text{height of cylinder})$
$= \pi d \times h$
$= 2\pi rh$

Thus curved surface area of a cylinder: $\boldsymbol{A = 2\pi rh}$.

(If the area of the circular ends are required too, we use the formula $\boldsymbol{A = \pi r^2}$ for each.)

Example The diagram shows the net of a cylinder.

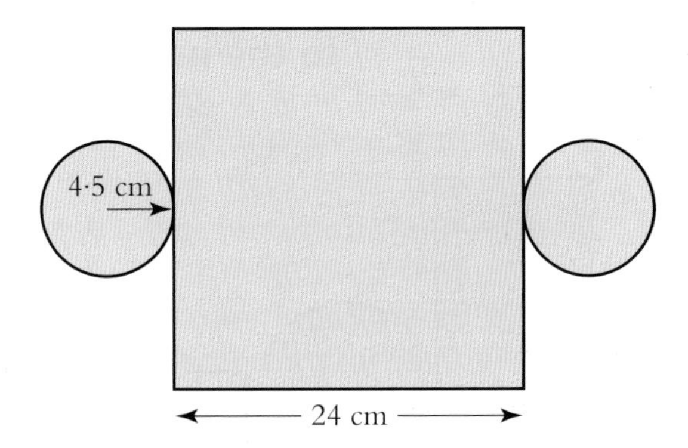

Calculate the area of the curved surface of the cylinder.

Solution Curved surface area $= 2\pi rh$

$= 2 \times \pi \times 4{\cdot}5 \times 24$

$= 679\,\text{cm}^2$.

Exercise 6

Give answers to all questions (except 9) to the nearest whole unit.

1 Calculate the area of the curved surface of each cylinder.

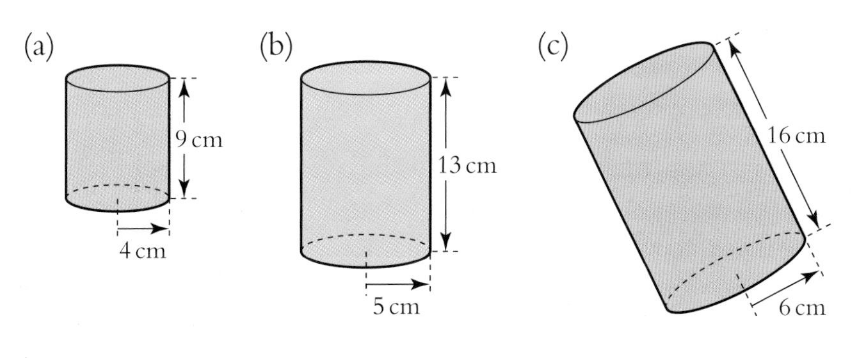

Exercise 6 continued

Give answers to all questions (except 9) to the nearest whole unit.

2 The diagrams show the nets of two cylinders.

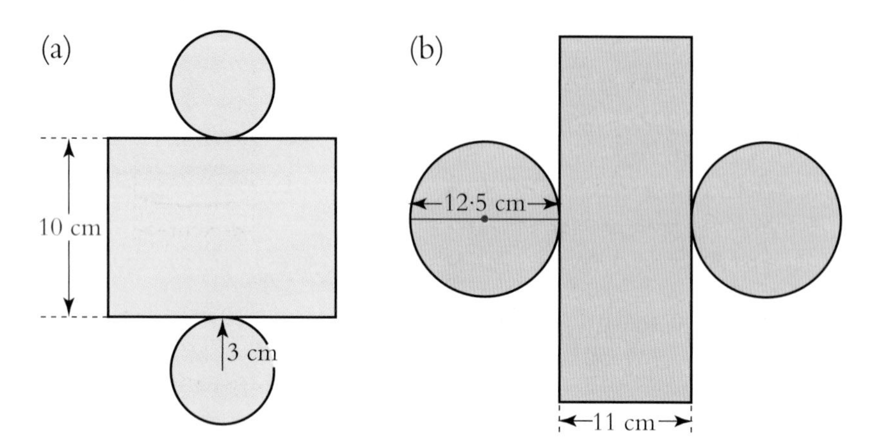

Calculate the area of the curved surface of each cylinder.

3 Calculate the area of the curved surface of the cylinder shown.

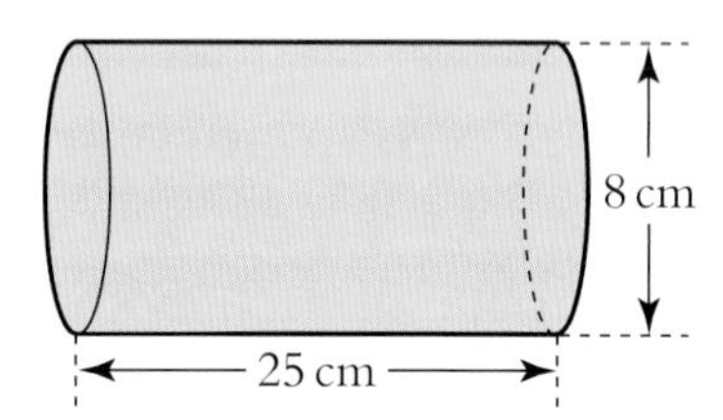

4 A label fits exactly around the curved surface of a cylinder.

If the cylinder has radius 8 cm and height 14 cm, find the area of the label.

5 Find the area of the curved surface of a cylinder of diameter 7·5 cm and height 9 cm.

(Do not use a calculator for questions 6 and 7)

6 A cardboard tube in the shape of a cylinder is used to hold wrapping paper.

It has radius 2 centimetres and length 50 centimetres.

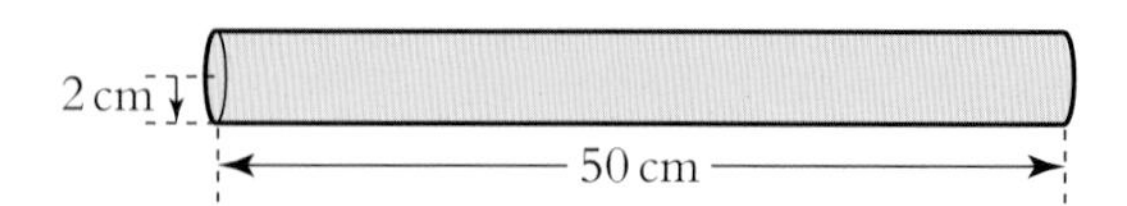

Calculate the area of the curved surface of this tube. (**Take π = 3·14.**)

Exercise 6 continued

7 A toilet roll holder is in the shape of a cylinder of diameter 5 cm and height 10 cm.

Calculate the area of the curved surface of the toilet roll holder. (**Take π = 3·14.**)

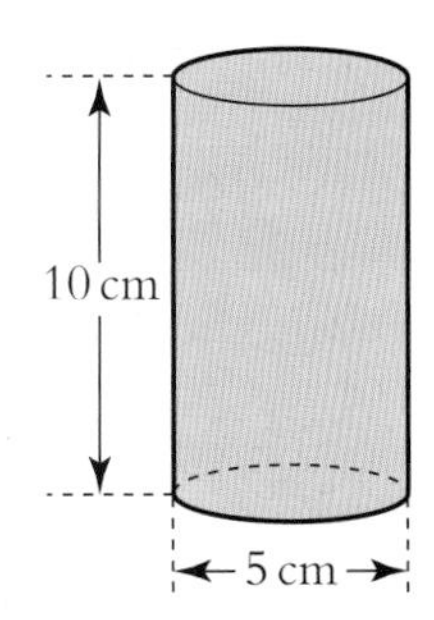

8 A can of beans is in the shape of a cylinder of diameter 7·5 cm and height 11 cm.

The label on the tin goes all the way round and with an extra 1 cm for overlap.

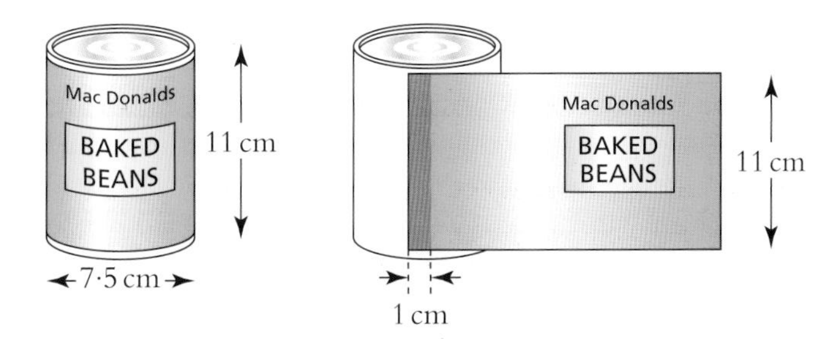

Calculate the area of the label.

9 A haystack is in the shape of a cylinder. It is 2 metres high and has a diameter of 1·8 metres. The haystack is to be completely wrapped in polythene for transport.

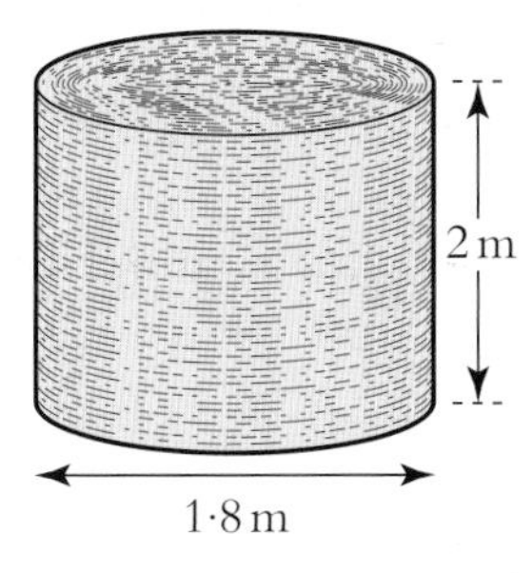

(a) Calculate the area of the top of the haystack.

(b) Write down the area of the base of the haystack.

(c) Calculate the area of the curved surface of the haystack.

(d) Hence calculate the total minimum area of polythene required for wrapping. (Give your answer correct to 1 decimal place.)

Exercise 6 continued

10 A cylinder has radius 12 cm and height 15 cm.

Calculate its **total** surface area.

11 The diagram shows the net of a solid shape.

(a) Name the solid shape made from this net.

(b) Calculate the total surface area of the solid shape.

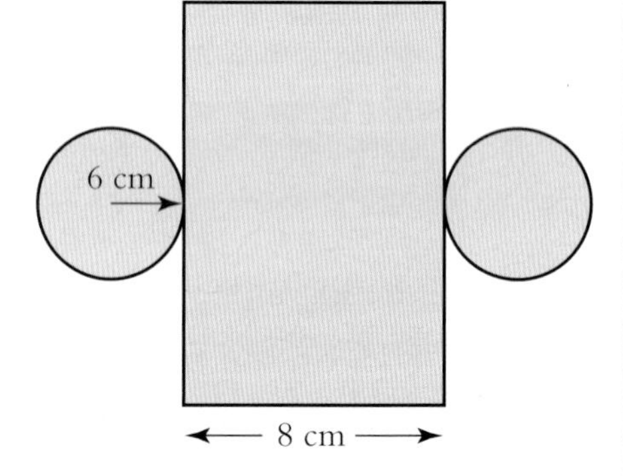

Exercise 7 – Revision of Chapter 16

1 What is the three-figure bearing of:

(a) North (b) South (c) East (d) south-east ?

2 What direction is indicated by a three-figure bearing of:

(a) 045° (b) 315° (c) 270° (d) 225° ?

3 Use a protractor to find the three-figure bearing of each direction shown.

(a)

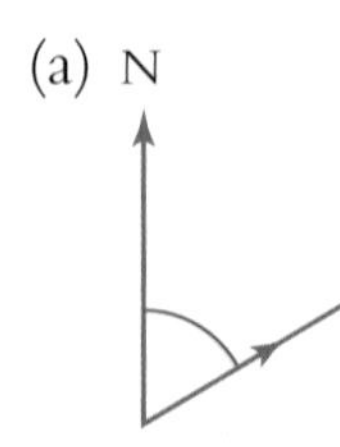

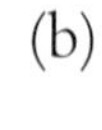

(b)

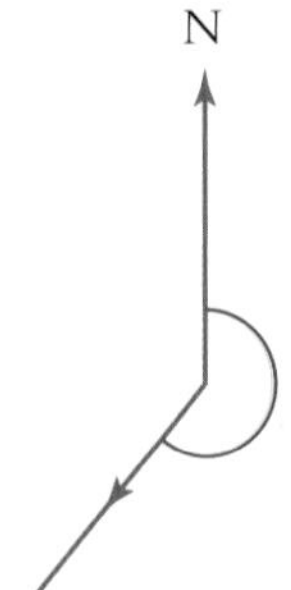

(c)

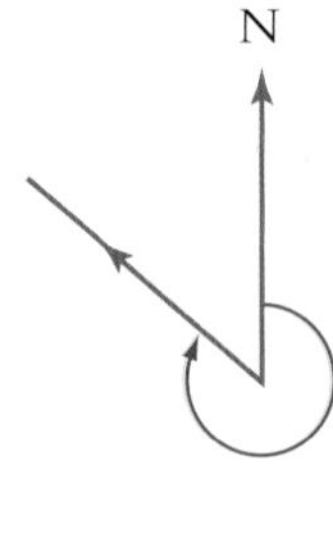

Exercise 7 – Revision of Chapter 16 continued

4 Angie observes the top of her school from two different positions A and B.

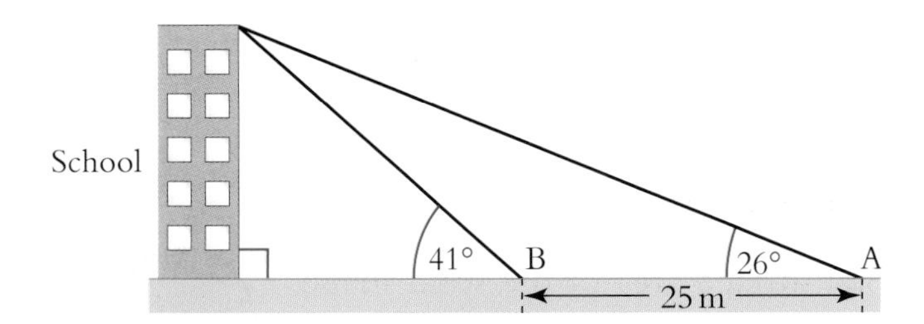

At A, the angle of elevation to the top of the school is 26°. At B, the angle of elevation is 41°. Positions A and B are 25 metres apart.

Make a scale drawing using a scale 1 cm to 5 m, and find the height of the school.

5 A fishing boat leaves harbour and sails for 30 kilometres on a bearing of 050°.

It then sails for 20 kilometres on a bearing of 160°.

(a) Using a scale of 1 cm to 5 km, make a scale drawing of the boat's journey.

(b) How far now is the journey back to harbour?

6 Town B is 60 km to the East of town A.

(a) Make a scale drawing using a scale of 1 cm to 10 km.

(b) Town C is on a bearing of 140° from A and 260° from B.
Plot the position of town C on your scale drawing.

7 The diagram shows the net of a solid shape.

(a) Name the solid shape made from this net.

(b) Calculate the total surface area of the solid shape.

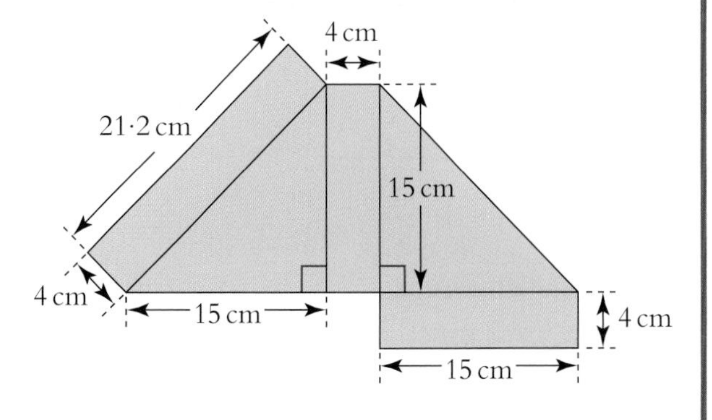

Exercise 7 – Revision of Chapter 16 continued

8 The diagram shows the net of a cylinder.

Calculate the area of the curved surface of the cylinder.

(Give your answer to the nearest cm^2.)

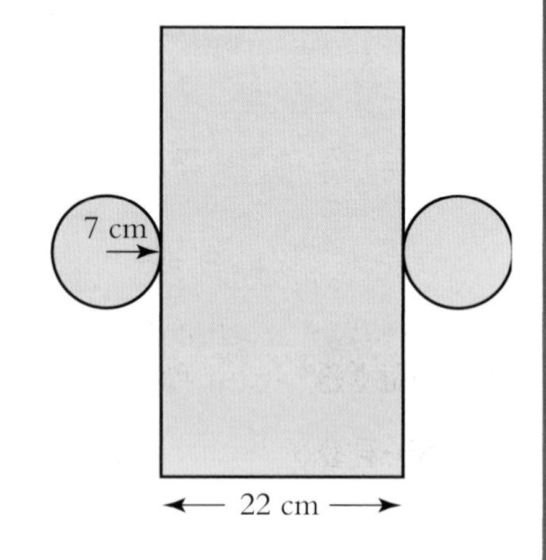

Summary

(In this short summary you are reminded of three-figure bearings, scale drawings, and scale drawings using bearings. In addition the calculations of surface area for triangluar prisms and cylinders are included.)

1 ***Directions*** may be given by

(i) the points of the compass;

(ii) three-figure bearings.

Three-figure bearings are measured clockwise from North. North is 000°, north-east is 045°, East is 090°, and so on.

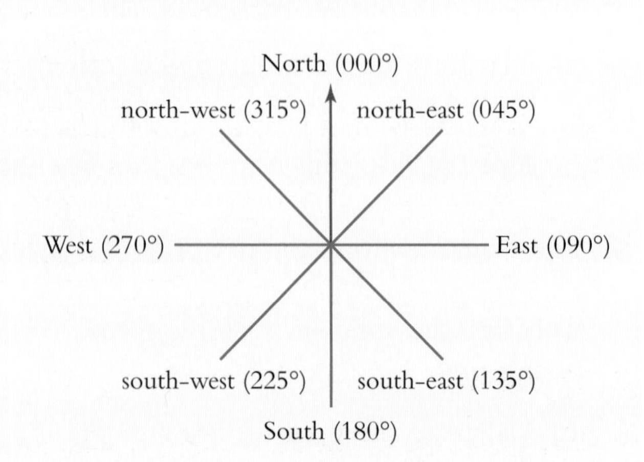

2 ***Scale drawings*** allow us to show very large distances or sizes on paper. Every scale drawing must include a clear scale, such as 1 centimetre represents 10 metres, 1 cm to 10 m or 1 : 1000.

You must know the terms *angle of elevation* and *angle of depression*.

Summary continued

3 Scale drawings may involve ***three-figure bearings***.

Example From town K, town L is on a bearing of 055° and at a distance of 50 kilometres.

Town M is on a bearing of 080° from K and 145° from L.

Show the positions of towns K, L and M on a scale drawing.

Solution

Scale: 1 cm to 10 km

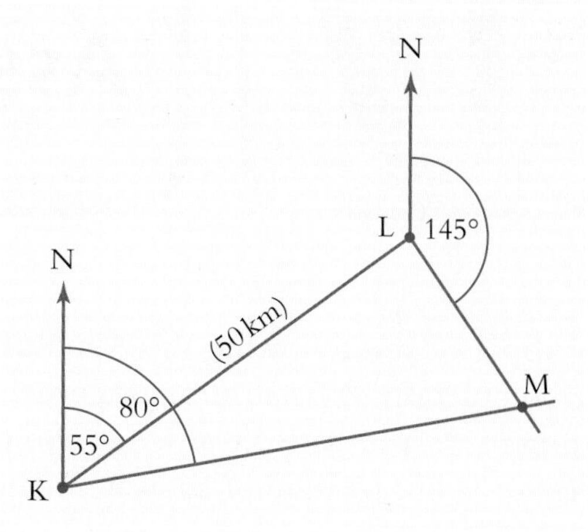

4 ***Triangular prisms***

The net of a triangular prism can be folded up to make a triangular prism.

You should try to prove that the total surface area of this prism will be $406{\cdot}8\ \text{cm}^2$.

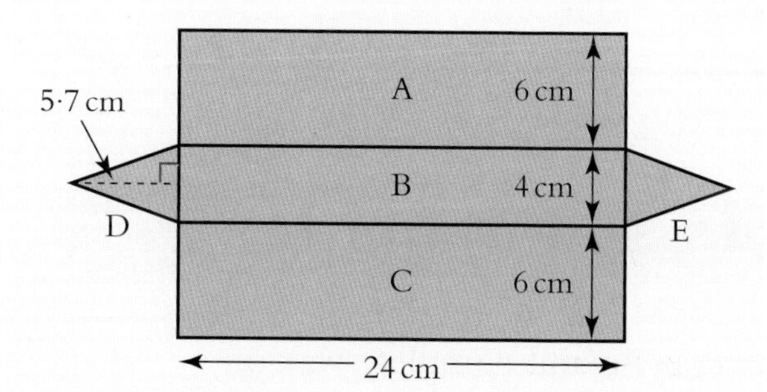

5 ***Cylinders***

The net of a cylinder can be folded up to make a cylinder.

The area of the curved surface of a cylinder is given by the formula $A = 2\pi rh$. If we wish to find the total surface area of the cylinder, including the top and the base, then this is given by the formula $A = 2\pi rh + 2\pi r^2$.

17 Statistical Assignment

For those of you following this Applications of Mathematics Unit the completion of a statistical assignment is an essential requirement. You have already met some calculated statistics in Chapter 9, but to attempt an assignment successfully, you must learn a few additional skills in data handling. We begin by looking at these additional skills before moving on to the assignment itself.

Five-figure Summaries

In this section we will study the preparation of a five-figure summary of a data set. This summary includes lowest and highest values, lower and upper quartiles, and also the median of the data.

Key words and definitions

A five-figure summary is a concise and complete way of summarising a data set.

It consists of:

- the lowest value of the data, L
- the lower quartile of the data, Q_1
- the median of the data, Q_2
- the upper quartile of the data, Q_3
- the highest value of the data, H

To produce a five-figure summary, we start by arranging the data set in order from lowest value to highest value.

The **median** is the middle number in a set of ordered numbers.

If there are n numbers in the ordered data set, we use the formula $(\boldsymbol{n} + \mathbf{1}) \div \mathbf{2}$ to find the position of the median. For example, if there are 15 numbers, $(15 + 1) \div 2 = 8$, so the median is the 8th number. If there are 16 numbers, $(16 + 1) \div 2 = 8{\cdot}5$, so the median is midway between the 8th and 9th numbers. To find such a median, we calculate the mean of the two numbers.

The median splits a data set in half.

The **lower quartile** is the median of the lower half of the data.

The **upper quartile** is the median of the upper half of the data.

Example Prepare a five-figure summary for the given data set.

4, 5, 5, 6, 7, 8, 9, 10, 10, 11, 11, 13, 16.

Solution The data is already in order and there are 13 items ($n = 13$).

Here lowest value, $L = 4$, highest value, $H = 16$.

Position of median $= (13 + 1) \div 2 = 7$ (the 7th number along).

4, 5, 5, 6, 7, 8, **9**, 10, 10, 11, 11 13, 16

Hence the median, $Q_2 = 9$.

(The median splits the data set into two halves with 6 numbers in each.)

For the lower half (4, 5, **5**, **6**, 7, 8), the median is between 5 and 6.

The mean of 5 and 6 is $\frac{5 + 6}{2} = \frac{11}{2} = 5{\cdot}5$.

Hence the lower quartile, $Q_1 = 5{\cdot}5$.

For the upper half (10, 10, **11**, **11**, 13, 16), the median is between 11 and 11.

Hence the upper quartile, $Q_3 = 11$.

The five-figure summary is therefore
$L = 4$, $Q_1 = 5{\cdot}5$, $Q_2 = 9$, $Q_3 = 11$, $H = 16$.

Example Prepare a five-figure summary for the following data set:

13, 9, 23, 14, 4, 6, 9, 10, 11, 3, 15, 19, 25, 8, 6, 5, 17, 20.

Solution First, we re-arrange the data in order:

3, 4, 5, 6, 6, 8, 9, 9, 10, 11, 13, 14, 15, 17, 19, 20, 23, 25.

Here, lowest value, $L = 3$, highest value, $H = 25$.

Since $n = 18$, the position of the median
$= (18 + 1) \div 2 = 19 \div 2 = 9{\cdot}5$.

Hence the median is between the 9th and 10th numbers in the list. These are 10 and 11, so the median $Q_2 = \frac{10 + 11}{2} = \frac{21}{2} = 10{\cdot}5$.
The median splits the data set into two halves with 9 numbers in each.

The lower quartile is the median of the lower half
(3, 4, 5, 6, **6**, 8, 9, 9, 10).

Hence lower quartile, $Q_1 = 6$.

The upper quartile is the median of the upper half
(11, 13, 14, 15, **17**, 19, 20, 23, 25).

Hence upper quartile, $Q_3 = 17$.

The five-figure summary is therefore
$L = 3$, $Q_1 = 6$, $Q_2 = 10{\cdot}5$, $Q_3 = 17$, $H = 25$.

Exercise 1

1 Prepare a five-figure summary for each data set:

(a) 3, 7, 8, 10, 11, 12, 13, 14, 16, 19, 20

(b) 4, 6, 8, 12, 13, 15, 16, 16, 21

(c) 7, 7, 9, 10, 11, 13, 14, 16, 17, 20

(d) 4, 8, 13, 14, 15, 17, 18, 19, 20, 20, 22, 29.

Exercise 1 continued

2 By putting each data set in order, prepare a five-figure summary for each:

(a) 4, 10, 15, 13, 9, 6, 1, 20, 12

(b) 12, 7, 6, 4, 2, 9, 14, 17, 19, 21

(c) 3, 6, 4, 10, 17, 4, 1, 12, 9, 13, 19, 15, 7

(d) 12, 27, 10, 7, 14, 19, 23, 30.

3 The marks of a class for an Intermediate 1 test are as shown:

38, 45, 53, 60, 39, 47, 71, 55, 62.

Prepare a five-figure summary for this data.

4 The marks of all students for another Intermediate 1 test are illustrated in an ordered stem and leaf diagram, as follows:

Intermediate 1 marks

3	3 5 8 9
4	1 3 5 5 9
5	2 2 7 8 9 9
6	0 1 4 7 8 9 9
7	0 2 3

($n = 25$) (4 | 1 means 41)

Prepare a five-figure summary for these results.

5 The results of the students in Class 5 in their Art exam are listed as follows:

34 36 70 47 48 37 50 53 60 52

45 49 50 61 58 60 41 58 48 55.

(a) Illustrate this data in an ordered stem and leaf diagram.

(b) Prepare a five-figure summary for the data.

Boxplots

A boxplot is a simple diagram which illustrates a five-figure summary of data. When we draw a boxplot, we always give it a title and include a suitable scale.

Example A sample of 24 students was asked how many times each had visited the cinema in the last two months. The results are as shown:

1 4 5 6 3 3 2 5
3 2 5 6 3 2 4 4
3 6 5 3 1 1 4 5.

(a) Prepare a five-figure summary for this data.
(b) Construct a boxplot for the data.

Solution

(a) First we arrange the results in order:

1, 1, 1, 2, 2, 2, 3, 3, 3, 3, 3, 3, 4, 4, 4, 4, 5, 5, 5, 5, 5, 6, 6, 6.

Here, lowest value, $L = 1$, highest value, $H = 6$.

Since $n = 24$, the position of the median
$= (24 + 1) \div 2 = 25 \div 2 = 12{\cdot}5$.

Hence the median is between 12th and 13th numbers along.

These are 3 and 4, so the median $Q_2 = 3{\cdot}5$.

The lower quartile is the median of the 12 numbers in lower half. Hence $Q_1 = 2{\cdot}5$.

The upper quartile is the median of the 12 numbers in upper half. Hence $Q_3 = 5$.

The five-figure summary is therefore
$L = 1, Q_1 = 2{\cdot}5, Q_2 = 3{\cdot}5, Q_3 = 5, H = 6$.

(b)

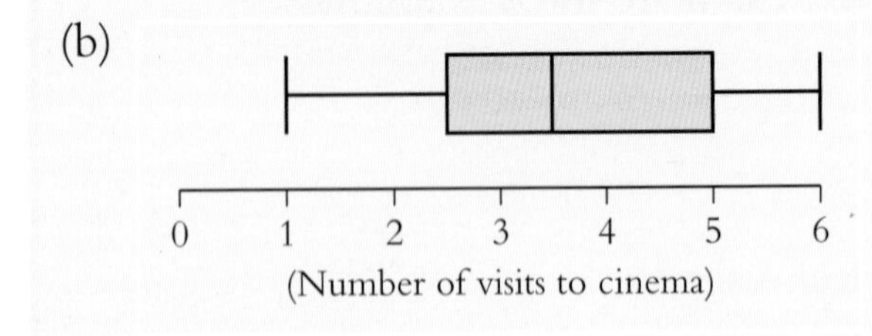

Exercise 2

1 The numbers of paperclips in a sample of boxes are counted.

The results are indicated in the boxplot.

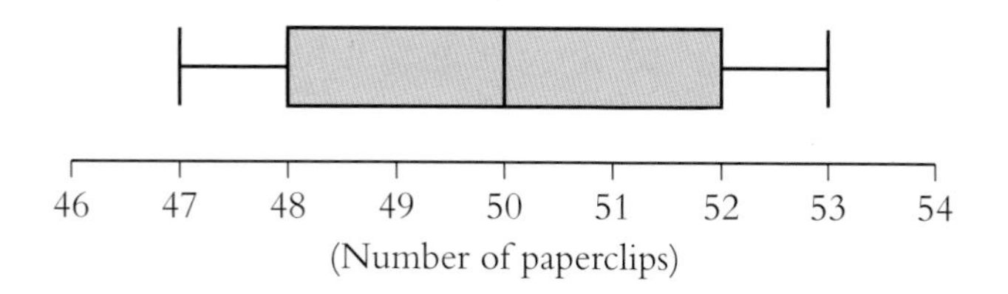

Make a five-figure summary of the data.

2 The number of buses which passed a bus stop between 9 am and 10 am was counted every day for three weeks. The results are shown in the boxplot.

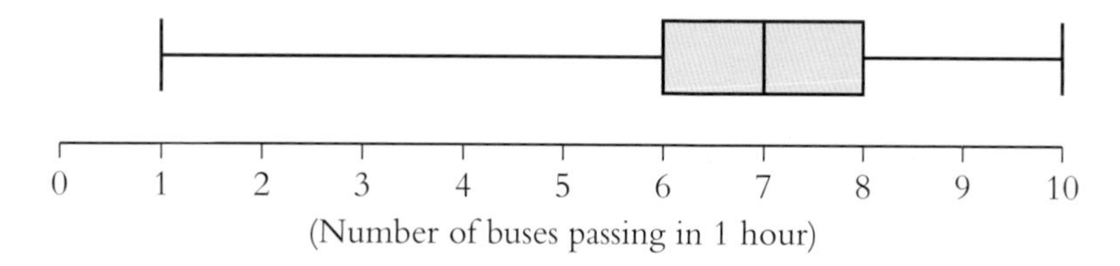

Make a five-figure summary of the data.

3 The numbers of tacks in a sample of boxes were counted. The results were as follows:

30 32 20 30 32 35 33 31 28.

(a) For this data, find (i) the median

(ii) the lower quartile

(iii) the upper quartile.

(b) Make a five-figure summary of the data.

(c) Construct a boxplot for the data.

4 A group of students was asked how many times each had attended a football match during the last month. The results are as shown:

1 3 2 0 1 1 1 0 6 7 0 4
1 5 2 0 0 8 1 0 2 3 3 1.

(a) Prepare a five-figure summary of this data.

(b) Construct a boxplot for the data.

Exercise 2 continued

5 Steve measured the heights (in millimetres) of some plants in his garden. They were:

45 52 39 48 58 50 61 30 24 38 49 40 36.

Copy and complete the boxplot to show the heights of the plants.

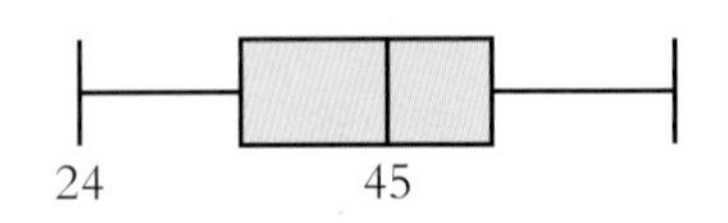

6 The numbers of goals scored by the teams in the SPL during one weekend were:

0 3 0 1 3 2 1 1 1 4 4 0.

(a) Prepare a five-figure summary of this data.

(b) Construct a boxplot for the data.

7 In a tournament, a group of golfers recorded the following scores:

70 74 68 69 70 66 75 72

71 69 68 70 71 77 73 70.

(a) Prepare a five-figure summary of this data.

(b) Construct a boxplot for the data.

8 The heights of a group of female students are shown (to the nearest centimetre).

156 148 178 188 160 175 164 174 174 168 169 171 172

(a) Prepare a five-figure summary for this data.

(b) Draw a boxplot to illustrate the data.

Measures of Spread

In this section, you should be able to

- calculate the range and interquartile range of a data set
- recognise the range and interquartile range as measures of spread.

Main points

In Chapter 9, we looked at three 'averages' – the mean, mode and median. These are sometimes called measures of *central tendency*. This means that they tell us about the centre or middle of a data set. However, these 'averages' tell us nothing about how the data is spread. To describe spread we need different measures, and one of these, the Range, we have already met. You will remember that for a given data set:

Range = Highest value – Lowest value

Another and perhaps more useful measure of spread is the Interquartile Range. For any data set:

Interquartile Range = Upper quartile – Lower quartile

That is, Interquartile Range = $Q_3 - Q_1$.

The greater the Range and Interquartile Range, the more spread out is the data.

In a boxplot, the wider the boxplot, the greater the Range. Similarly the wider the box, the greater the Interquartile Range, and the greater the spread of data.

Example Ali kept a record of how much 15 customers spent in his newsagent shop. The amounts (in pence) are as shown:

30 15 28 55 35 15 15 30 55 50 45 35 28 36 60.

(a) Prepare a five-figure summary of this data.
(b) Construct a boxplot for the data.
(c) Calculate the Interquartile Range.

Solution (a) First we arrange the numbers in order. Thus:

15 15 15 28 28 30 30 35 35 36 45 50 55 55 60.

Since $n = 15$, the position of the median is $(15 + 1) \div 2 = 8$.

Hence the median is the 8th number. Thus $Q_2 = 35$.

The lower quartile, $Q_1 = 28$, and the upper quartile Q_3, = 50.

The five-figure summary is therefore
$L = 15$, $Q_1 = 28$, $Q_2 = 35$, $Q_3 = 50$, $H = 60$.

(b)

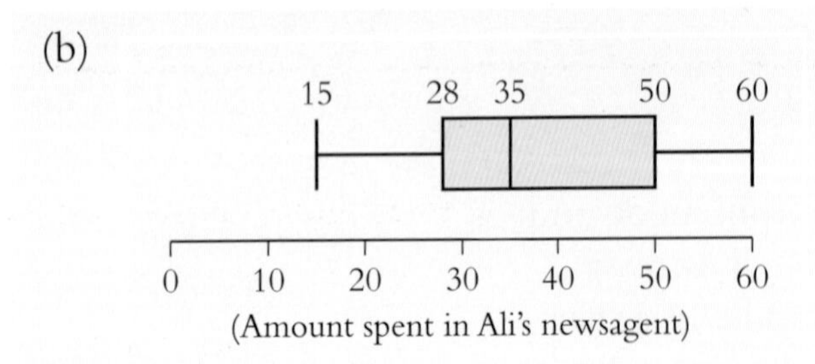

(c) Interquartile range $= Q_3 - Q_1 = 50 - 28 = 22$.

Example A record was also kept of how much customers spent in Joe's newsagent shop around the corner from Ali. The boxplot for this data along with that for Ali is as shown.

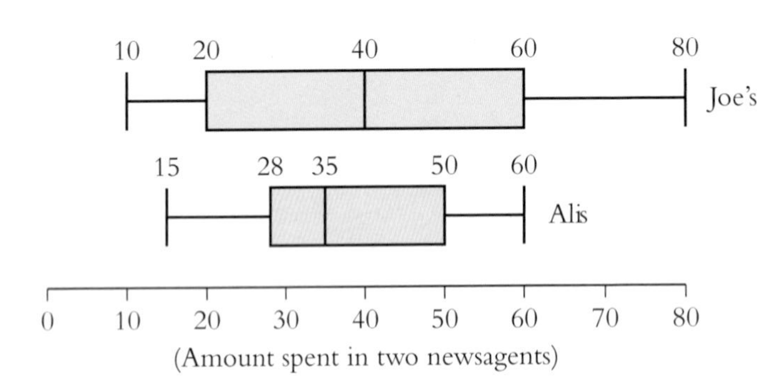

Compare and comment on the two boxplots.

Solution The interquartile range for Joe's newsagent is $Q_3 - Q_1 = 60 - 20 = 40$. This is greater than the interquartile range for Ali (22), therefore the amounts spent in Joe's are more spread than in Ali's.

The median for Joe is 40p compared with 35p for Ali. This suggests that customers spent more in Joe's shop.

Note. If you are asked to compare two boxplots, you should always comment on the medians, saying which is greater **and** also comment on the spread of the data either by comparing the Ranges or Interquartile Ranges.

Exercise 3

1 Calculate the Range and Interquartile Range for each of these five-figure summaries:

(a) $L = 12, Q_1 = 16, Q_2 = 18, Q_3 = 21, H = 25$

(b) $L = 6, Q_1 = 7, Q_2 = 8{\cdot}5, Q_3 = 9{\cdot}5, H = 12$.

2 The numbers of toffees in nine packets are counted, as follows:

20 21 23 25 25 26 28 28 29.

(a) Calculate the Range of toffees.

(b) Calculate the Interquartile Range of toffees.

3 A survey was carried out to find how many hours of sleep some children had on a particular night.

The results are shown in the boxplot.

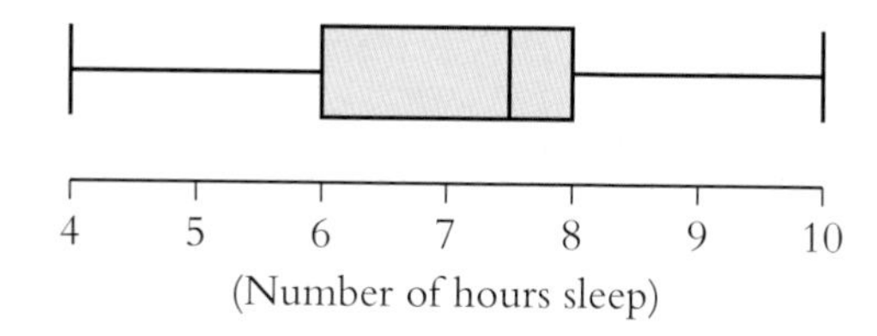

(a) Calculate the Range for the group.

(b) Calculate the Interquartile Range for the group.

4 The numbers of points gained by the teams in the Scottish Premier League in season 2005–06 is as shown.

91 74 73 56 55 54 58 49 33 33 33 18.

(a) Calculate the Range of points.

(b) The Range in season 2004–05 was 60 points. Comment on the two seasons.

5 The volumes in litres, of some refrigerators were as follows:

78 82 92 76 78 99 72 90.

(a) Calculate the Range of volume.

(b) Calculate the upper and lower quartiles of volume.

(c) Calculate the Interquartile Range of volume.

Exercise 3 continued

6 These boxplots show the sales of DVDs in a shop during the months of July and December.

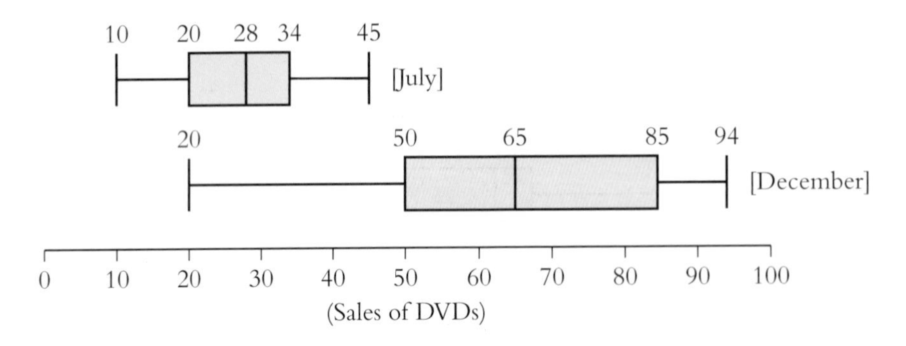

(a) Calculate the Interquartile Range for the December sales.

(b) In his report, the shop manager stated that 'Sales are generally higher in December than in July, but they are also more variable'.

How do the boxplots show that this statement is correct?

7 Mr Smith waits for the number '38' bus at 2 pm every weekday. He kept a record of how many minutes late the bus was over a two-week period.

The results were as follows:

2 4 0 0 5 6 10 6 1 8.

For these results calculate (a) the median

(b) the Range.

Mrs. Jones kept a similar record for the number '38' bus at 5 pm each day.

The median of her results was 8 minutes, and the Range was 19 minutes.

(c) Make two comments comparing the lateness of the number '38' bus at 2 pm and 5 pm.

8 A survey was conducted to find how many times 19 families visited a local restaurant over a period of two months. The results are as shown:

5 5 8 2 10 13 4 5 11 8 0 1 2 7 0 8 7 0 1.

(a) Find the lower quartile.

(b) Calculate the Interquartile Range.

Exercise 3 continued

9 Mr Cameron teaches two first year classes. He recorded their results in the October test and produced the following boxplots.

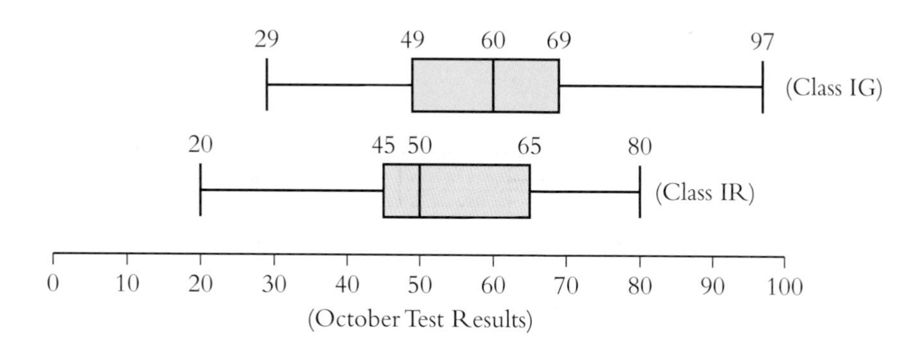

(a) Write down the median for each class.

(b) Calculate the Interquartile Range for each class.

(c) Compare the class results and comment on them.

10 The ages of the players in Sparta Tranent FC are as follows:

27 25 29 35 19 37 26 22 28 29 31 20 17.

(a) Prepare a five-figure summary of this data.

(b) Illustrate the data in a boxplot.

(c) Calculate the Interquartile Range of ages.

A boxplot was drawn to show the ages of the players in Atletico Dunbar FC.

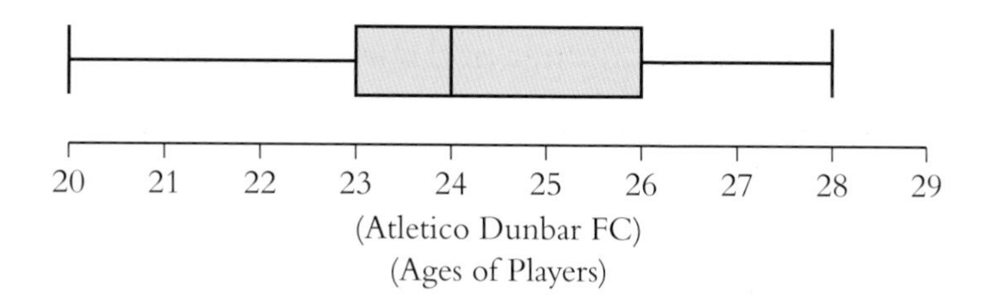

(d) Compare the boxplots for the two teams and comment.

Exercise 4 – Revision of Chapter 17

1 Prepare a five-figure summary of each set of data:

(a) 5, 7, 9, 9, 11, 13

(b) 6, 4, 7, 8, 10, 12, 15, 2, 12.

2 The heights of some male students were measured to the nearest centimetre, as follows:

170 177 175 158 160 175 176 184 190 165 182 171 172.

(a) For these heights find:

(i) the median

(ii) the lower quartile

(iii) the upper quartile.

(b) Draw a boxplot to illustrate this data.

3 The numbers of cars passing the school gate was counted in 5 minute intervals between 8 am and 9 am on Monday morning. The results are as shown:

3 5 8 11 6 4 5 16 17 19 22 10.

Calculate (a) the median

(b) the Range.

A similar survey was conducted on Saturday morning at the same times. On Saturday morning, the median was 4 and the Range was 7.

(c) Make two comments comparing the numbers of cars passing the school gate on Monday morning and Saturday morning.

4 Mr Gascoigne gives his class a written general knowledge test of twenty questions.

The number of correct answers given by each student is as shown:

13 10 19 7 10 9 15 18 7 11 14 18 10 13 18.

(a) Prepare a five-figure summary of this data.

(b) Calculate the Interquartile Range.

Exercise 4 – Revision of Chapter 17 continued

5 Gail is comparing the temperatures (in °C) of 10 cities in Europe and 10 in North America.

One day in July, she gathered the following results:

Europe	18	23	26	19	20	27	34	30	25	22
North America	24	35	30	23	21	37	30	28	27	25.

(a) Prepare a five-figure summary for Europe.

(b) Prepare a five-figure summary for North America.

(c) Construct boxplots for each set of data.

(d) Compare the boxplots and comment.

Statistical Assignment

As part of the Intermediate 1 course, you must complete a statistical assignment. This assignment should give you the opportunity to demonstrate all of the skills you have learned in this chapter along with those you learned earlier.

Your assignment requires you to do the following:

- collect two data sets
- prepare a five-figure summary for each of your data sets
- illustrate your data sets in boxplots
- calculate the Ranges and Interquartile Ranges for your data
- compare the two data sets from the boxplots.

The assignment must also include an introduction explaining how you collected your data and a conclusion summarising your findings.

The following 'completed assignment' will serve as a useful specimen example!

Numbers of Goals Scored in Serie A and the Premiership – A Comparison

Since I am interested in football, I have decided to do a project comparing two football leagues – The 'Premiership' in England and 'Serie A' in Italy.

I collected my data from the Internet and also from completed league tables for season 2005-06. I have listed the teams in each league in order of the position in which they finished along with the number of goals scored by each.

Fortunately for me, both leagues contain 20 teams, and each team played 38 games.

Premiership 2005-06		Serie A 2005-06	
Team	Goals	Team	Goals
1. Chelsea	72	1. Inter Milan	68
2. Manchester United	72	2. Roma	70
3. Liverpool	57	3. AC Milan	85
4. Arsenal	68	4. Chievo Verona	54
5. Tottenham Hotspur	53	5. Palermo	50
6. Blackburn Rovers	51	6. Livorno	37
7. Newcastle United	47	7. Empoli	47
8. Bolton Wanderers	49	8. Parma	46
9. West Ham United	52	9. Fiorentina	66
10. Wigan Athletic	45	10. Ascona	43
11. Everton	34	11. Udinese	40
12. Fulham	48	12. Sampdoria	47
13. Charlton Athletic	41	13. Reggina	39
14. Middlesbrough	48	14. Cagliari	42
15. Manchester City	43	15. Siena	42
16. Aston Villa	42	16. Lazio	57
17. Portsmouth	37	17. Messina	33
18. Birmingham City	28	18. Lecce	30
19. West Bromwich Albion	31	19. Treviso	24
20. Sunderland	26	20. Juventus	71

(Originally Juventus were first in the Serie A league and AC Milan were second. However after an investigation into corruption, five teams in the league had points deducted and Inter Milan became the Champions.)

From these results, stem and leaf diagrams were drawn for each league to show the numbers of goals scored in order.

Premiership - Goals Scored

Stem	Leaves
2	6 8
3	1 4 7
4	1 2 3 5 7 8 8 9
5	1 2 3 7
6	8
7	2 2

($n = 20$) (2|6 means 26)

Serie A - Goals Scored

Stem	Leaves
2	4
3	0 3 7 9
4	0 2 2 3 6 7 7
5	0 4 7
6	6 8
7	0 1
8	5

($n = 20$) (2|4 means 24)

From these stem and leaf diagrams, five-figure summaries were prepared for each league, as follows.

Premiership

$L = 26,\ Q_1 = 39,\ Q_2 = 47{\cdot}5,\ Q_3 = 52{\cdot}5,\ H = 72.$

Serie A

$L = 24,\ Q_1 = 39{\cdot}5,\ Q_2 = 46{\cdot}5,\ Q_3 = 61{\cdot}5,\ H = 85.$

From these five-figure summaries, two measures of spread were calculated for each set of data – the Range and the Interquartile Range. These were as follows:

Premiership: Range $= 72 - 26 = 46$, Interquartile Range $= 52{\cdot}5 - 39 = 13{\cdot}5$.

Serie A: Range $= 85 - 24 = 61$, Interquartile Range $= 61{\cdot}5 - 39{\cdot}5 = 22$.

The data was then displayed in boxplots, as follows.

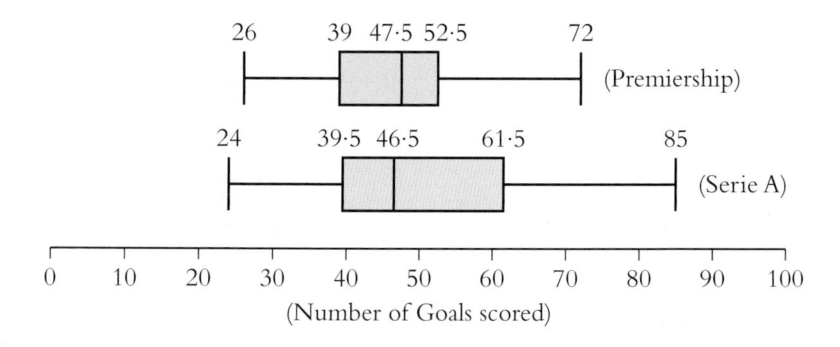

Conclusion

On average, it would appear that the numbers of goals scored by the teams in the Premiership and Serie A leagues is very similar. This is because the medians are very close to each other, 47·5 and 46·5 respectively.

It is interesting to look at the means too:

Premiership: Mean $= \frac{944}{20} = 47{\cdot}2$

Serie A: Mean $= \frac{991}{20} = 49{\cdot}55.$

This indicates that there were 47 more goals scored in Serie A than in the Premiership (991–944). Also, on average, the Italian teams scored over 2 goals more than teams in the Premiership during the season. The mean is perhaps a preferable average to the median in this case since it takes *every* goal scored into account.

As for spread, the boxplots clearly show that the pattern of goals in Italy is more spread out. This is shown too in the calculations of the Range (61 in Italy compared with 46 in England) and the Interquartile Range (22 to 13·5).

The larger Range in Serie A is partly due to AC Milan who scored 85 goals, 13 more than the leading goalscorers in the Premiership. There were five teams in Serie A who scored over 60 goals compared with only three in the Premiership. This accounts for the higher upper quartile. The lowest scores and lower quartiles are very similar in both leagues.

Finally, although there were many similarities, more goals *were* scored in Italy, particularly by the better teams, and the higher spread might indicate that results in Italy are less predictable.

Summary

(The specimen assignment has already summarised most of the work in this chapter but if you have any remaining doubts, you should read the following items carefully.)

1 ***Five-figure Summary***

A five-figure summary of a data set consists of the lowest value (L), the lower quartile (Q_1), the median (Q_2), the upper quartile (Q_3) and the highest value (H).

Summary continued

(If there are n entries in the data set, the formula $(n + 1) \div 2$ enables you to find the **position** of the median. If the median is between two numbers, then find their mean to find the median.)

The median divides the whole data set into two equal halves. The lower quartile is then the median of the lower half, and the upper quartile is the median of the upper half.

Thus for the data set 7, 10, 6, 5, 8, 11, 6, 3, 8, 12, (ten items), the correct order is 3 5 6 6 7 8 8 10 11 12.

Since $n = 10$, position of median $= (10 + 1) \div 2 = 5{\cdot}5$ (between 5th and 6th numbers).

Hence median $Q_2 = \dfrac{7 + 8}{2} = 7{\cdot}5$.

Also lower quartile $Q_1 = 6$, and upper quartile $Q_3 = 10$.

Hence the five-figure summary is $L = 3$, $Q_1 = 6$, $Q_2 = 7{\cdot}5$, $Q_3 = 10$, $H = 12$.

2 ***Boxplots***

A boxplot is used to illustrate the results of a five-figure summary.

The data in 1 is shown in this boxplot.

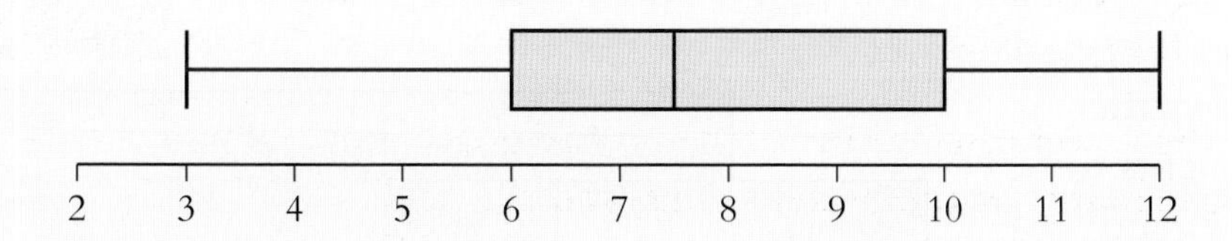

3 ***Measures of Spread***

Range = Highest Value – Lowest value $(= H - L)$.

Interquartile Range = Upper Quartile – Lower Quartile $(= Q_3 - Q_1)$.

For the data in 1, Range $= 12 - 3 = 9$, and Interquartile Range $= 10 - 6 = 4$.

The greater the Range and Interquartile Range, the more spread the data is.

4 ***Comparing two data sets***

If you are asked to compare boxplots, compare both the medians and the spread.

End Of Unit Tests

Test One (Non-Calculator)

1 Flora has completed her first week in her new job. Her basic rate of pay is £6 per hour. She is paid at time and a half for working overtime. During the week, she worked 40 hours at the basic rate and 4 hours of overtime.

How much did Flora earn?

2 Mark works in a shop and is paid £7·50 per hour. He works a 40-hour week.

Copy and complete his payslip.

Name: Mark Swann	Week ending: 13/4/07	
Hours Worked 40	Hourly Rate £7·50	Gross Pay
Income Tax £39·74	National Insurance £23·76	Total Deductions
		Net Pay

3 The table shows the monthly payments to be made, with (w) and without (w/o) payment protection, when money is borrowed from a bank.

	1 year		3 years		5 years	
Amount Borrowed	W/O	W	W/O	W	W/O	W
£3000	£262	£283	£98	£109	£66	£76
£5000	£426	£462	£154	£171	£100	£116
£7000	£597	£646	£216	£239	£139	£162
£10 000	£853	£923	£308	£341	£200	£231

(a) John Paul borrows £5000 over three years without payment protection. State his monthly payment.

(b) How much extra would John have paid over the three years if he had taken the loan with payment protection?

4 The network diagram shows the streets to be visited by a wheely bin lorry.

(a) How many arcs are there in the network diagram?

(b) The lorry starts at the depot. It goes along each street once, and it does not need to finish at the depot. List the streets, in order, for one possible route.

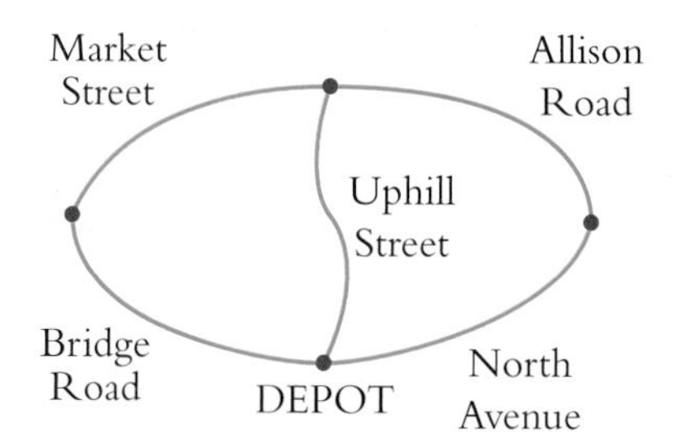

5 Xavier sells furniture. The flowchart shown is used to calculate his monthly salary.

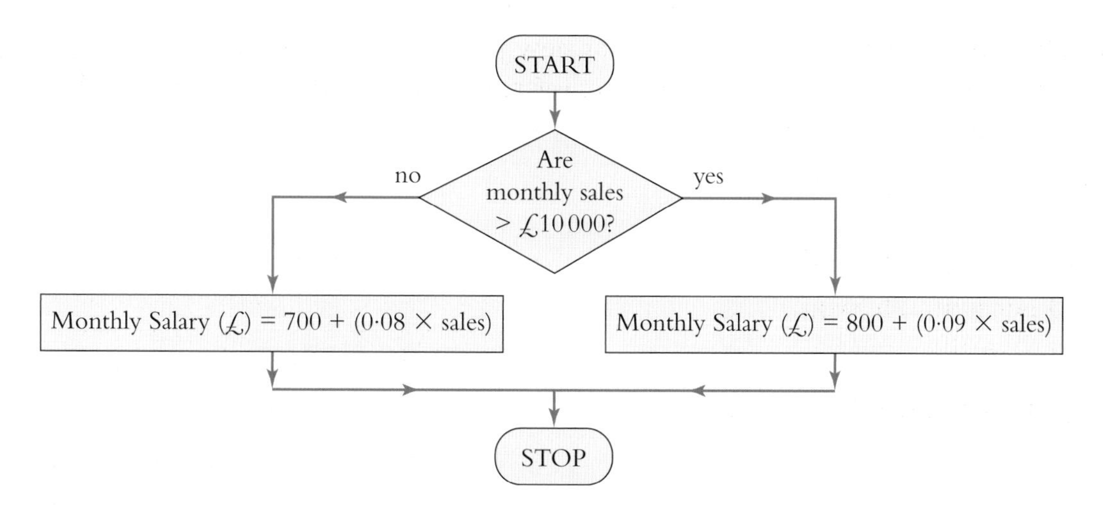

Calculate Xavier's salary for May when he sold £9000 worth of furniture.

6 A spreadsheet is used to keep a record of sales in a chip shop.

	A	B	C	D	E	F
1			**Sales of Suppers in Chip Shop**			
2		**Pie**	**Sausage**	**Fish**	**Black Pudding**	**Haggis**
3	Monday	10	7	15	10	7
4	Tuesday	8	12	20	7	8
5	Wednesday	12	8	20	6	11
6	Thursday	11	13	35	12	13
7	Friday	20	15	40	15	14
8	Saturday	25	15	56	18	19
9						
10	Totals					
11	Averages					

(a) The result of the formula =SUM(D3..D8) is to be entered in cell D10. What value would appear in cell D10?

(b) What formula would be used to enter the average daily sale of fish suppers in cell D11?

7 The scale drawing shows the position of a ship and a harbour.

Use this scale drawing to find the distance and bearing of the harbour from the ship.

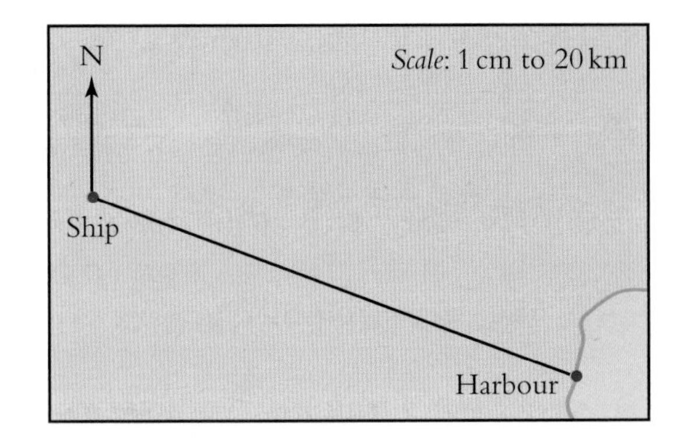

8 The scale of a map is 1 : 2000.

Two buildings on the map are 8 cm apart. What is the actual distance between the buildings? Give your answer in metres.

9 The diagram shows the net of a solid shape.

(a) Name the solid shape formed from this net.

(b) Calculate the surface area of the solid shape.

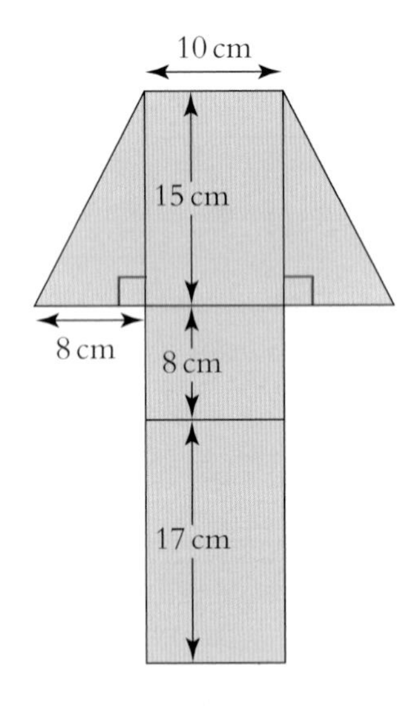

10 A cylinder has height 10 centimetres and radius 5 centimetres.

Calculate the area of its curved surface. (Take $\pi = 3{\cdot}14$.)

11 The ages of the players in a school netball club are:

12 12 14 13 18 17 17 14 12 13 15 14 15.

(a) Find the median, lower quartile and upper quartile for this data.

(b) Illustrate the data on a boxplot.

(c) Calculate the Interquartile Range.

Test Two (Calculator)

1 Niamh is paid £7·60 per hour for working a basic 40-hour week. Overtime is paid at time and a half. How much will Niamh earn in a week in which she works 45 hours?

2 Oliver works as a hairdresser. His payslip for the week ending 20th April is as shown.

Oliver Stephens		Week Ending: 20/04/07	
Basic Pay £236·00	Overtime £35·40	Bonus £26·40	Gross Pay
Income Tax £39·25	National Insurance	Superannuation £17·87	Total Deductions
			Net Pay £217·16

Calculate (a) Gross Pay

(b) Total Deductions

(c) National Insurance.

3 The table shows the monthly repayments to be made on a loan of £1000.

(Monthly repayments on £1000)			
Period of Loan	Annual Interest Rate		
	(10%)	(12%)	(14%)
12 months	£86·84	£87·67	£88·51
24 months	£45·49	£46·32	£47·14
36 months	£31·75	£32·59	£33·44

Louie borrows £6000 to be repaid over 24 months. If he is charged an annual interest rate of 12%, find the total repayments to be made on the loan.

4 The network diagram shows seven towns A, B, C, D, E, F, and G, and the distances between them in miles.

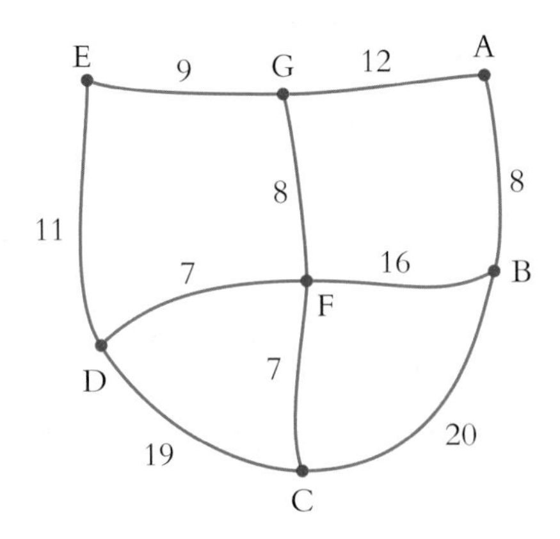

(a) State the degree of the vertex at B.

(b) What is the shortest distance from A to C?

5 The flowchart shows how a shop assistant's weekly pay is calculated.

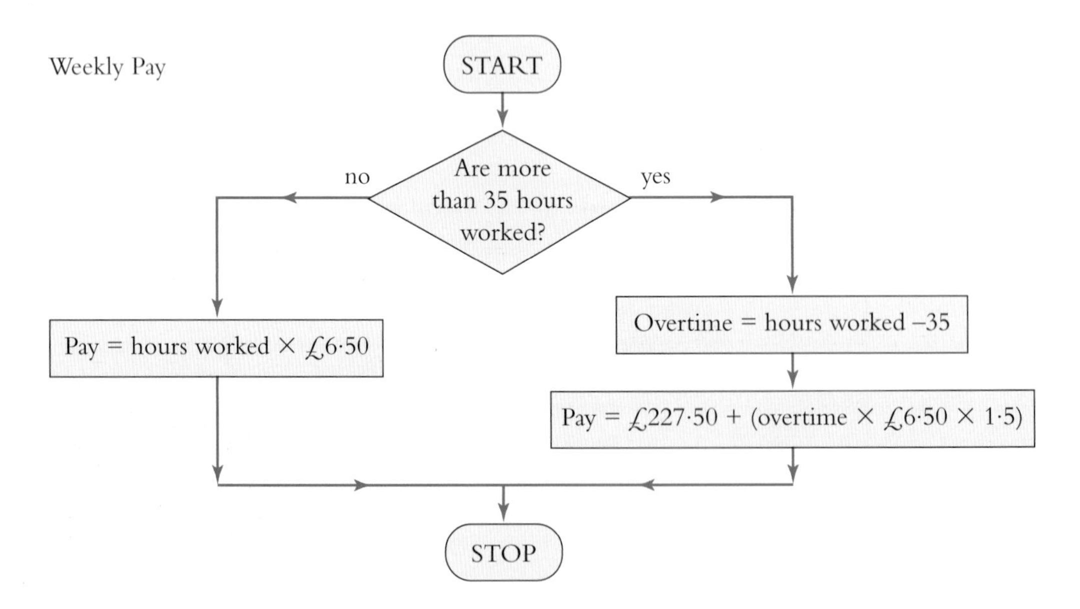

Calculate the weekly pay of an assistant who works 41 hours.

6 The box office manager in a theatre records the ticket sales on a spreadsheet.

(a) The formula =C7*D7 is entered in cell E7.

What value will appear in cell E7?

(b) What formula would be entered in cell E13 to find the total takings at the box office?

	A	B	C	D	E
1	**Theatre Box Office**		Takings for Concert on 14/04/2007		
2					
3	**Area of Theatre**		**Price per ticket**	**Number of tickets sold**	**Total proceeds**
4					
5	Stalls		£73·00	120	£8760·00
6	Dress Circle Front		£67·00	30	£2010·00
7	Dress Circle Back		£57·00	40	
8	Upper Circle Front		£47·00	50	
9	Upper Circle Back		£36·00	50	
10	Balcony Front		£26·00	120	
11	Balcony Back		£16·00	130	
12					
13		**Total takings at the Box Office**			

7 From a distance of 40 metres, the angle of elevation to the top of a cliff is 59°.

(a) Using a scale of 1 centimetre to 5 metres, make a scale drawing.

(b) Hence estimate the actual height of the cliff.

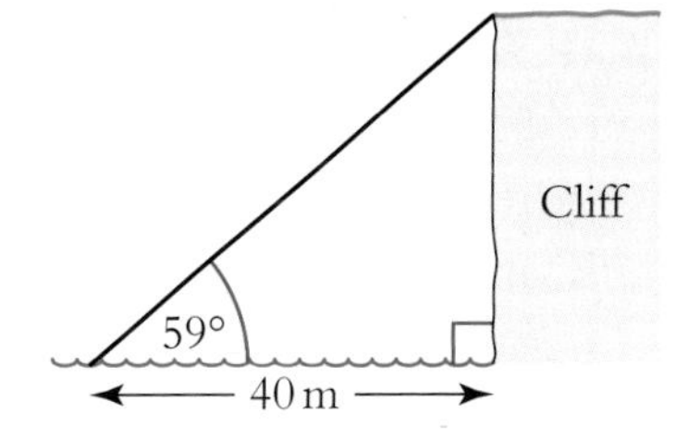

8 The diagram shows the net of a triangular prism.

Calculate the total surface area of the triangular prism.

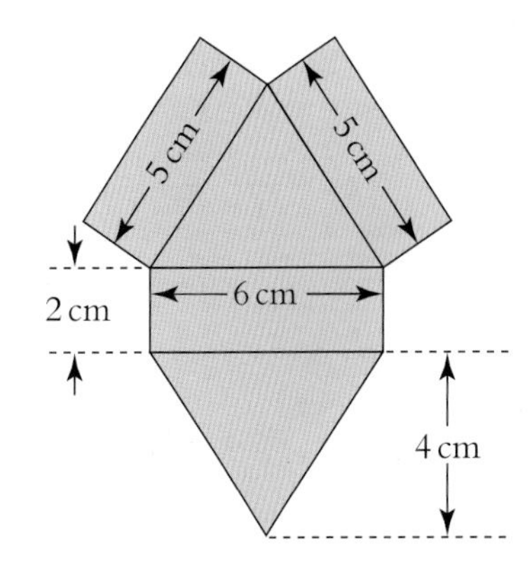

9 A cylinder has radius 12 cm and height 35 cm. Calculate the area of its curved surface. (Give your answer to the nearest square centimetre.)

10 The marks scored by 16 students in a test were:

59 63 75 63 64 66 70 74 75 68 68 70 71 72 68 77.

Copy and complete the boxplot to show these marks.

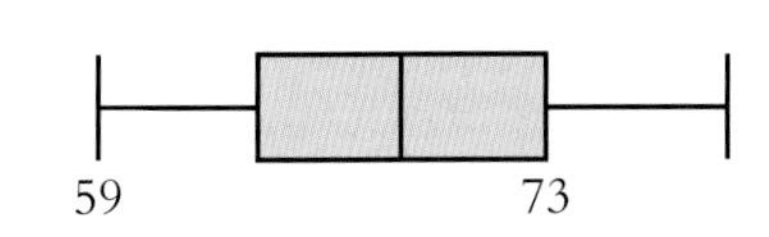

11 The activity diagram shows how a meal of mince with potatoes and peas is cooked. There are four rings on the cooker, so all the items can cook at the same time. All times are in minutes.

Find the minimum time required to complete the meal.

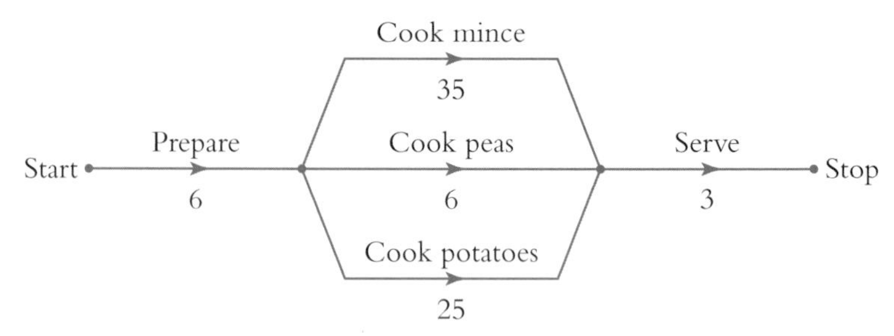

Test Three (A/B Content)

1 Anders Back works in a store. He works a basic 8-hour day at a rate of £6·90 per hour. Extra hours worked are paid at time and a half.

During one week he works 8 hours on each of Monday, Tuesday, and Wednesday, and 11 hours both on Thursday and Friday. Calculate his total gross wage for the week.

2 The flowchart shows how to calculate interest when an amount of money is invested for 1 year in the Bank of Strathclyde.

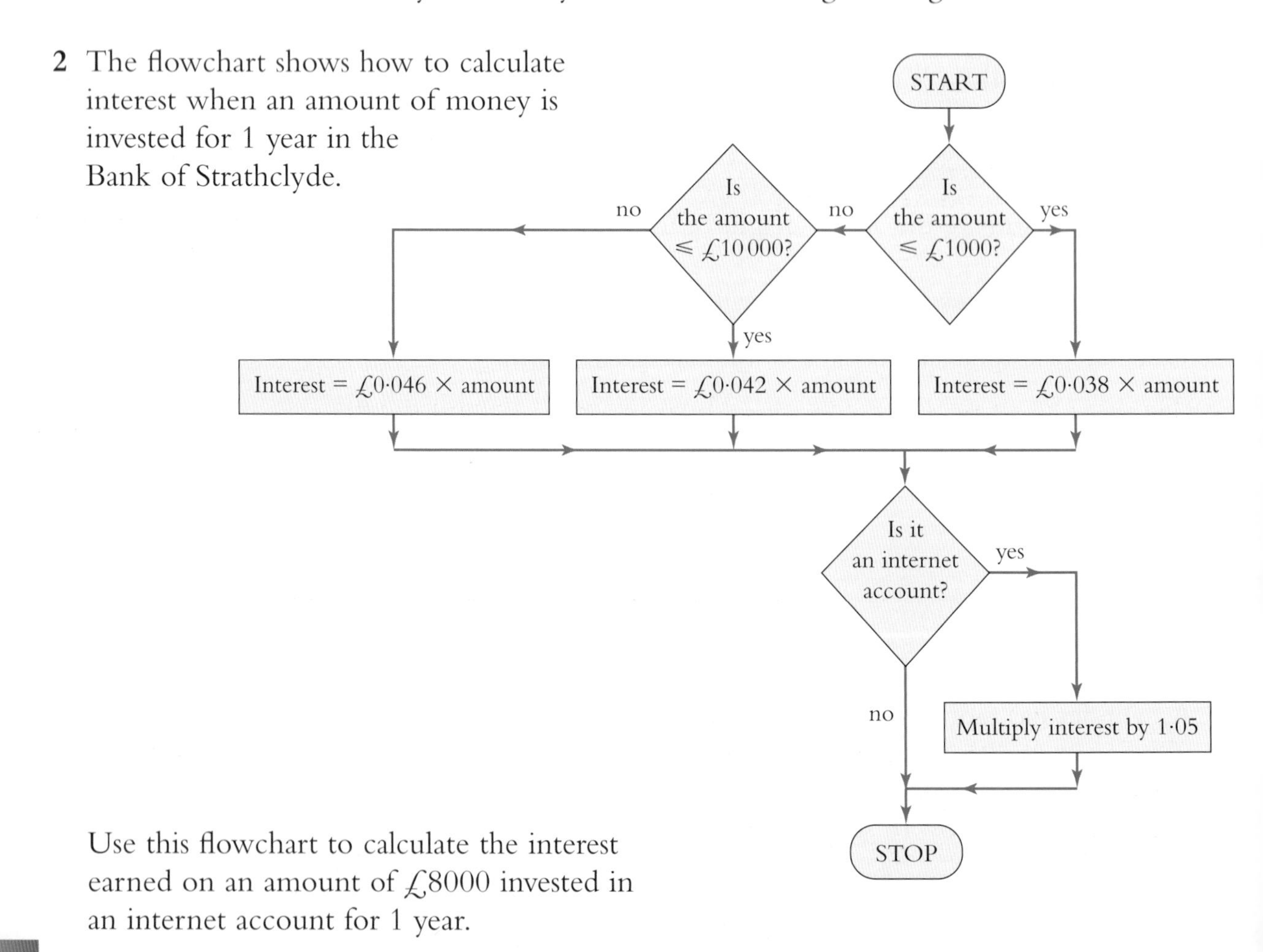

Use this flowchart to calculate the interest earned on an amount of £8000 invested in an internet account for 1 year.

3 Ship A is 50 kilometres due North of ship B.

(a) Use a scale of 1 cm to 5 km, and show this in a scale drawing.

(b) A third ship C is on a bearing of 220° from A and 330° from B. Show ship C on your drawing.

4 The diagram shows the net of a triangular prism

Calculate the surface area of this triangular prism.

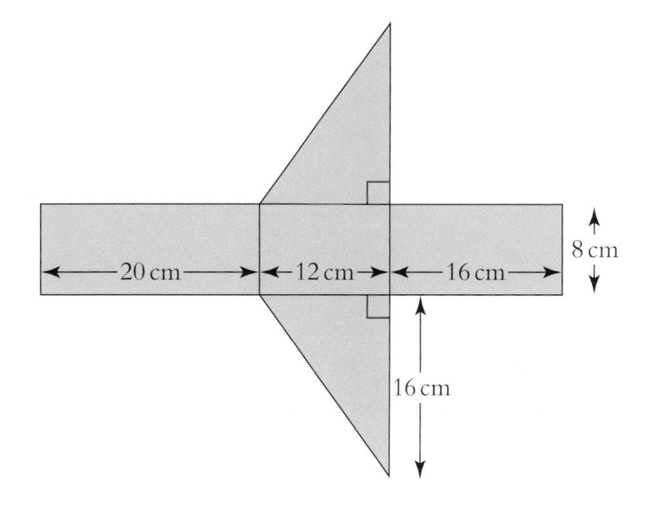

5 A label is designed to fit all of the way round a cylindrical can of soup.

The can has diameter 12 centimetres and height 16 centimetres. There is an extra 2 centimetres of length in the label for overlap.

Calculate the total area of the label.

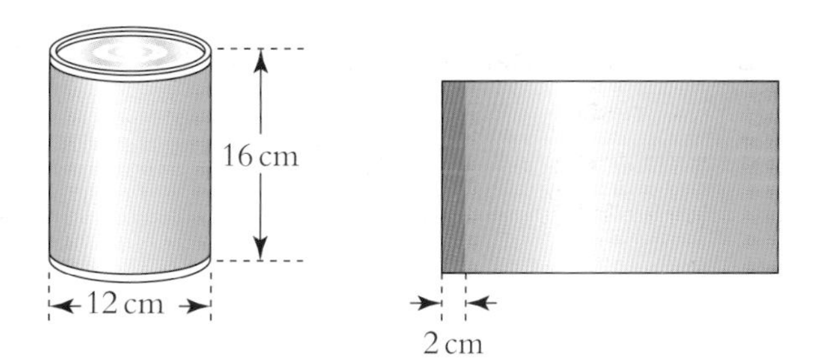

6 The boxplot shows the marks of the boys in the S4 English exam.

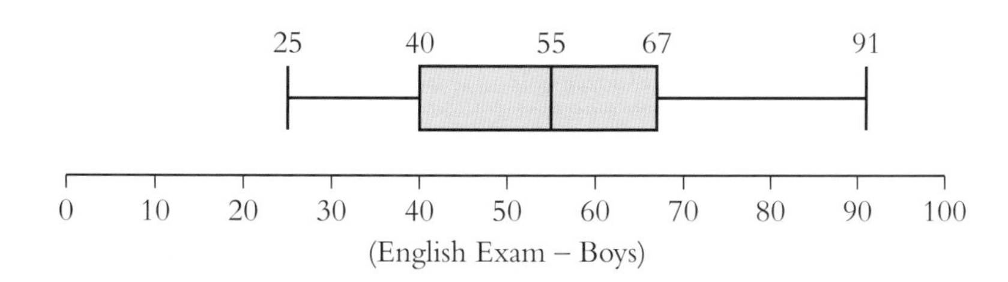

(a) The marks of the girls were analysed. The median for the girls was found to be 63 and the Range was 42.

Make two comments comparing the results of the boys and girls.

7 The diagram shows the net of a solid shape.

(a) Name the solid shape formed from this net.

(b) Calculate the total surface area of this net.

(Give your answer correct to one decimal place.)

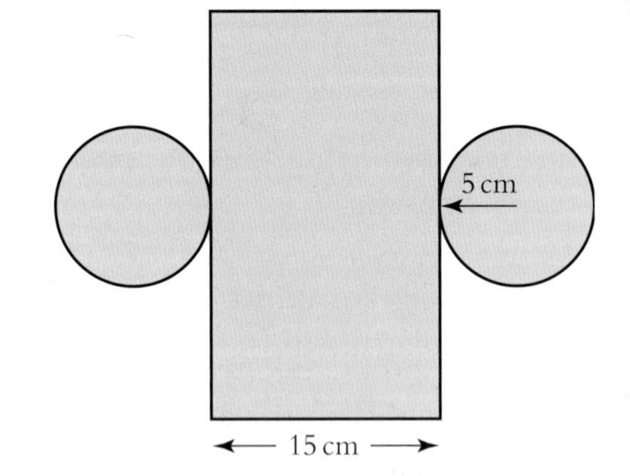

Practice Exam Paper

Now you should attempt practice examination papers in which the standard of questions is the same as that in the actual exam, and the papers are the same length.

In these papers all students should do Section A, covering Units 1 and 2.

Those studying Unit 3 should do Section B.

Those studying Applications of Mathematics should do Section C.

Paper 1 is the non-calculator paper which is worth 30 marks and for which 35 minutes is allowed.

Paper 1 (Non-Calculator)

Section A – All students do this section.

1 (a) Find $0{\cdot}486 \div 9$. **(1)**

(b) Find $\frac{5}{8}$ of 120. **(1)**

(c) Find 70% of £150. **(1)**

2 The following rule may be used to convert miles to kilometres.

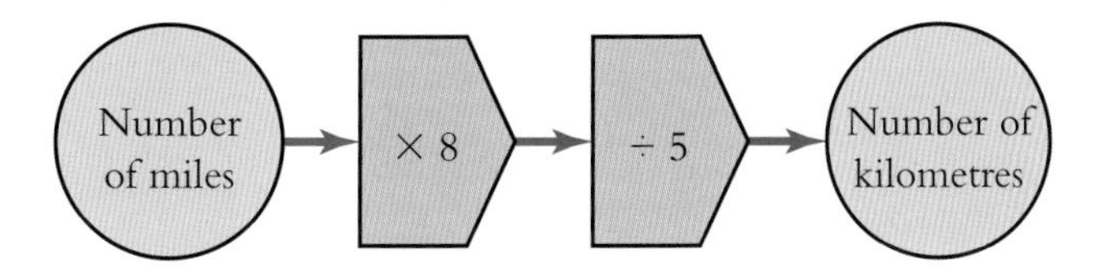

Convert 75 miles to kilometres. **(2)**

3 The number of visitors to a tourist office was counted each day for three weeks, as follows:

25 29 35 47 50 59 18

28 57 21 16 45 42 22

58 32 28 36 42 57 21.

(a) Illustrate this data in an ordered stem and leaf diagram. **(3)**

(b) State the median for this data. **(1)**

4 Douglas drives from Oban to Fort William.

The graph shows his journey.

(a) He stops on the way for coffee. How long does his coffee break last? **(1)**

(b) Does Douglas drive faster before or after the coffee break? **(1)**

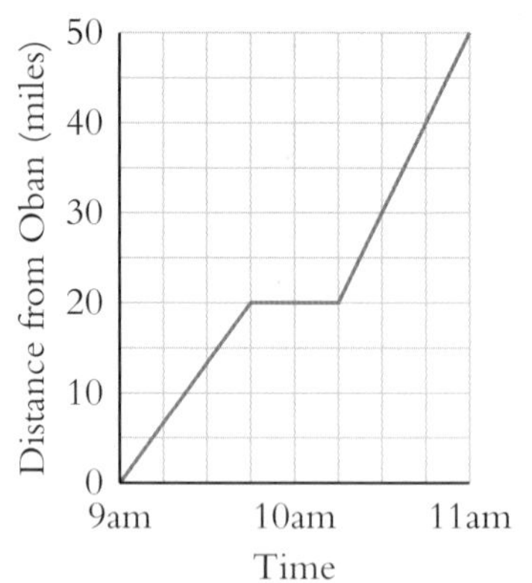

5 The number of residents in each house in Home Street is shown in this frequency table.

Copy and complete the table and calculate the mean number of residents in each house. **(3)**

Number of Residents	Frequency	Number of Residents × Frequency
1	12	12
2	23	46
3	25	75
4	19	76
5	15	
6	6	
	(Total = 100)	

6 Twenty people work in a primary school. Three of them are left-handed.

What percentage is this? **(3)**

7 Sam is buying some accessories for his computer. He sees the following items in a sale.

Headphones £20; Mouse £15; Keyboard £40; Printer £35; Scanner £30.

Sam only has £85 and he wants to buy **three** of these items.

One combination of items is shown in the table.

Copy and complete this table to show all the possible combinations of items Sam can afford. **(3)**

Headphones (£20)	Mouse (£15)	Keyboard (£40)	Printer (£35)	Scanner (£30)	Total Value
	✓		✓	✓	£80

Section B – Only do this section if you are studying Unit 3

8 Factorise $21p - 28$. **(2)**

9 (a) Complete the table for $y = 1 + 2x$. **(2)**

x	−4	0	3
y			

(b) Hence draw the line $y = 1 + 2x$ on a piece of squared paper. **(2)**

10 (a) Solve the inequality $6y - 5 > 19$. **(2)**

(b) Multiply out the brackets and simplify:

$5(2a - 3b) + 6b$. **(2)**

Section C – Only do this section if you are studying Applications of Mathematics

8 A spreadsheet is used to calculate the gross wages of the workers in a shop.

	A	B	C	D
1				
2				
3	**Name**	**Hourly rate of pay**	**Number of hours worked**	**Gross pay**
4	W. Allison	£7·50	40	£300·00
5	J. Christie	£9·20	40	
6	G. Kerr	£5·80	35	
7	R. Lorimer	£5·40	40	
8	J. Mitchell	£6·00	20	
9	W. Smith	£7·50	20	
10				
11	**Average gross wage per worker =**			

(a) The formula =B9*C9 is entered in cell D9.

What value will appear in cell D9? **(1)**

(b) Write down the formula to enter in cell D11 to show the average gross wage per worker. **(1)**

9 This network diagram shows four towns and the routes connecting them.

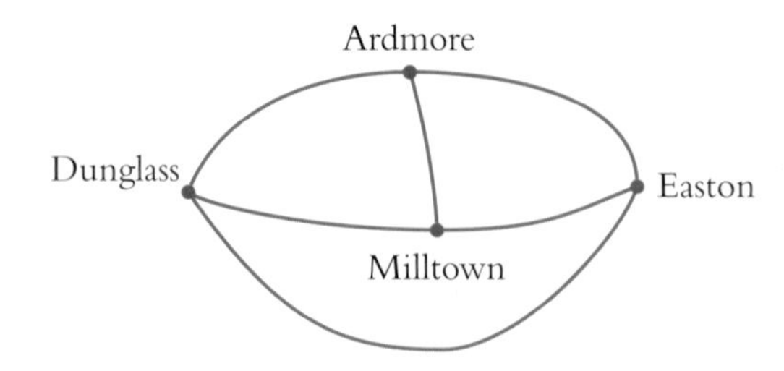

(a) How many arcs are there in the network diagram? **(1)**

(b) What is the order of the node at Ardmore? **(1)**

10 The number of absentees in a factory was counted each Monday for ten weeks. The results were as follows:

13 19 17 11 10 8 19 16 8 11.

(a) Copy and complete the boxplot to show the number of absentees.

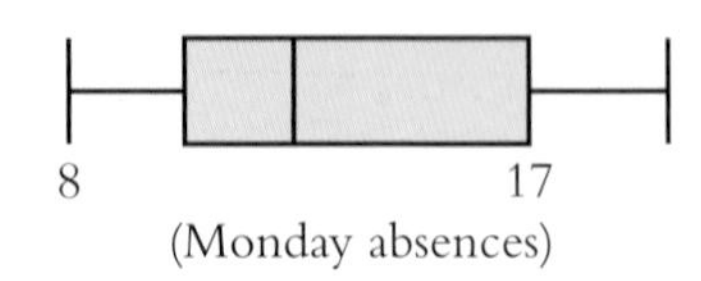

(4)

The management introduces a bonus payment for good attendance.

The following boxplot shows the number of absentees for ten weeks **after** the bonus payment is introduced.

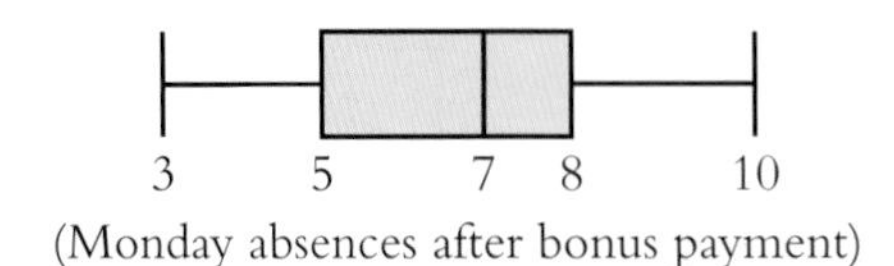

(b) Make two comments comparing the number of absentees before and after the bonus payment is introduced. **(2)**

Paper 2 is the Calculator paper. It is worth 50 marks and the time allowed for it is 55 minutes.

Paper 2 (Calculator)

Section A – All students do this section.

1 A company delivers wine in boxes each of which is in the shape of a cube.

Calculate the volume of each box.
(Give your answer in litres.)

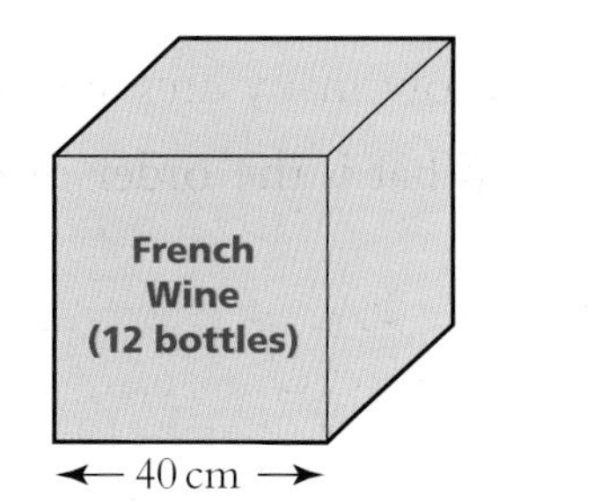

(3)

2 A driver leaves Aberdeen at 10.55 pm and reaches Edinburgh arriving at 1.25 am the following morning.

If the distance from Aberdeen to Edinburgh is 210 kilometres, calculate the average speed of the journey. **(3)**

3 Lydia has invested £5400 in the bank. If the rate of interest is 4·5% per annum, how much interest will she earn in 8 months? **(3)**

4 The numbers of yellow cards issued during 12 weekend league matches were as follows:

5 2 3 4 5 6 1 0 7 3 3 9.

(a) Find the modal number of yellow cards. **(1)**

(b) Find the median number of yellow cards. **(2)**

(c) Find the probability that there were less than 2 yellow cards issued at a match that weekend. **(1)**

5 Margaret is going to France on holiday. She changes £800 to euros.

(a) If the exchange rate is £1 = 1·46 euros, how many euros will she receive? **(2)**

(b) In France she spends 1150 euros. How many euros has she left? **(1)**

(c) She changes the remaining euros back to pounds after her holiday.

If the exchange rate is now £1 = 1·43 euros, how much will she receive?

(Give your answer to the nearest penny.) **(2)**

6 Some shoppers were asked which brand of toothpaste they preferred.

Their responses are shown in the pie chart.

There were 900 shoppers in the group. How many of them preferred Gleam?

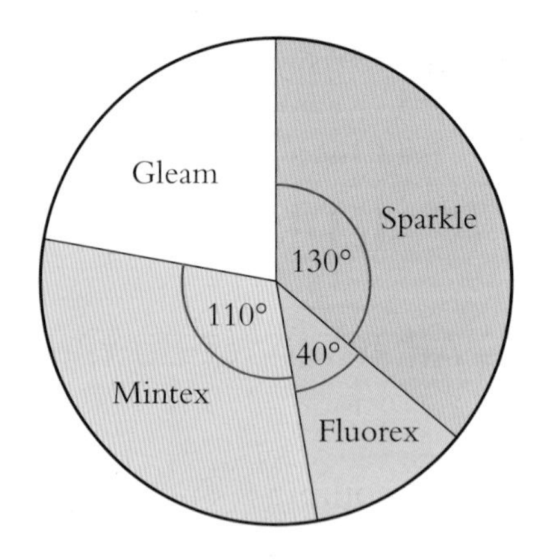

(3)

7 The diagram shows a circle with centre O.

Triangle ABC has been drawn with AB as diameter.

The diameter AB is 15 centimetres long. Angle ACB is 90°.

AC is 13 centimetres long. Find the length of BC.

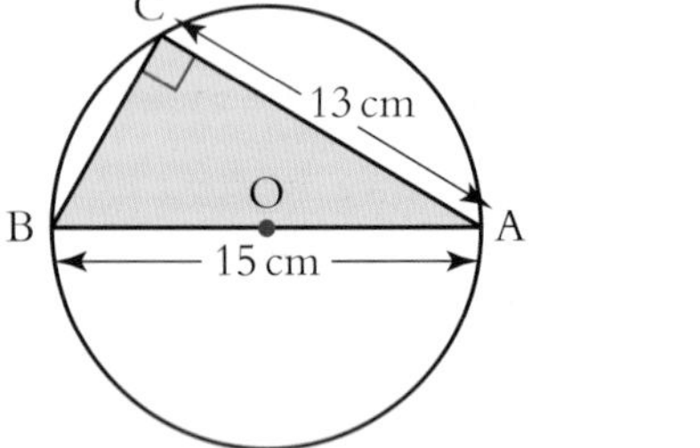

(3)

8 Norman bought a new bed on hire purchase. The cash price of the bed was £850. He paid a deposit of 20% of the cash price and then 24 equal monthly instalments of £31·25. How much extra did Norman pay to buy the bed on hire purchase? **(4)**

9 The trophy for winning a tennis competition is designed as a shield.

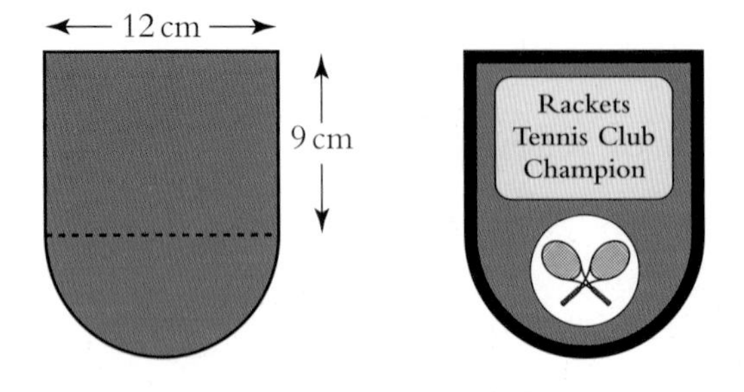

This shield is in the shape of a rectangle with a semi-circular base.

Black edging is to placed around the outside of the shield.

Find the length of the edging required.

(Give your answer correct to one decimal place.) **(5)**

10 (a) Calculate $-7 \times (-12)$. **(1)**

This is a number cell.

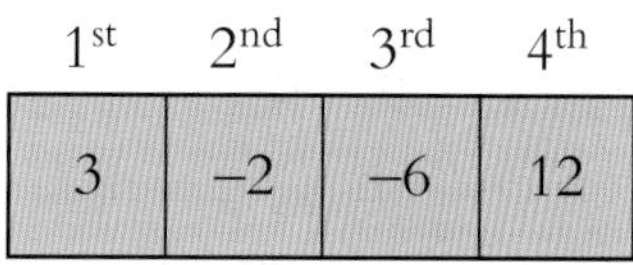

- 1st number × 2nd number = 3rd number — $3 \times (-2) = -6$
- 2nd number × 3rd number = 4th number — $(-2) \times (-6) = 12$

(b) Copy and complete this number cell

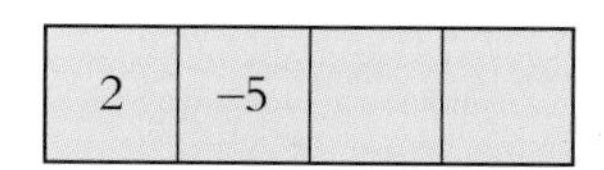

(1)

(c) Copy and complete this number cell

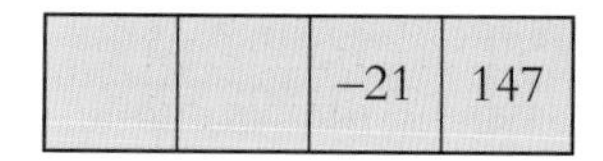

(2)

Section B – Only do this section if you are studying Unit 3

11 Solve algebraically the equation $7y + 8 = 44 - 2y$. **(3)**

12 Use the following formula to calculate the value of h when $D = 576$ and $k = 8$.

$$h = \sqrt{\frac{D}{k}}$$

(3)

13 The diagram shows the side view of a new house.

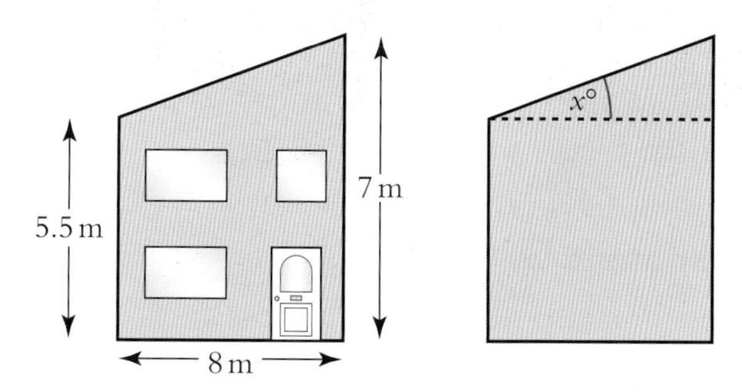

Calculate the angle, $x°$, the sloping edge of the roof makes with the horizontal. **(4)**

14 Old Trafford stadium in Manchester holds $7 \cdot 6 \times 10^4$ spectators.

During one season the stadium was full to capacity for all 19 Premiership games played there.

How many spectators attended all the games?
(Give your answer in standard form.) **(3)**

Section C – Only do this section if you are studying Applications of Mathematics

11 The table shows the monthly repayments to be made when money is borrowed from the Bank of Ecosse.

Loan \ Term	Monthly Repayments 20 years	25 years	30 years
£20 000	£161·81	£149·32	£139·72
£30 000	£244·05	£220·05	£210·24
£40 000	£326·43	£296·71	£279·20

Frances needs to borrow £40 000 to buy a house.

(a) She decides to repay the loan over 25 years.

How much will she pay each month? **(1)**

(b) How much will she pay altogether? **(2)**

12 The scale drawing shows the positions of a church and a tree which are 75 metres apart.

(a) Find the scale of this drawing. **(1)**

(b) Buried treasure is on a bearing of 080° from the church and 150° from the tree.

Copy and complete the diagram to show the position of the buried treasure. **(2)**

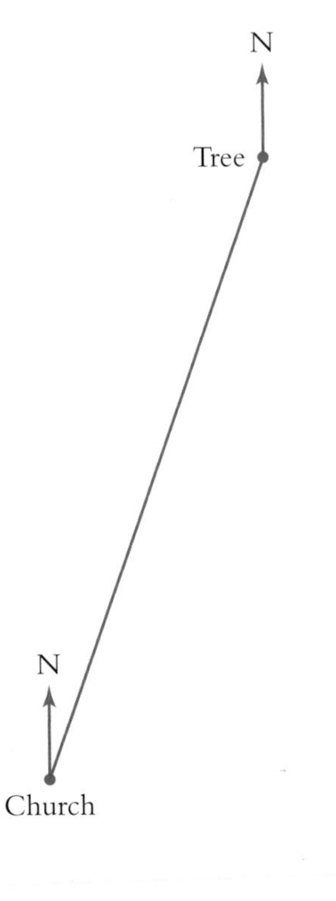

13 Kamran earns £5·60 per hour. During one week he works 4 hours overtime at time and a half.

(a) Calculate his overtime pay. **(1)**

(b) Copy and complete his payslip. **(2)**

Kamran Raja	Week Ending: 09/02/2007	
Basic Pay £212·80	Overtime	Gross Pay
Income Tax £20·07	National Insurance £12·96	Total Deductions £33·03
		Net Pay

14 The diagram shows the net of a solid shape.

(a) Name the solid shape formed from this net. **(1)**

(b) Calculate the area of the curved surface of the solid shape. **(3)**

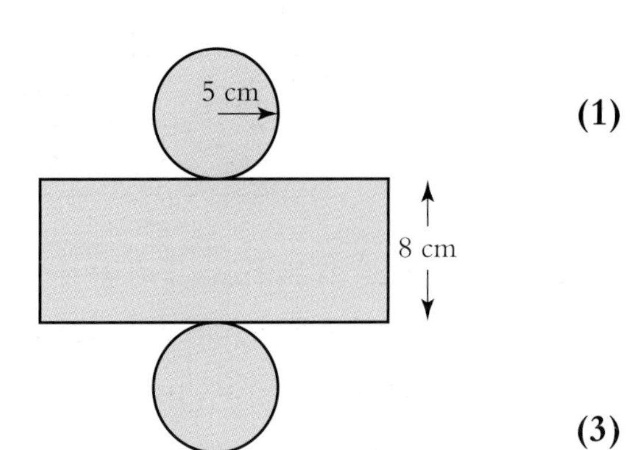